T0175778

Cyber Ecosystem and Security series

Machine Learning for Computer and Cyber Security
Principles, Algorithms, and Practices

Editors

Brij B. Gupta
National Institute of Technology, Kurukshetra, India

Michael Sheng
Department of Computing, Macquarie University, Sydney, Australia

CRC Press
Taylor & Francis Group
Boca Raton London New York

CRC Press is an imprint of the
Taylor & Francis Group, an **informa** business

A SCIENCE PUBLISHERS BOOK

CRC Press
Taylor & Francis Group
6000 Broken Sound Parkway NW, Suite 300
Boca Raton, FL 33487-2742

First issued in paperback 2020

ISBN-13: 978-1-138-58730-4 (hbk)
ISBN-13: 978-0-367-78027-2 (pbk)

Library of Congress Cataloging-in-Publication Data

Names: Gupta, Brij, 1982- editor. | Sheng, Quan Z. editor.
Title: Machine learning for computer and cyber security : principles,
    algorithms, and practices / editors Brij B. Gupta, National Institute of
    Technology, Institute of National Importance, Kurukshetra, India, Michael
    Sheng, Department of Computing, Macquarie University, Sydney, Australia.
Description: Boca Raton, FL : Taylor & Francis Group, [2019] | "A science
    publishers book." | Includes bibliographical references and index.
Identifiers: LCCN 2018054743 | ISBN 9781138587304 (acid-free paper)
Subjects: LCSH: Computer networks--Security measures--Data processing. |
    Computer security--Data processing. | Machine learning. | Artificial
    intelligence.
Classification: LCC TK5105.59 .M3284 2019 | DDC 006.3/1--dc23
LC record available at https://lccn.loc.gov/2018054743

Visit the Taylor & Francis Web site at
http://www.taylorandfrancis.com

and the CRC Press Web site at
http://www.crcpress.com

Dedicated to my wife Varsha Gupta and daughter Prisha Gupta for their constant support during the course of this book.

B.B. Gupta

Dedicated to my mum for her love.

Michael Sheng

Foreword

"A baby learns to crawl, walk and then run.
We are in the crawling stage when it comes to applying machine learning."

—Dave Waters

Whether you are a student in any school or a research fellow in any university or faculty, or an IT professional or for that matter, any person who is drawn to cyber security, you will find something useful and enjoyable in this book. This book caters to the needs of the global research community. On going through the book, you shall discover the profoundness of the influence which machine learning and data-mining techniques have in developing defense solutions against attackers. It explores some of the most exciting solutions developed and the future opportunities for current researchers. A lot of effort has been put into the designing of this book. It is written and compiled in a lucid, visually distinct and illustrative way to provide insights to the user about the manner in which upfront technologies, like machine learning and data mining, have integrated with cyber security. The inspiration behind this book was to enlighten the researchers and to gather greater appreciation for a difficult, but fascinating subject-machine learning. This book is exemplary of infinite applications that machine learning has.

The emerging technologies, like machine learning and data mining, are the gift of many tedious endeavours; a gift that possesses immense power to shape the futuristic secure cyberspace. We hope this book will contribute to the imagination of the students and young researchers in this field to explore further as there is so much yet to be discovered and invented by man. But this is only a first foot in this direction; many more volumes will follow. I congratulate the authors as I think there are going to be many grateful readers who shall gain broader perspectives on the discipline of machine learning as a result of their efforts.

Gregorio Martinez Perez
Department of Computer Science
University of Murcia (UMU), Spain
E-mail: gregorio@um.es
Homepage: http://webs.um.es/gregorio/

Acknowledgement

Many people have contributed greatly to this book on *Machine Learning for Computer and Cyber Security: Principles, Algorithms, and Practices*. We, the editors, would like to acknowledge all of them for their valuable help and generous ideas in improving the quality of this book. With our feelings of gratitude, we would like to introduce first the authors and reviewers of the chapters for their outstanding expertise, constructive reviews and devoted effort; secondly the CRC Press, Taylor and Francis Group staff for their constant encouragement, continuous assistance and untiring support and finally, the editor's family for being the source of continuous love, unconditional support and prayers not only for this work, but throughout our life. Last but far from the least, we express our heartfelt thanks to the Almighty for granting us the courage to face the complexities of life and complete this work.

B.B. Gupta
Michael Sheng

Series Preface

The cyber world is up against security and privacy related issues in every field including finance, healthcare, transportation, retail, and so forth; due to the extensive use of internet-based applications and services. Analyzing the impact and prioritizing, industry-specific requirements has become a prime concern. Moreover, the risks associated with cyber-crimes is making the study of forensics, essential.

The series will present emerging aspects in the cyber ecosystem with respect to various applications, underlying technologies, stakeholders involved, etc. and review the security and privacy related challenges. It will cover the role of digital forensics in cyber-crimes, and review the ongoing research to protect the misuse of technology. This series is targeted at researchers, students, academicians, and business professions in the field.

Brij B. Gupta
Dharma P. Agrawal
Gregorio Martinez Perez

Preface

Computer security pertains to the use of technology and policies to assure confidentiality, integrity and availability by adopting the activities of prevention, detection and recovery. While researching and teaching cyber security for many years, we noticed an increasing trend towards machine learning-based solutions, most of which revolve around machine learning and data-mining techniques. Being unable to find any appropriate compilation of these modern techniques, we decided to come up with this book titled, *Machine Learning for Computer and Cyber Security: Principles, Algorithms, and Practices*.

The cyberspace continues to face challenging threats and vulnerabilities, exposing us to different kinds of cyber attacks of varied severity levels and to counter which, machine learning is an ideal solution. We believe learning and understanding of any technical matter demands studying it multiple times in different manners, which offer an ever new perspective of the problem. Thus, for ensuring an acceptable and satisfactory level of security, it is required that researchers clearly understand the new technologies and their working mechanism. This book serves as a valuable reference point for researchers, educators and engineers who want to explore different machine-learning techniques and tools in cyber security. It is also of general interest to one using machine-learning techniques in software development, particularly in developing cyber security applications. It is our hope that the work presented in this book will stimulate new discussions and generate original ideas that will further develop this important area.

We would like to express our heartfelt gratitude and acknowledgement to the authors for their contributions and the reviewers for their expertise to improve the manuscript. We are grateful to CRC Press, Taylor & Francis Group for giving the opportunity to publish this book. We would like to thank them for their support and professionalism during the entire publication process.

August 2018

B.B. Gupta
Michael Sheng

Contents

CHAPTER 1

A Deep Learning-based System for Network Cyber Threat Detection

Angel Luis Perales Gomez,* Lorenzo Fernandez Maimo and *Felix J. Garcia Clemente*

New network technologies and paradigms are rendering existing intrusion detection and defense procedures obsolete. In particular, the upcoming fifth generation (5G) mobile technology with large volumes of information and high transmission rates is posing new challenges on cybersecurity defense systems. In this regard, this chapter proposes a system for network cyberthreats detection in 5G mobile networks by making use of deep learning techniques with statistical features obtained from network flows. Since these features are payload-independent, they can be computed even with encrypted traffic. The system analyzes network traffic in real time and, additionally, is able to adapt in order to manage traffic fluctuation. This proposal has been evaluated in a botnet context, reaching reasonable classification accuracy. Moreover, the model prediction runtime has also been evaluated with a variety of deep learning frameworks and a wide range of traffic loads. The experimental results show that the model is suitable in a real 5G scenario.

1. Introduction

Over the years, cyber-security researchers have used diverse techniques and developed multiple solutions to protect assets of organizations from malicious attackers. Cyber-security solutions provide encryption, rights management and inspection capabilities for protecting network data. However, cyber security threats, such as trojans, viruses, worms and botnets, among others [1] require cyber security solutions to be constantly updated. In this sense, solutions based on Intrusion Detection Systems (IDS) include proactive techniques to anticipate vulnerabilities and trigger reactive actions.

Departamento de Ingeniería y Tecnología de Computadores, University of Murcia, 30100 Murcia, Spain.
E-mails: lfmaimo@um.es; fgarcia@um.es
* Corresponding author: angelluis.perales@um.es

The large volume of traffic carried by modern networks make IDSs ineffective to collect and analyze every network packet. As an illustration, a deep packet inspection (DPI) tool like Snort [2] begins to discard packets from 1.5 Gbps [3] due to overheads. Recently, intensive experiments were conducted in order to obtain a thorough performance evaluation of Snort and the application of machine learning techniques on it [4]. These experiments covered a wide range of bandwidths, resulting in 9.5 per cent of packets dropped at 4 Gbps, while packet drop rose to 20 per cent at 10 Gbps. In order to improve performance, advanced parallelization techniques based on hardware accelerators were proposed. Among them, techniques based on Field Programmable Gate Array (FPGA) support speeds of up to 4 Gbps without loss [5] while the ones based on Application-Specific Integrated Circuit (ASIC) reach speeds close to 7.2 Gbps [6].

Consequently IDS-based solutions based on DPI adopted new ways of detection, evolving towards the use of innovative AI-based techniques [7]. A complete survey dealing with solutions to quickly classify gathered network flows and detect attacks, can be found in [8]. However, upcoming networks with even higher transmission rates will make these solutions insufficient. That is the case with the Internet of Things [9] or the new fifth generation (5G) mobile technology, where its new advanced features will be a challenge for the existing detection procedures; therefore, these procedures need to be adapted accordingly to the new requirements.

The 5G-PPP consortium has identified a pool of Key Performance Indicators (KPI) that have a high impact when analyzing and inspecting traffic network flows [10]. This inspection computes certain characteristics of the incoming network flow needed for the detection procedure in an efficient and quick manner. Among these KPIs, the following four make detection procedures an even greater challenge in 5G mobile networks:

- 1000 times higher mobile data volume per geographical area.
- 10 to 100 times more connected devices.
- 10 times to 100 times higher typical user data rate.
- End-to-end latency of < 1 ms.

The large number of User Equipments (UE) belonging to 5G subscribers, the large volumes of data traffic produced by them and the reduced latency in connectivity make us face new challenges to be solved without losing cyber threat detection accuracy in real-time scenarios.

To overcome this challenge, this chapter presents a 5G-oriented architecture to identify cyber threats in 5G networks by means of deep learning techniques. This approach is based on a feature vector made up of statistical measures obtained from the network flows within a given time period. These features are payload-independent, making them suitable to be used even with encrypted traffic. In our proposal, a deep learning model is trained in a supervised way with the feature vectors computed from a well-known botnet dataset. Subsequently, this model is used to classify the network traffic as normal or anomalous within this botnet context [11]. Our experimental results show the classification accuracy of our proposal as well as its prediction runtime. This runtime is shown to be suitable for the time restrictions imposed by the 5G networks. Moreover, this chapter first presents a detailed study of different deep learning techniques and how they can be applied to network anomaly detection.

2. Deep Learning

For years, machine learning has been used in a wide range of applications, such as spam filters or recommendation systems. Nowadays, machine learning is being replaced with more advanced deep learning techniques. Both are closely related and share the same philosophy—they build a model from an input dataset and use it to make predictions on unseen data. Indeed, deep learning is considered as a subfield of machine learning (Fig. 1). However, machine learning needs feature engineering to generate the input features, thus requiring more domain expertise, whereas deep learning obtains their own features from raw data [12]. This fact, together with the increase of computation power in modern hardware and the availability of public datasets to be used in a wide domain of problems, are the main reasons of deep learning success.

In this section, a brief introduction to machine learning is presented, describing the typical problems where machine learning is used and the different training methods available. Next, the main concepts behind deep learning are explained, starting from a basic feed-forward artificial neural network and ending with more complex architectures.

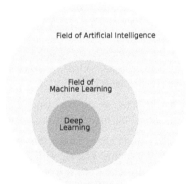

Fig. 1: Relationship among artificial intelligence, machine learning and deep learning.

2.1 *Machine Learning*

Machine learning is an artificial intelligence field that allows a computer to learn from data without being explicitly programmed. Machine learning algorithms (also called 'models') need a training phase before being used to make predictions. During the training stage, the algorithm uses the input data to find a model that allows predictions about unseen data belonging to the same context. If the model accurately predicts unseen data, the model can generalize and, hence, can be used to make future predictions. Conversely, if the model predicts accurately the data used during the training but has much less accuracy when used with new data, then the model has overfitted the training data. Overfitting means that it has also learned the noise, and thus, it cannot generalize.

Additionally, the model may have parameters that cannot be optimized from the data; they are called 'hyperparameters'. In many cases, their default values are sufficient to obtain an acceptable result. However, reaching state-of-the-art requires the tuning of these model hyperparameters. Frequently this is the more time-consuming part because it implies fitting and testing the model for a wide range of hyperparameter values.

We can consider two main machine-learning tasks regarding their desired output: classification and regression (Fig. 2).

Classification: We want to determine the most probable class of an element, for example, distinguishing between pictures of cats and dogs. This typically consists in finding a boundary that separates the different classes. When we have only two classes, it is called binary classification. If we have more than two classes, it is called categorical classification. A typical classification problem in machine learning is to recognize handwritten digits. One approach to solve this task is to train a model by using the well-known MNIST dataset. In this problem, the input data is an image that represents a handwritten digit and the output is the digit itself. This can be seen as a 10-class classification task.

Regression: Attempt to model the relationship between a dependent variable and one or more independent variables. It is used to make both prediction and inference. A prediction example is when we want to predict the house pricing from variables such as number of bedrooms, number of bathrooms, square meters, city, and district. Conversely, if we want to know how much increase in house price is associated with a given city, we are making inference.

In addition to categorizing machine-learning algorithms regarding their desired output, we can also classify them as supervised or unsupervised, depending on whether or not they need labeled data respectively:

Supervised: Each sample has its label, so we can compute the error between the real and the predicted values and adjust the model parameters accordingly.

Unsupervised: In this case, the dataset has no labels, thus the approach to this problem is mainly based on clustering algorithms. The algorithms are based on grouping data in clusters, that is, the samples of each cluster are more similar to each other than to those in the other cluster.

A non-exhaustive list of supervised machine-learning algorithms includes linear regression, logistic regression, support vector machine, artificial neural networks, random forest, or tree decision just to name a few. Some of those algorithms, such

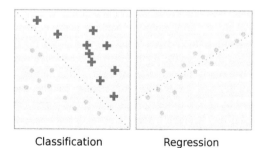

Classification   Regression

Fig. 2: Classification versus regression. In classification the dotted line represents a linear boundary that separates the two classes; in regression, the dotted line models the linear relationship between the two variables.

as logistic regression, focus on classification tasks; while others focus on regression tasks (e.g., linear regression). Some of them can be used both for regression and classification; for example, support vector machines, artificial neural networks or random forests. Regarding unsupervised algorithms, an example is the well-known k-means clustering method.

2.2 Artificial Neural Networks

Feedforward neural network is one of the basic artificial neural network architectures and it is made up of neurons organized in layers. Artificial neurons are inspired by biological neurons in the sense that they receive an input signal from different connections, perform a weighted sum of the inputs and apply an activation function to produce an output.

A detailed artificial neuron scheme can be seen in Fig. 3. The inputs are $x_1 \ldots x_n$ and they are multiplied by the weights labeled as $w_0 \ldots w_n$. The weighted sum is performed and, finally, the activation function is applied, producing an output.

From a mathematical point of view, a neuron performs a well defined operation given by Eq. 1. We can represent the weighted sum operator as a matrix multiplication, where both x and w have an $n \times 1$ shape, and n is the number of inputs of the neuron.

$$g(x) = \sigma(\sum_{i=0}^{n} w_i x_i + b) = \sigma(w^T x + b) \tag{1}$$

The activation function σ is used to introduce non-linearity to the output. Three of the most common activation functions are sigmoid, hyperbolic tangent (*tanh*) and REctifier Linear Unit (*ReLU*). The output of the sigmoid activation function takes values between 0 and 1. A special case of sigmoid, called logistic function, is frequently used as final step in binary classification problems (Eq. 2). These problems are rather common; for example, in supervised anomaly detection, the goal is to predict between two classes, normal and abnormal. Therefore, a sigmoid can be applied as a last step to limit the output between 0 and 1. Finally, a threshold decides the class (e.g., class one if *sigmoid*(x) > 0.5 or class two if *sigmoid*(x) ≤ 0.5).

$$sigmoid(x) = \frac{1}{1+e^{-x}} \tag{2}$$

Hyperbolic tangent activation function can be expressed as a function of hyperbolic cosine and hyperbolic sine. The output values goes from −1 to 1 (Eq. 3).

$$tanh(x) = \frac{sinh(x)}{cosh(x)} = \frac{e^x - e^{-x}}{e^x + e^{-x}} \tag{3}$$

ReLU, in turn, is based on a maximum function and it takes the maximum value between 0 and the input (Eq. 4). *ReLU* is simpler than *sigmoid* or *tanh* because any negative value is set to 0; hence, it does not need to compute any operation, such as multiplication or exponential. An interesting property of *ReLU* is that it helps to solve the vanishing gradient problem [13].

$$ReLU(x) = max(0, x) \tag{4}$$

The three activation function are plotted in Fig. 4.

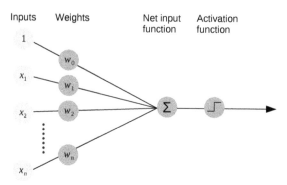

Fig. 3: One-layer artificial neural network—the perceptron.

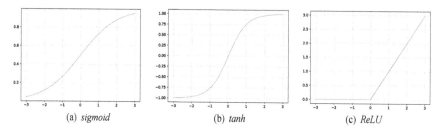

(a) *sigmoid*   (b) *tanh*   (c) *ReLU*

Fig. 4: The most common activation functions used in neural networks.

In general, as shown by Fig. 5, a feed-forward neural network is made up of a first layer called 'input layer', one or more 'hidden layers', and a final layer called 'output layer'. In the training phase, the input layer receives the training data; then, each neuron computes its output and propagates it to the next layer. This is repeated for each hidden layer until the output layer is reached. This is called 'forward propagation'. When the output layer is reached, a loss function computes the error between the output and the ground truth. Common loss functions are *mean square error* (*MSE*) for regression and *binary cross-entropy* (two classes) or *categorical cross-entropy* (three or more classes) for classification. The *MSE* function is computed as the mean of the difference between the observed values y and the predicted values $\hat{y}$, where y and $\hat{y}$ are vectors of size n (Eq. 5).

$$MSE = \frac{1}{n}\sum_{1}^{n}(y_i - \hat{y}_i)^2 \qquad (5)$$

The *binary cross-entropy* and *multi-class cross-entropy* are closely related. Indeed, the last one is a generalized case of the first. *Binary cross-entropy* and *multi-class cross-entropy* are computed as shown in Eqs. 6 and 7 respectively. In these formulas, $\hat{y}$ is the prediction vector, y is the observed values vector and i and j the i-th element in the vector and the j-th class respectively.

$$-\frac{1}{n}\sum_{1}^{n}[y_i - \log(\hat{y}_i) + (1 - y_i)\log(1 - \hat{y}_i)] \qquad (6)$$

$$-\frac{1}{n}\sum_{1}^{n}\sum_{1}^{m}[y_{ij}\log(1-\hat{y}_{ij})] \tag{7}$$

This error measure is used to update the weight of each layer by means of a technique called 'backpropagation' [14]. This technique is an efficient way to compute the gradient of the loss functions with respect to each optimizable parameter of the network. When the backpropagation process ends, the weights are updated by a version of gradient descent. This procedure is repeated until a minimum is reached.

There exists a wide variety of neural network architectures. Historically only shallow models were considered suitable; however, the modern hardware has achieved enough computational power to deal with millions of parameters of the deeper architectures. This compute power, together with enhanced regularization methods, initialization heuristics and larger datasets have substantially improved the convergence and performance of the deeper models.

In the following sections, two of the most popular architectures are described—convolutional neural networks (CNN), which have achieved state-of-the-art results in fields such as computer vision or natural language processing; and recurrent neural networks (RNN), including Long Short-Term Memory (LSTM), a deep architecture that is capable of detecting complex temporal patterns.

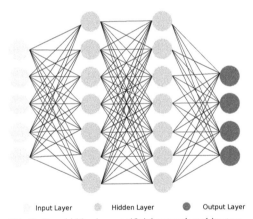

Input Layer   Hidden Layer   Output Layer

Fig. 5: Two-hidden layer artificial network architecture.

2.3 Convolutional Neural Networks

Feed-forward neural networks are based on fully-connected layers, in which every neuron is connected to every other in adjacent layers. By contrast, CNNs utilize local receptive fields (also called filters, which enforce a local connectivity pattern between neurons of adjacent layers, that is, a neuron is connected to a small region of the previous layer), shared weights (the weights are associated to the filter, so every output uses the same weights) and pooling (which simplifies the output).

Although CNNs have achieved state-of-the-art results in computer vision tasks (e.g., image classification or object detection), they are neither limited to image processing nor to two-dimensional inputs. In addition, they have proved their effectiveness in fields such as natural language processing or time series prediction

Three main types of layers are commonly used to build convolutional networks—convolutional layer, pooling layer and fully-connected layer (the same as in feed-forward networks).

2.3.1 Convolutional Layer

This is the core layer in the convolutional neural networks architecture and it is based on the convolution operator. Convolution is a linear operation widely used in computer vision and image processing applications. For images, convolution is represented as a matrix called kernel, filter or feature detector. The kernel matrix is moved across the image and applied to each image position. The resulting matrix is called feature map, activation map or convolved feature.

Figure 6 graphically shows how the operation is performed. It is important to notice that it is not a traditional matrix multiplication, but the dot product of the kernel and the underlying pixels. Different kernels result in different activation maps. In the figure, the output is computed as follows:

$$1 \times 0 + 2 \times 1 + 4 \times 0 + 2 \times 1 + 4 \times -4 + 6 \times 1 + 3 \times 0 + 2 \times 1 + 7 \times 0 = -4$$

The purpose of a convolutional layer is to extract features from the input image, where the filter weights determine the feature that is going to be detected. These weights are the parameters that the network will learn during the training phase.

Fig. 6: Convolution operation—The result is stored in the output cell corresponding to the center of the filter.

2.3.2 Pooling Layer

Frequently, a pooling layer is used just after a convolutional layer. The purpose is to reduce the dimensionality of the output by selecting the most representative feature in a patch. Pooling operation consists in a subsampling of its input by means of an aggregation operation (typically maximum or average) applied in a given stride. One interesting property of the pooling operation is that it also introduces translation invariance.

2.3.3 Fully Connected Layer

Fully connected layer is similar to the layers used in feed-forward neural networks. In CNN context, this layer is used as a final classification stage. The last convolutional layer outputs a given number of feature maps. This output is flattened into a single vector before being passed to the fully connected layer.

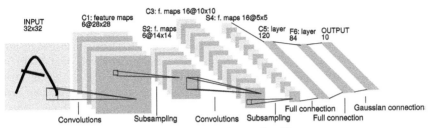

Fig. 7: Convolutional neural network architecture.

In Fig. 7, we can see a well-known convolutional neural network architecture called LeNet-5 [15] used in handwritten digit classification. LeNet-5 accepts images of 28 × 28 pixels as input. The network is made up of two convolutional layers, each of them with a corresponding max pooling layer, and two fully connected layers, followed by a final 10-output classification layer. Each output represents a class-membership estimation. This neural network architecture can be used in conjunction with the above-mentioned MNIST dataset to classify handwritten digit.

2.3.4 Recurrent Neural Networks

A recurrent neural network (RNN) is an architecture of artificial neural networks where connections between units form a directed graph along a sequence. Unlike feed-forward neural networks, RNNs can use their internal state to process sequences of inputs. This makes RNNs suitable for use in tasks that require data sequences, such as speech recognition, language modeling, translation or image captioning.

The input to the recurrent neural networks is a sequence of length n, denoted by $x_0, x_1, \dots, x_{n-1}$. The internal state can be represented by Eq. 8, where h_t is the hidden state after processing x_t input, h_{t-1} is the previous hidden state. W and U are the corresponding weights. The output function (y_t) can be seen in the Eq. 9, where V are the weights.

$$h_t = \sigma(W_t x_t + U_t h_{t-1}) \tag{8}$$

$$y_t = \sigma(V_t h_t) \tag{9}$$

The key concept in RNN architecture is the use of the previous state h_{t-1} that gives the network information about the past. This provides the network with a temporal context that can be used to predict the next element of a temporal series, or to translate a given sentence considering a number of previous words. However, RNN suffers from vanishing gradient; hence has trouble dealing with long-term dependencies. Long short-term memory neural networks (LSTM) solve this drawback by using a more complex architecture.

2.3.5 Long Short-Term Memory Neural Network

Long short-term memory neural network (LSTM) [16] is a special RNN architecture capable of learning long-term dependencies. It can remember information for a long period of time. An LSTM architecture uses the LSTM unit as a base block to build the network.

An LSTM unit is made up of a cell state c_t and three different gates: the forget gate (f_t), the input gate (i_t), and the output gate (o_t). In this model, the hidden state (h_t) is also the cell output and there is a new internal cell state (c_t) that is managed by the forget and input gates. The forget gate provides a forgetting coefficient computed from the input data (x_t) and the previous hidden state (h_{t-1}). Finally a sigmoid function (σ) is applied. The operations to compute the forget gate, the input gate and the new cell state are shown in Eq. 10. Here, W and U are the weights associated with the current input data and the previous hidden state respectively; and b is the bias.

$$f_t = \sigma(W_f x_t + U_f h_{t-1} + b_f)$$
$$i_t = \sigma(W_i x_t + U_i h_{t-1} + b_i) \tag{10}$$
$$c_t = f_t * c_{t-1} + i_t * tanh(W_c x_t + U_c h_{t-1} + b_c)$$

The input gate decides how the cell state should be updated. The next step is to obtain a candidate value to update the cell state. This candidate value is generated by applying *tanh* to a function of the previous hidden state and the input data. The new cell value is computed as a sum of the former cell state and the candidate, entry-wise multiplied (*) by the forget and input values respectively.

The output gate is responsible for deciding what information will be passed to the next unit. Its value is calculated from the input data and the previous hidden state. Finally, the output (or hidden state) is computed by entry-wise multiplying the output gate and the *tanh* function applied to the cell state (Eq. 11).

$$o_t = \sigma(W_o x_t + U_o h_{t-1} + b_o)$$
$$h_t = o_t * tanh(c_t) \tag{11}$$

3. Deep Learning Applied to Anomaly Detection

Anomaly detection is the identification of items, events or observations which do not conform to an expected pattern or behavior. A simple anomaly detection procedure could be to define a region representing normal behavior and label any sample in the data that does not belong to this normal region as an anomaly. According to [17], there are several challenges in this simple approach—difficulty in defining such a normal region; anomalies produced by an adaptive malicious attacker; normal behavior evolution; context dependent definition of anomaly (e.g., in medical domain, a small fluctuation in body temperature can be an anomaly, whereas similar deviation in stock market domain may be normal); or lack of availability of labeled datasets.

Given the variety of anomalies, the general anomaly detection problem is considered as extremely hard. Moreover, even in a given context, the concept of anomaly can change with time. As an illustration, let us assume a network where P2P traffic is not allowed. In this case, P2P traffic would be classified as anomalous. However, if the company decides to take advantage of some P2P platform to provide a new service, the anomalous P2P traffic would be extremely difficult to differentiate from the allowed one.

In computer networks, anomaly detection has been a subject of study for decades and many approaches have been explored [17, 18]. The deep learning perspective, however, has received special attention in recent years [19].

Deep-learning anomaly detection techniques typically act as classifiers that distinguish between normal and anomalous classes. These techniques operate in one of the following three modes:

Supervised detection requires a dataset with traffic labeled as normal or anomalous. The main issue is how to build a really comprehensive training set with all the anomalous traffic properly labeled. These datasets are not usually available and they can be difficult to collect and maintain [20]. There are several reasons for this, such as the excessive effort needed to collect such data, the level of expert knowledge that would be necessary for the analysis and even the issues with the users' privacy. In fact, there are different legal aspects in collecting all the traffic from the network of a company or institution.

Unsupervised detection does not require any labeled training. The goal in this technique is typically to discover groups of similar samples within the data (clustering). A question that arises is how to be certain that the identified classes correspond to the desired ones. Even if we have a large amount of data covering all the different scenarios and traffic patterns, we cannot blindly trust the results. However, when used for dimension reduction, these methods are well suited to extract discriminative higher-level features that can improve the classification performance of a supervised or semi-supervised algorithm.

Semi-supervised detection tries to estimate the probability distribution of the normal traffic from a sufficient amount of collected samples. This defines a tight boundary around the region (not necessarily convex) where a sample is classified as normal. The difference with respect to the supervised method is that there is no information about the shape of the anomalous region in the sample space. The new traffic is classified as anomalous in case it exceeds a threshold distance.

Among the unsupervised learning methods, Deep Belief Networks (DBN) [21] and Stacked Autoencoders (SAE) [22] have proved to be effective in learning invariant features from complex and high-dimensional datasets. The Restricted Boltzmann Machine (RBM) is the building block of a DBN, where each layer is separately trained in a greedy way as a RBM, that takes the input from the feature layer learned in the previous layer. SAE use the same idea to train stacked autoencoders in an unsupervised way, one at a time, to obtain a set of more descriptive low-dimension features. Both can be fine-tuned by means of backpropagation or Support Vector Machine (SVM) layers in a supervised way. They can also be configured as a semi-supervised one-class method, for example, adding a one-class SVM as a last layer [23, 24]. These semi-supervised one-class algorithms are well suited in anomaly detection, where the set of anomalous traffic captured is usually much smaller than the set of normal traffic. They can also be used in a prior phase to detect background traffic outliers, which could give us some useful insights into the traffic characteristics.

Training deep learning methods is a costly process because they need a great amount of data and iterations to converge. However, these methods usually outperform other classic algorithms. Additionally, they exhibit highly parallel computation patterns in prediction mode that take great advantage of GPU's computing power. We are interested in time-effective solutions that offer sufficient accuracy with a low evaluation runtime, to process the features coming from the large volumes of input information expected in 5G mobile networks

4. Proposal for Network Cyber Threat Detection

To achieve effective network anomaly detection, we use a distributed system that integrates in 5G networks [25] following the ETSI NFV architecture [26], which has also been used as the basis for other proposals [27, 28].

The system is depicted in Fig. 8, which is made up of two Virtualized Network Functions (VNF)—Anomaly Symptom Detection (ASD) and Network Anomaly Detection (NAD). The former is located within the Radio Access Network (RAN) infrastructure. It focuses on the quick search of anomaly symptoms, where a symptom is basically a sign of anomaly detected in the traffic. The latter is a collector of timestamped and RAN-associated symptoms, where a central process analyzes the timeline and the relationship among these symptoms to identify any network anomaly. Once an anomaly is detected, it is immediately communicated to the monitoring and diagnoser module.

Our proposal is flexible because it can deploy new virtualized resources to continue detecting anomaly symptoms when network traffic increases. It is also extensible, since the symptom detection is performed in a distributed way in the RANs, while the anomaly detection is performed in a centralized way in the core network, known as Evolved Packet Core (EPC).

Regarding ASD and NAD functions, we propose using deep-learning techniques to analyze network flows. We divide flow accounting in two-step process: flow export and flow collection. The former is responsible for generating the flows, while the latter aggregates flows to compute a feature vector, suitable for the anomaly detection task. However, a prime design decision is to determine the set of network flow features to be computed in order to train our system.

The anomaly detection problem can be approached from a wide range of methods. In particular, we propose to use deep learning because it has been shown to be a useful choice in other similar problems, and because it has a set of features that are well aligned with the new requirements based on the KPIs identified for 5G networks.

In our architecture, the anomaly detection is arranged in two levels, as shown in Fig. 9. At the low level, the flow collector gathers all the different flows during a given period of time and calculates a vector of features that the ASD module will classify as anomalous or normal. This initial classification has to be made as quickly as possible, even at the expense of sacrificing accuracy for a lower response time. If an anomaly is suspected, a symptom packet composed of the feature vector involved, a time stamp and the type of anomaly detected (when a multi-class method is being used), is sent to the next level, the NAD module.

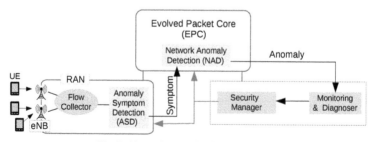

Fig. 8: Network anomaly detection system.

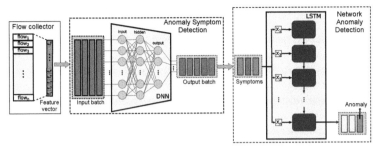

Fig. 9: Detail of the low-level (ASD) and high-level (NAD) modules.

The NAD receives several streams of symptoms from all the ASDs, sorts them by their time stamps and assembles their time sequence. The task of deciding whether this sequence belongs to any of a set of attack categories can be seen as a sequence-of-symptoms classification problem.

4.1 Suitable Deep Learning Models

Focusing on NAD, the model selected to detect anomalies was LSTM. The main reason is that the NAD module has less strict time restrictions if the ASDs have sufficient classification performance. In our research we are mainly interested in the ASD module, as its runtime performance is critical.

Regarding the ASD module, the deep learning models selected to be used were DBN and SAE. Both models share the same structure and can be used in supervised and unsupervised learning. We can use a DBN or SAE model followed by a classification layer if a labeled dataset is available. Otherwise, we propose to use DBN as a semi-supervised method, training with the normal network traffic and using DBN as a sort of Discriminative RBM without any further back-propagation layer [29].

Using a DBN or SAE followed by an SVM as a final layer is proposed by some authors [23]. This model can be used as a supervised two-class method or as a semi-supervised one-class method. This approach is discarded because it is not well suited to large-scale high-dimensional datasets.

We evaluated three different supervised DBN architectures. They have one, three and six hidden layers, respectively. As a final layer we chose a classification layer with a sigmoid activation function that gives the probability of the flow being normal or abnormal. These models are assumed to be trained with a labeled dataset, having two classes—*normal* or *anomalous*.

4.2 Network Flow Features

As mentioned in previous sections, machine-learning features need engineering by hand. In order to avoid the domain expertise, we try to automatically compute discriminative features from an input vector with many metrics.

Flow export requires NetFlow that defines how flow records are transported to the flow collector. NetFlow records contain for each flow—source and destination IP addresses and ports, start and end timestamps, type of service, level 3 protocol, TCP flags, next hop router, input and output SNMP interfaces, source and destination autonomous systems and network masks. In addition, each flow record carries aggregated information about the number of packets and bytes exchanged.

Each flow record could be directly sent from the flow collector to the symptom detection module. However, the flow record is too simple and only allows extracting of a few features. A step forward is to obtain aggregated views from a window of flow records [24]. In this case, a timer and/or accounting function triggers the aggregation process after receiving a certain number of flows (offset). The aggregated view is computed, taking one or more time periods in a multi-scale way (e.g., three feature vectors could be obtained considering 5-minute, 1-minute and 1-second periods).

The required features can be different, depending on the anomaly that we want to detect; for example, if we want to detect TCP anomalies, we only need TCP features. In our experiments we identified 288 features that were calculated from the features presented in Table 1. In order to generate this features we considered two subset of flows—one whole set of TCP/UDP flows and another with TCP/UDP flows filtered so that their source/destination IPs belong to the flows within the offset.

Among the selected TCP/UDP features, we highlighted the features related to calculate the entropy and variance of NetFlow records. Firstly, the entropy of source IPs for incoming flows and the entropy of destination IPs for outgoing flows allow features regarding how the corresponding IPs appear in the set of flows. The entropy of source ports for incoming flows and the entropy of destination ports for outgoing flows allow similar features related to network ports. Secondly, the variances of the number of IP packets and total bytes for incoming, outgoing and total flows allow features regarding how different the flows are in the analyzed set.

Table 1: Selected TCP/UDP features computed from the network flows for the classification test.

Features (t)
Number of flows, number of incoming flows, number of outgoing flows.
% of incoming and outgoing flows over total.
% of symmetric and asymmetric incoming flows over total.
Sum, maximum, minimum, mean and variance of IP packets per incoming, outgoing and total flows.
Sum, maximum, minimum, mean and variance of bytes per incoming, outgoing and total flows.
Sum, maximum, minimum, mean and variance of source bytes per incoming, outgoing and total flows.
Number of different source IPs for incoming flows and destination IPs for outgoing flows.
Number of different source and destination ports for incoming and outgoing flows.
Entropy of source IPs for incoming flows and destination IPs for outgoing flows.
Entropy of source and destination ports for incoming and outgoing flows.
% of source and destination ports > 1024 for incoming and outgoing flows.
% of source and destination ports ≤ 1024 for incoming and outgoing flows.

5. Experimental Results

The rather complex 5G scenario imposes important constraints on the runtime consumed in evaluating the model. Therefore, the experiments are aimed at not only finding a deep learning model with sufficient classification accuracy, but also an architecture with a low evaluation runtime. In this section, the ASD module depicted in the Fig. 9 is analyzed from two different perspectives—its classification accuracy and its model evaluation runtime.

In Section 3 the difficulty of dealing with the general concept of anomaly was discussed. Bearing this in mind, the scope of the proposed anomaly detection method needed to be constrained. It is well known that anomalies in communication networks are commonly related to malicious behavior of computers infected by malware forming botnets. This fact, together with the availability of recent public datasets with realistic labeled traffic, made us focus on botnets attack detection. Therefore, for this experimental section, a supervised method has been chosen, leaving to future work the discussion of an unsupervised approach.

5.1 The CTU Dataset

CTU [30] is a publicly available dataset, captured in the CTU University, Czech Republic, in 2011 and suitable to be used in this context. It tries to accomplish all the good requirements to be considered as a good dataset—it has real botnets attacks and not simulations, unknown traffic from a large network, ground-truth labels for training and evaluating as well as different types of botnets. The CTU dataset comprises 13 scenarios with different numbers of infected computers and seven botnet families. The traffic of each scenario was captured in a pcap file that contained all the packets and was post-processed to extract other types of information. The number of packets and total size of the captured traffic and the bot present in each scenario can be seen in Table 2.

Although a detailed description of the different scenarios is available at [31], we briefly present some of them. Scenario 1 is based on the Neris botnet running for 6.15 hours. This botnet used an HTTP-based C&C channel and tried to send SPAM. In the scenario 3, the Rbot botnet ran for several days and scanned large network ranges. The Murlo botnet was run in the scenario 8 for 19.5 hours and tried to connect to several remote C&C systems. The botnets in the rest of scenarios did other actions, like UDP and ICMP DDoS, scan web proxies or P2P connections.

Table 2: CTU botnet scenarios.

Dataset	Packets	Size	Bot
Scenario 1	71 971 482	52 GB	Neris
Scenario 2	71 851 300	60 GB	Neris
Scenario 3	167 730 395	121 GB	Rbot
Scenario 4	62 089 135	53 GB	Rbot
Scenario 5	4 481 167	37.6 GB	Virut
Scenario 6	38 764 357	30 GB	Menti
Scenario 7	7 467 139	5.8 GB	Sogou
Scenario 8	155 207 799	123 GB	Murlo
Scenario 9	115 415 321	94 GB	Neris
Scenario 10	90 389 782	73 GB	Rbot
Scenario 11	6 337 202	5.2 GB	Rbot
Scenario 12	13 212 268	8.3 GB	NSIS.ay
Scenario 13	50 888 256	34 GB	Virut

5.2 Classification Performance

In our classification experiments that support our decision to use a simple deep learning model, we need a model to be tested. Any of the models mentioned in 4 could have been selected; however, we decided to use a DBN followed by a classification layer.

First, feature vectors are created from the CTU dataset by taking aggregated views of 30 and 60 seconds after receiving every network flow. In this way, if the last received network flow is anomalous, then the corresponding feature vector will be labeled as anomalous as well.

Unfortunately, this sort of dataset is highly unbalanced as the percentage of anomalous traffic is usually smaller in a real network. This makes the learning process more challenging because, in this case, neural networks tend to classify all traffic as normal and consider anomalies as noise, achieving an accuracy close to 100 per cent. In this case, we are using a weighted loss function to compensate for the different label frequencies. In addition, we use more appropriate metrics than accuracy: the pair (*precision, recall*), or more concisely the *F1 score*. The entries in the confusion matrix are denoted as:

True Positive (TP): This entry refers to the number of positive examples which are correctly predicted as positives.

True Negative (TN): It denotes the number of negative examples correctly classified as negatives.

False Positive (FP): This entry is defined as the number of negative examples incorrectly classified as positives.

False Negative (FN): It is the number of positive examples incorrectly assigned as negatives.

And the proposed error metrics are:

Precision: Indicates what per cent of positive predictions were correct.

$$Precision = \frac{TP}{TP + FP}$$

Recall or sensitivity: Defines what per cent of positive cases a classifier predicted correctly.

$$Recall = \frac{TP}{TP + FN}$$

F1 score: Shows the trade-off between the precision and recall regarding the positive class.

$$F1\ score = 2 \times \frac{Precision \times Recall}{Precision + Recall}$$

In our experiment, we use two different training/test partitions. The first partition is built from the whole dataset splitting it into training, validation and testing sets, while the second dataset is built using the same partition suggested by the authors of the CTU dataset [30]. The method they used divides the whole dataset in two subsets: training and testing. The division process considers the following restrictions: the training and cross-validation subset should be 80 per cent approximately of the whole dataset; the testing subset should be 20 per cent of the whole dataset and none of the botnet families used in the training and cross-validation dataset should be used in the testing dataset.

Bearing all this in mind, we wanted to find a simple deep-learning model with sufficient classification performance. We restricted the search domain to three DBNs with up to six hidden layers and *ReLu* as activation function, an input vector of 288 features, batch normalization and a binary classifier as output layer (anomaly vs. normal). The activation function of the output was a sigmoid and cross-entropy was the cost function used. The hyperparameters of each of the given networks were: learning rate ∈ [0.001, 0.5], dropout ∈ [0, 0.4] and L2-regularization ∈ [0, 0.2]. Hyperparameter tuning was carried out by means of 10-fold cross-validation and randomized search.

When trained and tested with all the botnets, the 128-64-32 model achieved the highest F1-score, with a precision of 0.9537 and a recall of 0.9954; that is, 95.37 per cent of the anomaly predictions were actually anomalies, and 99.54 per cent of the actual anomalies were correctly classified (Table 3). Regarding the models tested with unknown scenarios, the 16-8-4 model obtained the better generalization results in spite of its simplicity. The evaluation of the trained model reached a precision of 0.6861 and a recall of 0.7096 on an average. Approximately 71 per cent of the anomalous traffic was correctly detected, even though the model had never seen those botnets before. The global scores of each model evaluated on the unknown scenarios are presented in Table 4. In addition, a breakdown by unknown scenario of the scores obtained by the 16-8-4 model is provided in Table 5.

Table 3: Classification results of the three selected deep models regarding known botnets.

Hidden Layers	LR	λ_{L2}	Precision	Recall	F1-score
16-8-4	0.01	0.00	0.8311	0.9947	0.9055
128-64-32	0.01	0.00	**0.9537**	**0.9954**	**0.9741**
128-128-64-64-32-32	0.1	0.01	0.9534	0.9904	0.9715

Table 4: Classification results of the three selected deep models regarding unknown botnets.

Hidden Layers	LR	λ_{L2}	Precision	Recall	F1-score
16-8-4	0.01	0.00	0.6861	**0.7095**	**0.6976**
128-64-32	0.01	0.01	0.6910	0.6775	0.6842
128-128-64-64-32-32	0.01	0.01	**0.8693**	0.4803	0.6187

Table 5: Breakdown of classification results by unknown scenario for the 16-8-4 model.

Test Set	Precision	Recall	F1-score
Scenario 1	0.742	0.902	0.814
Scenario 2	0.631	0.544	0.584
Scenario 6	0.724	0.937	0.812
Scenario 8	0.410	0.756	0.531
Scenario 9	0.941	0.300	0.610

Although the precision obtained is not too high, it is expected that the second-level detector in the NAD module substantially improves it. This module is in charge of detecting complex patterns in the temporal series of symptoms received from all the ASDs.

5.3 Execution Performance

There exist a variety of commercial deep-learning frameworks and libraries. To train the neural network proposed, the selected framework must support LSTM Recurrent Networks (the model used in NAD module) and DBNs, besides having multiple-GPU support. The selected deep-learning libraries to test our performance benchmark were: TensorFlow 1.4 [32], Caffe2 0.8.1 [33, 34], Theano 1.0.0rc1 [35], PyTorch 0.2.0 4 [36], MXNet 0.11.0 [37] and CNTK 2.2 [38]. All of them have been widely used in different kinds of R&D and commercial projects and in different scenarios. In addition, the selected frameworks are open-source projects and all of them except Theano are supported by prominent high-tech companies. Caffe2 and Pytorch are sponsored by Facebook; Tensorflow is supported by Google; CNTK, by Microsoft; and MxNet, by Amazon. Theano is considered one of the first deep-learning frameworks and it was developed by a research group of the University of Montreal.

A common feature of all the selected frameworks is to provide developers with a Python API, allowing to prototype quickly. Another relevant feature is that most of the frameworks also offer a C API that enables to compile to machine code ready for deployment in production. We use both Python and C API in the deployment of our experiments.

In these experiments, the performance of the above frameworks has been evaluated by using a DBN. We have included in the evaluation our own implementation on cuBLAS [39] as a baseline solution with minimal overhead. The DBN model keeps the same input and output size for each experiment (i.e., the input layer has the same size as the feature vector, and the output layer size is 2). Softmax is then used to estimate the likelihood of belonging to the normal or anomalous class.
The chosen architectures are:

- One hidden full layer of 128 floats.
- Three hidden full layers of 128, 64 and 32 floats.
- Six hidden full layers of 128, 128, 64, 64, 32 and 32 floats.

The model can be evaluated on either the CPU or the GPU, and the batch of feature vectors is assumed to arrive at the system's memory at maximum speed. Therefore, if the model is evaluated by the CPU, the batch is directly accessible. Conversely, if the model is evaluated by the GPU, it is necessary to transfer the batch to the GPU's memory.

Several factors have to be taken into account in order to reach the best performance. The batch size has a great influence on the execution time, regardless of which processor is running the model. If the model is evaluated by the CPU, a big batch takes advantage of the Advanced Vector Extensions. However, above a certain batch size, there will probably be more cache/TLB misses and page faults, resulting in a worse throughput. Conversely, if the model is evaluated by the GPU, increasing the batch size can decrease the execution time as a greater percentage of the available processing units can be used simultaneously. However, in this latter case, the time spent on transferring the batch to the GPU's memory also increases. In this experimental work, we used only one GPU so that the batch size was limited by the GPU's memory size.

We carried out the performance evaluation using a workstation with 32 GB of RAM, a six-core Intel i7-5930K at 3.5 GHz with hyper-threading running Linux, and one NVIDIA GeForce GTX 1080 with 8 GB RAM.

5.3.1 Determining the Optimum Batch Size and Framework

Although most of deep-learning frameworks are designed to work as a computation graph, we have decided to take a deep-learning framework as a black box and determine the more suitable framework by using a benchmark executed during the installation procedure of our software on each RAN. This benchmark runs several tests to find out the best batch size for a given framework. Performance is measured for all the feasible combinations of the tuple (*model, framework, vector size, batch size*). Here, *vector size* is the number of components of the feature *vector; batch size* is the number of feature vectors in the batch; *model* is the trained deep learning model; and *framework* is the deep learning framework used to train and evaluate the model. These models and frameworks have been described above.

For the shake of clarity, Fig. 10 only shows the resulting performance for the six hidden-layer neural network running on the GPU. The performances of the one and three hidden-layer neural networks have no significant behavior change except scale. The figure shows that the speedup, as a function of the vector size when the batch size is fixed, is not linear. There might be several reasons for this fact—memory transfer time, computing time, number of GPU threads or CUDA's grid and block dimensions, among others. However, our goal is not to provide an insightful description of this behavior but tuning our software to maximize its performance. Specifically, Fig. 10

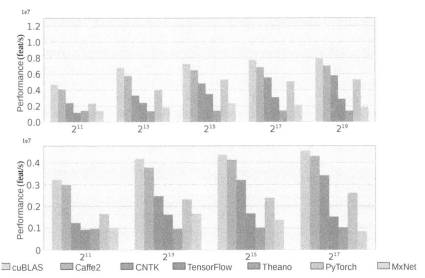

Fig. 10: Six hidden layer neural network (GPU) performance for a wide range of batch sizes and two selected feature vector sizes when running on the GPU. (*Top*) 128-feature vector (*Bottom*) 256-feature vector.

presents the marks obtained by every framework for an input vector of 128 and 256 features, and a variety of batch sizes.

As seen in Fig. 10, the best performance for all vectors and batches sizes is our own implementation using cuBLAS, followed closely by Caffe2. Given that all the analyzed frameworks use cuBLAS for matrix operation, our ad hoc implementation of the models can be considered as an upper bound of the achievable throughput. Our version does not suffer from the overhead imposed by the abstraction layers that the frameworks need in order to provide flexibility and extended functionality.

Examining the behavior of each framework, cuBLAS's best marks are 0.8 and 0.46 million feature vectors per second for 128 and 256 vector size, respectively. It is followed closely by Caffe2, with a small performance difference that becomes even smaller when batch size is increased. In contrast, MxNet and Theano obtain the worst marks; Theano achieves its best result (0.15 million feature vectors per second) with a batch size of 2^{17} and a vector size of 128; while MxNet obtains 0.21 million feature vectors per second as its best marks for a batch size of 2^{15} and a vector size of 128. It should be noticed that Theano is a special case because the performance remains almost unchanged for all batch sizes with the vector size fixed.

Although the number of features can be potentially unlimited, we set the feature vector to 256 elements. This number was chosen because it is a representative value considering the previous ASD classification performance (see Section 4.2). With this input vector size and running the models on our workstation, the best batch sizes in the three DBN architectures are shown in Table 6.

The cuBLAS implementation can be considered as the maximum throughput achievable by using the frameworks. However, fixing the vector size to 256 and taking into account only commercial frameworks, the best performance is offered by Caffe2 for all batch sizes. For one-layer architecture and a batch size of 2^{17} it reaches 5.1 million feature vectors per second, while the second best option is CNTK offering 3.8 million feature vectors per second. On the other hand, MxNet and Theano compete for the worst position in the ranking, providing poor results in almost every configuration when compared with Caffe2 and PyTorch. Particularly, in this configuration MxNet reaches its best mark of 1.54 million feature vectors per second for the batch size 2^{13}. Theano obtains 1.64 million feature vectors per second for a batch size of 2^{17}.

We should bear in mind that these results have been obtained by running the benchmark on the specific workstation described above. Therefore, the marks could

Table 6: Optimum batch size.

Framework	One Hidden Layer	Three Hidden Layers	Six Hidden Layers
TensorFlow	2^{15}	2^{17}	2^{15}
Caffe2	2^{17}	2^{17}	2^{17}
Theano	2^{17}	2^{17}	2^{17}
PyTorch	2^{11}	2^{17}	2^{17}
MxNet	2^{15}	2^{13}	2^{13}
CNTK	2^{17}	2^{17}	2^{17}
cuBLAS	2^{17}	2^{17}	2^{17}

be rather different with other existing hardware configurations, where the virtualized computational resources available in each RAN can result in a different optimal combination of deep-learning framework and batch size. This makes it crucial to run the benchmark suite on each RAN at installation stage and determine the best framework that suits each particular RAN in the 5G network. It should also be noted that, although Caffe2 running on a GPU provides the best performance, it could not be the most suitable option. As a case in point, a system with low traffic can decide not to use a GPU due to power consumption policy. In this case, TensorFlow on a CPU would be the best option. Additionally, other system configurations could be considered, such as frameworks using more than one GPU. In any case, our architecture allows self-adaptation to every hardware context.

5.3.2 CPU vs GPU Performance

GPUs have become an essential tool for machine-learning computation, because deep-learning models are actually well suited to be executed on them. The more complex and independent the computations are, the more impressive the performance gain usually is.

In our particular case, the time spent in transferring the batch from main memory to the GPU cannot be disregarded when compared with the computing time. If the batch is small, the number of memory transfers increases, limiting the throughput. Conversely if the batch is large, the memory becomes the most common limitation. Figure 11 illustrates how some of the frameworks improve as the batch size increases. In contrast, some others even get worse in the same case. GPU shows for every

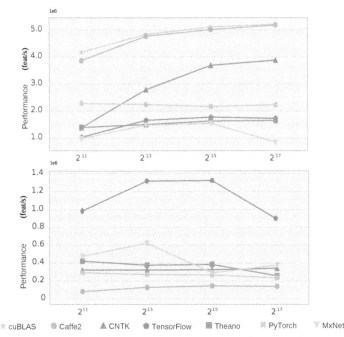

Fig. 11: Comparison between GPU (*Top*) and CPU (*Bottom*) performance with a vector of 256 features for one hidden layer architecture.

framework a significant performance improvement compared to CPU. It is important to note that cuBLAS is a library running on GPU, so the CPU benchmark does not include it.

It is worth mentioning that TensorFlow running on CPU gets a better mark than some frameworks running on GPU. In particular, for one hidden layer architecture with 2^{13} and 2^{15} batch sizes, TensorFlow behaves better than MxNet and similar to Theano. In contrast, Caffe2, which gets the best mark on GPU, obtains the worst mark on CPU. In general, all the frameworks running on CPU show a performance lower than 1 million feature vectors per second, with the TensorFlow exception, which reaches up to 1.3 million feature vectors per second in the one-layer architecture.

5.4 Discussion

In order to contextualize the experimental results, let us consider a real 5G-scenario of a city with a 5 RAN and 50 eNBs per RAN (elements of the LTE Radio Access Network for connectivity purpose) and 10 millions of habitants with 1 UE each of them. Let us assume that each RAN handles 2 millions inhabitants and each connected UE is generating 10 flows per second on an average and that there is a perfect balancing of 40000 UEs per eNB. In this configuration, each eNB will manage 400000 flows per second. If we assume an ASD per eNB and an offset of one flow will result in having 400000 feature vectors per second sending to each ASD. According to the experiment results, this load can be managed on CPU by TensorFlow using any batch size or MxNet using 2^{11} or 2^{13} batch size. Similarly, if we assume that there are five eNBs sending data to the ASD, instead of only one, we obtain 2 million feature vectors per second. In this case, the system should select a framework running on GPU as any framework running on CPU can handle the traffic load. The suitable frameworks are: PyTorch, CNTK, Caffe2 and the implementation based on cuBLAS as shown in Fig. 11.

The worst scenario would be when 13 or more eNBs are simultaneously sending data to the ASD. In this situation, a framework change wouldn't be enough even if the GPU is used. As a solution, two mechanisms are suggested: (a) to deploy dynamically more computational resources through new virtualized components as VNFs (for example, using an additional available GPU); and (b) to adapt the flow aggregation frequency in order to reduce the rate of collected feature vectors. This can be made by increasing the offset value.

These realistic scenarios illustrate the capacity of the system to adapt to the traffic circumstances. When the traffic load is low, that is, few UEs are connected to the eNBs and sending data simultaneously, the system can select a framework with a CPU performance sufficient to handle the traffic. When the traffic increases for any reason (for example, the course of an event, such as a football match or a music festival) and more UEs are connected and send data simultaneously in a short period of time, the system can switch to another framework that offers best performance. Conversely, if there is no suitable framework, the system can decide to use the GPU, either keeping the same framework or substituting it with another more suitable. Finally, when the traffic load is too high and no framework can handle it, the system can deploy more virtualized resources, if available.

It should be noted that the system behavior strongly depends on the available hardware in the RAN. Therefore, the marks obtained by each framework could be totally different with other hardware configuration. This makes it crucial to run the benchmark suite on each RAN at installation phase in order to measure the performance of each framework with a wide range of batch sizes.

6. Conclusion

A system for anomaly detection in the context of a 5G mobile-network architecture has been proposed. This proposal is based on a novel two-level deep machine-learning model where the first level is a supervised or semi-supervised learning method implementing a DBN or a SAE running on every RAN as quickly as possible. This sacrifices some accuracy due to the amount of network traffic that a RAN handles. Its task is to detect symptoms, that is, local anomalous traffic conditions happening during a configurable short time period. All the collected symptoms are sent to the NAD component, where they are assembled and used as input for a LSTM Recurrent Network, trained in a supervised way to recognize temporal patterns of cyber attacks.

A number of experiments were also conducted to realize a comprehensive comparative performance evaluation of deep-learning frameworks. The aim was to determine the best suited one for reaching the highest processing performance. The experimental results showed that each system in the 5G network were able to use one framework or another depending on, for example, the availability of physical resources (GPU or CPU) in a given RAN or the network flow reception rate.

As future work, several aspects can be considered. First, it is necessary to train the ASD and NAD module using real traffic from a 5G network and test the anomaly detection accuracy, taking the system as a whole. Second, another important issue is the efficiency of the feature vector creation which is carried out in the flow collector from a window of network flows. Some of the statistical features cannot be computed incrementally as the new flows arrive, making their calculation computationally expensive. Therefore, we need to develop an efficient implementation that enables us to generate the feature vector in real time. Finally, a last issue is the integration of our anomaly detection method based on deep-learning techniques, with deep packet inspection (DPI) tools, such as Snort. When the system detects an anomaly, the network flows involved should be stored in an efficient way, e.g., using an high-performance database. The DPI tool can then analyze them to determine more precisely the type of anomaly.

References

[1] Gupta, B., Agrawal, D.P. and Yamaguchi, S. (2016). Handbook of Research on Modern Cryptographic Solutions for Computer and Cyber Security. IGI Global.

[2] Sourcefire, Inc. (2018). Snort: An open source network intrusion detection and prevention system. http://www.snort.org.

[3] Richariya, V., Singh, U.P. and Mishra, R. (2012). Distributed approach of intrusion detection system: Survey. International Journal of Advanced Computer Research 2(6): 358–363.

[4] Raza Shah, S.A. and Issac, B. (2018). Performance comparison of intrusion detection systems and application of machine learning to Snort system. Future Generation Computer Systems 80: 157–170.

[5] Yu, J., Yang, B., Sun, R. and Chen, Y. (2009). FPGA-based parallel pattern matching algorithm for network intrusion detection system. pp. 458–461. *In*: 2009 International Conference on Multimedia Information Networking and Security.

[6] Hsiao, Y.M., Chen, M.J., Chu, Y.S. and Huang, C.H. (2012). High-throughput intrusion detection system with parallel pattern matching. IEICE Electronics Express 9(18): 1467–1472.

[7] Buczak, A.L. and Guven, E. (2016). A survey of data mining and machine learning methods for cyber security intrusion detection. IEEE Communications Surveys & Tutorials 18(2): 1153–1176.

[8] Gardiner, J. and Nagaraja, S. (2016). On the security of machine learning in malware C&C detection: A survey. ACM Computing Surveys 49(3): 59:1–59:39.

[9] Stergiou, C., Psannis, K.E., Kim, B.G. and Gupta, B. (2018). Secure integration of iot and cloud computing. Future Generation Computer Systems 78: 964–975.

[10] The 5G Infrastructure Public Private Partnership. (2016). Key Performance Indicators (KPI). http://5g-ppp.eu/kpis.

[11] Alomari, E., Manickam, S., Gupta, B., Karuppayah, S. and Alfaris, R. (2012). Botnet-based distributed denial of service (ddos) attacks on web servers: classification and art. arXiv preprint 1208.0403.

[12] Goodfellow, I., Bengio, Y. and Courville, A. (2016). Deep Learning. MIT Press.

[13] Glorot, X., Bordes, A. and Bengio, Y. (2011). Deep sparse rectifier neural networks. pp. 315–323. *In*: Gordon, G., Dunson, D. and Dudk, M. (eds.). Proceedings of the Fourteenth International Conference on Artificial Intelligence and Statistics, PMLR, Fort Lauderdale, FL, USA, Proceedings of Machine Learning Research, Vol. 15.

[14] LeCun, Y., Bottou, L., Orr, G.B., Müller, K.R. (1998). Efficient backprop. pp. 9–50. *In*: Neural Networks: Tricks of the Trade, This Book is an Outgrowth of a 1996 NIPS Workshop, Springer-Verlag, London, UK.

[15] Lecun, Y., Bottou, L., Bengio, Y. and Haffner, P. (1998). Gradient-based learning applied to document recognition. pp. 2278–2324. *In*: Proceedings of the IEEE.

[16] Hochreiter, S. and Schmidhuber, J. (1997). Long short-term memory. Neural Comput. 9(8): 1735–1780.

[17] Chandola, V., Banerjee, A. and Kumar, V. (2009). Anomaly detection: A survey. ACM Computing Surveys 41(3): 15:1–15:58.

[18] Garcia-Teodoro, P., Diaz-Verdejo, J., Maciá-Fernández, G. and Vázquez, E. (2009). Anomaly-based network intrusion detection: Techniques, systems and challenges. Computers & Security 28(1-2): 18–28.

[19] Hodo, E., Bellekens, X., Hamilton, A., Tachtatzis, C. and Atkinson, R. (2017). Shallow and deep networks intrusion detection system: A taxonomy and survey. arXiv preprint 1701.02145.

[20] Aviv, A.J. and Haeberlen, A. (2011). Challenges in experimenting with botnet detection systems. pp. 6. *In*: 4th Conference on Cyber Security Experimentation and Test.

[21] Salakhutdinov, R. and Hinton, G.E. (2009). Deep Boltzmann machines. pp. 448–455. *In*: Artificial Intelligence and Statistics.

[22] Hinton, G.E. and Salakhutdinov, R.R. (2006). Reducing the dimensionality of data with neural networks. Science 313(5786): 504–507.

[23] Erfani, S.M., Rajasegarar, S., Karunasekera, S. and Leckie, C. (2016). High-dimensional and large-scale anomaly detection using a linear one-class SVM with deep learning. Pattern Recognition 58: 121–134.

[24] Niyaz, Q., Sun, W. and Javaid, A.Y. (2016). A deep learning-based DDoS detection system in Software-Defined Networking (SDN). arXiv preprint 1611.07400.

[25] Fernández Maimó, L., Perales Gómez, A., García Clemente, F., Gil Pérez, M. and Martínez Pérez, G. (2018). A self-adaptive deep learning-based system for anomaly detection in 5G networks. IEEE Access 6: 7700–7712.

[26] ETSI NFV ISG. (2017). Network Functions Virtualisation (NFV); Network Operator Perspectives on NFV Priorities for 5G. Tech. rep.

[27] Siddiqui, M.S. et al. (2016). Hierarchical, virtualised and distributed intelligence 5G architecture for low-latency and secure applications. Transactions on Emerging Telecommunications Technologies 27(9): 1233–1241.

[28] Neves, P. et al. (2017). Future mode of operations for 5G-The SELFNET approach enabled by SDN/NFV. Computer Standards & Interfaces 54(4): 229–246.

[29] Fiore, U., Palmieri, F., Castiglione, A. and De Santis, A. (2013). Network anomaly detection with the restricted Boltzmann machine. Neurocomputing 122: 13–23.

[30] Garcia, S., Grill, M., Stiborek, J. and Zunino, A. (2014). An empirical comparison of botnet detection methods. Computers & Security 45: 100–123.

[31] Czech Technical University ATG Group. (2018). Malware capture facility project. https://mcfp.weebly.com.

[32] Abadi, M. et al. (2016). Tensorflow: Large-scale machine learning on heterogeneous distributed systems. arXiv preprint 1603.04467.

[33] Jia, Y., Shelhamer, E., Donahue, J., Karayev, S., Long, J., Girshick, R., Guadarrama, S. and Darrell, T. (2014). Caffe: Convolutional architecture for fast feature embedding. pp. 675–678. *In*: 22nd ACM International Conference on Multimedia.

[34] Facebook Open Source. (2017). Caffe2: A new hightweight, modular, and scalable deep learning framework. http://caffe2.ai (accessed April 17, 2018).

[35] Theano Development Team. (2016). Theano: A python framework for fast computation of mathematical expressions. arXiv preprint 1605.02688.

[36] Collobert, R., Kavukcuoglu, K. and Farabet, C. (2011). Torch7: A Matlab-like environment for machine learning. *In*: BigLearn, NIPS Workshop, EPFL-CONF-192376.

[37] Chen, T., Li, M., Li, Y., Lin, M., Wang, N., Wang, M., Xiao, T., Xu, B., Zhang, C. and Zhang, Z. (2015). Mxnet: A flexible and efficient machine learning library for heterogeneous distributed systems. arXiv preprint 1512.01274.

[38] Seide, F. and Agarwal, A. (2016). CNTK: Microsoft's open-source deep-learning toolkit. pp. 2135–2135. *In*: 22nd ACM SIGKDD International Conference on Knowledge Discovery and Data Mining.

[39] NVIDIA Corporation. (2016). Dense Linear Algebra on GPUs. http://developer.nvidia.com/cublas.

CHAPTER 2

Machine Learning for Phishing Detection and Mitigation

Mohammad Alauthman,[a,*] *Ammar Almomani,*[b]
Mohammed Alweshah,[c] *Waleed Omoush*[d] and
Kamal Alieyan[e]

1. Introduction

Phishing e-mails refer to fraudulent e-mails sent with the aim of deceiving people and tricking them into revealing confidential information, such as passwords, usernames and credit card information. Such information may be related to bank accounts. That is done to obtain an illegal financial gain. Sending such e-mails is criminal. Phishing attacks include unknown 'zero-day' phishing attack.[1] Such attacks can cause severe losses. Although there are many defense mechanisms used for fighting against 'zero-day' phishing attacks, phishers are developing methods for getting past these mechanisms.

The present study aims at identifying and discussing machine-learning approaches for fighting against and detecting phishing e-mails. These approaches are used for filtering the phishing attacks. Researchers too have carried out assessment of several phishing e-mails filtering methods besides comparing them. Such a comparison was

[a] Department of Computer Science, Faculty of Information Technology, Zarqa University, Zarqa, Jordan.
[b] Department of Information Technology, Al-Huson University College, Al-Balqa Applied University, P.O. Box 50, Irbid, Jordan; ammarnav6@bau.edu.jo
[c] Prince Abdullah Bin Ghazi Faculty of Information Technology, Al-Balqa Applied University, Salt, Jordan; weshah@bau.edu.jo
[d] Dept. of CS, Imam Abdulrahman Bin Faisal University, Dammam, Saudi Arabia.
[e] National Advanced IPv6 Centre (NAv6), Universiti Sains Malaysia, 11800 Gelugor, Penang, Malaysia; kamal_alian@nav6.usm.my
* Corresponding author: malauthman@zu.edu.jo

[1] They refer to unknown attacks through which the attackers bypass security defenses. These attacks occur by exploiting the vulnerability existing in the targeted system.

conducted in terms of advantages and limitations. Thus, the present study aims to deepen one's understanding of matters related to phishing attacks besides creating awareness about the methods used for fighting against and filtering phishing e-mails.

2. Background and Overview of Phishing E-mails

Through this section, the researchers identified the types, life cycle and analyses of phishing attacks. They also identified the several options available to detect such e-mails. It is necessary to shed light on e-mail-related problems because e-mails are used by all governments, organizations, people and sectors for several goals. For instance, they may be used for sharing, distributing and organizing data [1–3]. Due to the significance of using e-mails, it is necessary to shed light on phishing e-mails and the approaches used for fighting against them.

Phishing e-mails refer to e-mail scams that represent part of a fraud scheme. Phishers usually disguise as a trustworthy entity (e.g., bank officer) and through e-mails, ask the targeted victims to open an embedded link that distributes malware. It should be noted that this problem is becoming more serious day by day. For instance, Gartner [4] suggests that 3.6 million users in USA lose money each year due to phishing e-mails. He also suggests that this problem leads to losing 3.2 billion US dollars per year in USA alone, In 2006, 2.3 million people fell victim to phishing e-mails and in 2007, the number rose to 6 million people.

According to e-Crime Trends Report [1], phishing attacks are increasing each year by a percentage of 12 per cent. Such an increase leads to increase in the losses incurred by companies and buyers. These e-mails also have a negative impact on electronic commerce because customers will no longer consider companies and on-line commerce websites trustworthy. Today, unknown 'zero-day' phishing e-mails are considered one of the biggest threats facing the world of business. This is because they are launched by unknown attackers by bypassing security defenses. These attacks occur by exploiting the vulnerability existing in the targeted system [2–5].

Phishing e-mails pose a serious problem. Thus, programmers have been developing several solutions to detect such e-mails, whiel phishers have been developing various methods to overcome such approaches and exploit the vulnerabilities of the targeted system [6]. Solutions so far have not proven to be effective [7]. Such solutions include communication-oriented approaches, such as authentication protocols. Such solutions also include content-based filtering approaches that sometimes rely on a computing (AI) technique [12]. Current algorithms discover phishing e-mails supported-fastened options and rules, whereas a number of machine-learning algorithms square measure to figure in online mode [8].

Through an online mode, data arrives one at a time. Upon the arrival of each datum, local learning is performed, while in offline mode, all the data set is available for global learning [8, 9]. Online mode is designed to work with internet connectivity. The error level in the classification process increases over time, particularly while addressing unknown zero-day phishing e-mails [10].

2.1 Phishing Attacks

Phishing attacks may be considered as a type of spam that employs two techniques— one is the deceptive phishing technique and the other is the malware-based phishing technique. Phishing e-mails refer to e-mail scams that represent part of a fraud scheme [11]. By using the deceptive technique, the phisher usually disguises as a trustworthy entity or person (e.g., bank officer) and through e-mail ask the targeted victim to open an embedded link. Through the link, the phisher seeks redirecting of the targeted victim to a fake website, in which the phisher asks the victim to provide confidential information related to bank account or credit card. This enables the phisher to obtain financial gain illegally.

Through use if the malware-based phishing technique, the phisher usually disguises as a trustworthy entity or person (e.g., bank officer) and through e-mail, asks the targeted victim to open an embedded link that distributes malware or deems a malicious code. Through such malware or code, the phisher can control the victim's system and obtain the confidential data directly. That is done by exploiting a vulnerability existing in the targeted system. Sometimes, the phisher tries to misdirect the user to a fake or legitimate website monitored by proxies [16]. Through this present study, researchers focus on deceptive phishing because it is one of the common phishing technique. Figure 1 presents the forms of phishing attacks [17].

Fig. 1: Types of phishing attacks.

2.2 Life Cycle of Phishing E-mails

The stages that represent the life cycle of phishing e-mails are presented in Fig. 2. By identifying such stages, one can understand the way the phishing attack occurs. It becomes clear that the phishing attack starts with someone sending the fraudulent e-mail that includes a link. This stage is similar to the fishing process because the criminal during this stage, sends e-mails to many people, hoping to find a vulnerable victim, who may visit the attached link. It should be noted that phishing e-mails usually look like a legitimate e-mail. A criminal disguises as a bank officer or corporate employee and usually provides the name and address of a legitimate company/bank. As for the attached link, it appears to be legitimate. All this is provided by phishers to trick people into taking the bite (i.e., opening the link) [12].

At the beginning, phishers used to disguise as representatives of legitimate well-known websites. On March 9, 2004, a phishing attack was reported. The phisher disguised as a representative of e-bay website. The e-mail of the phisher is presented in Fig. 3 [18]. The phisher claimed that the e-mail receiver had provided invalid information. The e-mail receiver was asked to update his information.

The phisher provided a link in the e-mail which facilitated access to a fake webpage appearing to be for e-Bay. This fake webpage asked the user to provide contact information, information about his social security and MasterCard. He was also asked to provide e-Bay username and countersign.

The e-mail appeared to be a legitimate e-mail sent by e-Bay and was represented through the e-mail of the sender (S-Harbor@eBay.com). The website appeared legitimate. For instance, the web address includes (http). The user found that the IP address of the fake webpage was 210.93.131.250 and belonged to the Republic of Korea. Thus, there was no relationship between this fake website and the actual website of e-Bay. Figure 4 presents a screenshot of this fake webpage [18]. However, there are different methods used for classifying e-mails into legitimate and phishing which enable the users to detect phishing e-mails. The researchers identified these methods as given below.

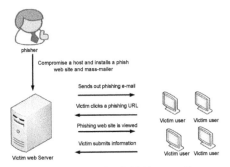

Fig. 2: Life cycle of phishing e-mail.

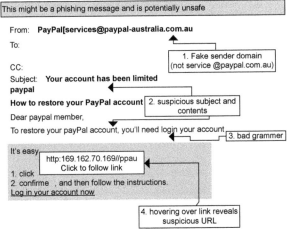

Fig. 3: An example of a phishing e-mail (the concerned phisher was disguised as a representative of e-bay website) [13].

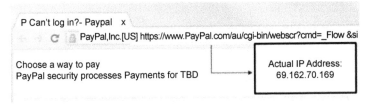

Fig. 4: Screenshot of the e-bay fake web page [13].

2.3 *Phishing E-mail Classification Methods (i.e., Methods for Filtering E-mails)*[2]

Filtering e-mails is a process through which e-mail messages are classified automatically into legitimate and phishing ones. This process is carried out by a phishing e-mail filter through which e-mail messages are analyzed and classified. E-mail messages may be processed separately which entails checking each e-mail to detect the presence of any distinct word. E-mail messages could also be filtered through a learning-based filter. This filter aims to analyze a set of labelled coaching information (pre-collected messages with upright judgments) [19]. The associate degree e-mails consist of 2 parts; the body and the header. E-mail headers include several fields (e.g., from, subject, to, etc.); in associate degree e-mails, the header lines are followed by the body. The header lines provide the routing data of the message explicitly besides supplying information about the subject, recipient and sender. As for the body, it provides the content of the intended message. Figure 5 presents the structure of the e-mail messages. As for Fig. 6, it presents an example of the structure of e-mail messages and are suggested for purposes related to feature extraction and selection [12].

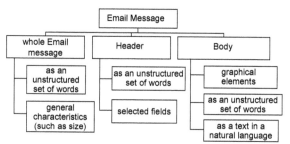

Fig. 5: Taxonomy of e-mail message structure [14].

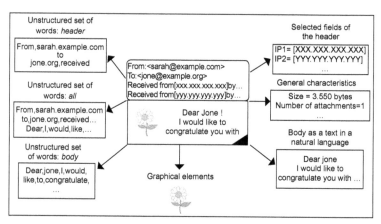

Fig. 6: An example of the structure of e-mail messages (suggested for purposes related to feature extraction and selection) [14].

---

[2] Methods for classifying e-mail messages into legitimate and phishing ones. Such methods enable the users to detect phishing e-mails.

In order to identify the steps of the pre-processing phase, the e-mail format must be identified. In addition, understanding the e-mail filtering techniques requires having an operating data about the e-mail format and structure. The structure of e-mail is defined in RFC 5322 [19]. Like ancient communicating mail, there is an associate degree envelope. However, users usually do not see the envelope because the e-mail systems throw away the envelope before delivering the e-mail message to the user. Figure 7 presents an example of an e-mail envelope. As for Fig. 8, it presents an example of an e-mail source code. Extra envelope recipients are defined with extra RCPT.

```
EHLO col0-omc4-s17.col0.hotmail.com
MAIL From:<rohanamin@live.com> SIZE=26103 BODY=8BITMIME
RCPT To:<rohan@rohanamin.com>
```

Fig. 7: An example of an e-mail envelope.

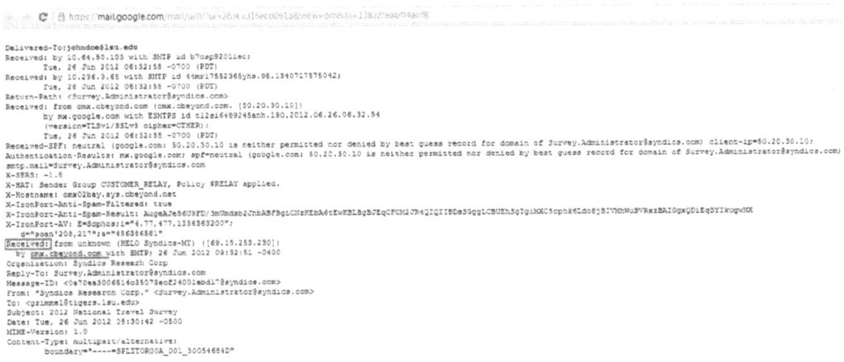

```
https://mail.google.com/mail/u/0/?ui=2&ik=315ec0091a&view=om&th=13823eab94acf8

Delivered-To:johndoe@lsu.edu
Received: by 10.64.50.103 with SMTP id b7csp92011ec;
        Tue, 26 Jun 2012 06:32:55 -0700 (PDT)
Received: by 10.236.9.65 with SMTP id 44mr17552368yhs.98.1340717575042;
        Tue, 26 Jun 2012 06:32:55 -0700 (PDT)
Return-Path: <Survey.Administrator@syndics.com>
Received: from omx.cbeyond.com (omx.cbeyond.com. [50.20.30.10])
        by mx.google.com with ESMTPS id t12si6489245anh.190.2012.06.26.06.32.54
        (version=TLSv1/SSLv3 cipher=OTHER);
        Tue, 26 Jun 2012 06:32:55 -0700 (PDT)
Received-SPF: neutral (google.com: 50.20.30.10 is neither permitted nor denied by best guess record for domain of Survey.Administrator@syndics.com) client-ip=50.20.30.10;
Authentication-Results: mx.google.com; spf=neutral (google.com: 50.20.30.10 is neither permitted nor denied by best guess record for domain of Survey.Administrator@syndics.com)
smtp.mail=Survey.Administrator@syndics.com
X-SBRS: -1.8
X-MAT: Sender Group CUSTOMER_RELAY, Policy $RELAY applied.
X-Hostname: omx02bay.sys.cbeyond.net
X-IronPort-Anti-Spam-Filtered: true
X-IronPort-Anti-Spam-Result: AugeAJe56U9FD/3mUmdsb2JnbABFBgiCNzKEbA6tEwKBLBgBJEqCFCM2JR4QIQIIBDeSGggLCBUEh3gIgiMXC5cphK6Ldo8jBIVMhWuBVRxzBAIGgxQDiEqBYIkugwNX
X-IronPort-AV: E=Sophos;i="4.77,477,1356363200";
   d="scan'208,217";a="456386581"
Received: from unknown (HELO Syndics-MT) ([69.15.253.230])
  by omx.cbeyond.com with SMTP; 26 Jun 2012 09:32:51 -0400
Organization: Syndics Research Corp
Reply-To: Survey.Administrator@syndics.com
Message-ID: <0a70ea3006614c35073ecf24001ebd17@syndice.com>
From: "Syndics Research Corp." <Survey.Administrator@syndics.com>
To: <gzimmel@tigers.lsu.edu>
Subject: 2012 National Travel Survey
Date: Tue, 26 Jun 2012 08:30:42 -0500
MIME-Version: 1.0
Content-Type: multipart/alternative;
        boundary="----=SPLITOR00A_001_30054684D"
```

Fig. 8: An example of e-mail source code.

HELO: The client sends HELO command to the SMTP server for identifying it and initiating the SMTP conversation.

From the above source code, the system shall extract 21 binary features. The e-mail message usually consists of two major elements—the body and the header. The body contains the message text and any additional attachment. Figure 8 presents an example of the associate degree of an e-mail. This is presented along with the body and the header. As for the e-mail source code, it consists of the following:

1. *Delivered-To*: It refers to the e-mail address that the message shall be delivered to.
2. *Received*: It refers to a list that includes all the servers through which the e-mail message travels through in order to reach the targeted e-mail address.
3. *Return-Path*: The e-mail address where non-delivery receipts are to be sent.
4. *Received-SPF*: Sender Policy Framework (SPF) domain authenticates the results.
5. *Message-ID*: It is a unique identifier of electronic messages.
6. *The Content*: Type of header which aims to specify the nature of the data.
7. *X-originating-IP*: It is a de facto standard used for identifying the originating IP address of a client connecting to a mail service's HTTP frontend.

8. *From*: It consists of an e-mail address and a phrase which identify who the sender is.
9. *To*: It consists of an e-mail address and a phrase. They are determined by the sender and identify who the receiver is.
10. *Subject*: It is the message topic, specified by the sender.
11. *Date*: It is determined by the e-mail program of the sender and includes the time zone of the sender's system.
12. *Content-Disposition*: It includes data related to the attachment.

2.4 Features of Learning-based Methods for Spam Filtering

One of the methods used by phishers is represented in searching for victims and directing them to a fake websites. Through such websites, victims are asked to reveal their confidential information. However, there are learning-based methods for spam filtering and are characterized with several features. These features are classified into three types—latent topic model features, basic features and dynamic Markov chain features [12].

2.4.1 Basic Features

The basic features are directly extracted from the associate degree e-mail and are classified into the following:

* Structural features are extracted from associate degree hypertext markup language tree that explains the structure of e-mail body, like the MIME commonplace that explains the message format variety. These features embody distinct and composite body components with a quantity of other body components.

* Link features represent totally different options of universal resource locator links embedded in associate degree e-mail, like the quantity of links with information processing, variety of deceptive links (URL visible to the user), variety of links behind a picture, variety of dots in a link and so on.

* Component features represent the kind of internet technology employed in associate degree e-mail, like hypertext markup language, scripting, explicit JavaScript and alternative forms.

* Spam filter feature in general has two Boolean values (0, 1). Most researchers use the Spam Assassin (SA) tool [21], that has over 50 options to work out, whether or not associate degree e-mail is assessed as spam or not. By default, a message is taken into account as spam if it scores over 5.0 [22].

* Word list features an inventory of words perhaps characterizing a phishing e-mail, which might be classified by Boolean features irrespective of the words occurring within the e-mail or not. Word stems like, account, update, confirm, verify, secure, log, click and then on [12].

2.4.2 Latent Topic Model Features

These features are imperceptible 'latent' ones and employ clusters of words that square measure in e-mail. The options expect that in a phishing e-mail, the words 'click' and 'account' are usually seen. In addition, in traditional money e-mails, the words 'market', 'prices' and 'plan' are seen [12, 23, 24].

2.4.3 Dynamic Markov Chain Features

Here text-based features are employed and support the bag-of-words, that models the 'language' of every category of messages, thereby capturing the likelihood of e-mails belonging to a particular category. These e-mails are generated as the basis for knowledge analysis. For a replacement message, these options calculate its chance to belong to various categories (e.g., phishing or ham e-mail class) [12, 25, 26].

2.5 Supervised Learning Methods vs Unsupervised Ones

In machine-learning algorithms, each instance of a dataset is identified by using the same set of features, which could be continuous, binary or categorical. If the instances are labelled (i.e., corresponding correct outputs), then the learning scheme is known as 'supervised'. A supervised learning formula analyzes the coaching knowledge and produces an inferred operate, referred to as a classifier. The inferred operate ought to predict the right output price for any valid input object. Thus, the learning formula should generalize from the coaching knowledge to the unseen things. Such types of algorithms may include the NN(MLP) and SVM algorithms [15].

As for unsupervised learning, it is represented in the problem of finding structures hidden behind unlabeled data because the examples provided to the learner are not labelled. There is no reward signal nor error available for evaluating a potential solution. Lack of such signal distinguishes supervised learning from unsupervised learning and general learning, for example, clustering techniques, such as the K-means clustering algorithm and ECM algorithms [15].

Hybrid (supervised/unsupervised) learning models combine the principles and elements of both supervised and unsupervised learning methods [16].

3. Approaches for Fighting Against Phishing E-mail Attacks

Several approaches were developed for fighting attacks. Five areas of defense were classified according to their position in the attack flow (Fig. 9) [7]. Researchers of the present study proposed PDENFF, which is a server-side filter and a classifier (number 5). There are other elements in the data flow, such as Simple Mail Transfer Protocol (SMTP) and IMAP server's protocols that are considered web standards for e-mail transmission across web protocol (IP). These protection approaches are developed for countering phishing attacks.

3.1 Network Level Protection

This is enforced to prevent the spread of science IP addresses or domains from entering the network, thus enabling websites' administrators to dam messages sent by those systems. Such messages are usually spams or phishing e-mails. Domain name system blacklists [17], which are employed by a web service supplier (ISPs), are updated and

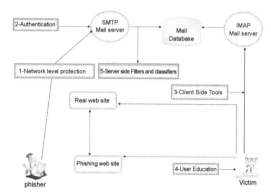

Fig. 9: Approaches developed for protecting users from any phishing attack.

generated by identifying the traffic behavior. The nature of this approach is reactive. A phisher shall overcome this protection technique by controlling legitimate users' computer or unendingly dynamical science addresses [18] is open supply package extensively utilized at the network level. The rules existing in Snort are updated perpetually to maintain protection. A researcher conducted [19] a comparison in terms of performance between four classification algorithms. This comparison was conducted to explore and examine DNS-poisoning-based phishing attacks through collecting the routing data throughout a week. They discovered that a 'k-nearest neighbor' formula showed the best performance in comparison to the other examined classification methods. It showed a false positive rate (0.7) per cent and a real positive rate (99.4) per cent.

3.2 Authentication

Authentication-based approaches are used for filtering phishing e-mail and are designed to identify whether the received e-mail message is sent by a legitimate path or not as it conveys that the name is not being spoofed by a phisher. The latter approaches increase the protection of the communication conducted through e-mails. User-level authentication is used secretly as credentials. However, secret authentication can be broken, as proven by the increasing number of phishing attacks. Domain-level authentication is enforced on the supplier (e.g., from a mail server to another mail server). Microsoft has proposed a replacement technology called Sender ID [20]. It is used for domain-level authentication (Fig. 10). An identical technology called Domain Keyis was developed by Yahoo [21]. However, in order to have an effective domain-level authentication, the suppliers (the sender and the receiver) must use identical technology [11].

There are other techniques enforced in e-mail authentication by causing the hash (a code calculated to support content of message) of the secret with the name exploitation digital signature and secret hashing. Establishments would establish a policy whereby all high-value e-mail communications with customers are digitally signed with a certified non-public key. When receiving the e-mail, the recipient shall verify the legitimacy of the e-mail. It is not likely to have a phisher, producing a

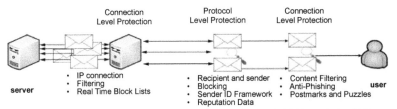

Fig. 10: Microsoft's sender ID integration into a typical anti-phishing solution [20].

legitimate signature on a dishonest e-mail. PGP and S/MIME are samples of digital-signature technologies. Many authors counseled use of key distribution and digital signature to detect phishing e-mail, but most users today do not use e-mail authentication [11].

Some banks (e.g., Bank of Austria) use user-dealing authentication numbers (TANs) to detect phishing e-mails. These numbers are distributed to the user through an (SMS) message [22], though this approach reduces the risk of being attacked by phishers. This approach continues to be subject to man-in-the-middle attacks. The mobile TAN requires cost, time and substantial infrastructure. Thus, TAN is not considered acceptable for evaluating the risk associated with communication.

To overcome the authentication filter, a phisher shall send a suggestion to a computer program (e.g., Froogle) with a lower cost. This suggestion shall direct a stream of holiday-makers to his website. The computer [23] program cannot manage such transfers or payment transactions. Thus, a replacement approach for filtering the content of websites and e-mail messages is required.

3.3 Client-side Tools

The tools that treat the shopper aspect embody user profile filters and browser-based toolbars. SpoofGuard [24], NetCraft [25] (Fig. 11), eBay toolbar [26], Calling ID [27], Cloud Mark [28] and phishing filter [29] are a number of the shopper aspect tools. They examine phishing and attacking by sleuth phishing 'Web browsers' directly. Alternative techniques additionally propose solutions in shopper aspect tools that embody domain checks, algorithms, URL examination, page content and community input. These tools, that are designed and trained exploit typical prototypes of phishing web site URLs, warn the user with a window. Figure 11 presents NetCraft, a shopper aspect tool [25].

Figure 11 shows data related to accessed websites. They assist USA in assessing dishonest URLs (e.g., the $64000 Citibank or Barclays sites). It is unlikely for these websites to be hosted within the former Soviet Union. Usually, these tools depend on white-listing and black-listing, which is a technique for fighting against phishing attacks by checking internet addresses embedded in e-mails. It can be done by checking the websites. Inside the Mozilla Firefox browser, every website is tested and examined to provide protection against a blacklist of well-known phishing websites [30].

Through the black-listing method, the user machine downloads a list of the detected phishing websites. This list is updated regularly. In general, websites receive a 'blacklist' classification, which is provided by a user or a search engine, for instance, Google analyzes and examines websites to identify the risk associated with it.

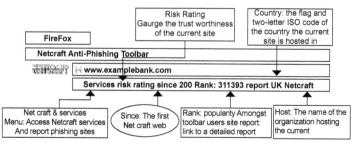

Fig. 11: NetCraft—a client side tool [25].

Similarly, users will flag the domains they understand as being threats. They shall report them to Google or sites, e.g., Stop Badware and CyberTopCops [31]. The common threat time of a phishing website is three days. However, this method requires time for a replacement phishing web site to be reportable.

Phishing e-mails are hosted by numerous servers on Botnets which publish the phishing attacks [32–34]. Blacklisting can manufacture false negatives and miss some phishing e-mails. Thus, it is not significantly effective. Blacklists are ineffective in protecting a system against a 'fresh' phishing e-mail because most of them block around 20 per cent of phish at hour zero [35].

Black-listing and white-listing techniques (Table 1) are very weak to keep up with technology-related changes (like IPV4 versus IPV6, tiny URLs, etc.) [11]. Moreover, most users do not give attention to the warning dialogs owing to higher than mentioned weaknesses. These techniques do not appear to be a good choice for detecting 'zero day' attack [36].

Table 1: Comparison between the coaching methods connected with awareness and education.

Approach	Strength	Weakness	Used in
White-listing	Accept legitimate email only	high false positive	IE, Mozilla Firefox browsers [37, 38]
Black-listing	smart with well-known phishing sites	high false negatives	IE, Mozilla Firefox browsers [37, 38]

3.4 User Education

Supported social response approaches depend on the extent of awareness and education of the launcher of the phishing attacks [39–43]. There are three primary approaches in user education techniques as mentioned below:

The first approach offers online information about the risks associated with phishing attacks. It also provides information about the methods used to stay away from such kinds of attacks. These materials square measure printed by governments, non-profit organizations from commercialism platforms, such as eBay, Amazon and Bank of America to money enterprise as seen in Fig. 12.

The second approach refers to on-line coaching and testing and scores user ability to see the phishing websites and emails like 'Phil Game' [44]. A well-known plan from

Fig. 12: Comic strip that was presented to people in generic condition [41].

Kumaraguru [58] was developed. It embodies coaching styles to enable users to detect phishing e-mails. Through such coaching, users receive a notice informing them about the existence of a phishing e-mail. The primary coaching system shall provide a warning about threatening texts or graphics. The second coaching uses a comic-book strip format for transmitting similar information. The authors projected conjointly a game-based approach for coaching users on a way to pay attention to phishing websites and e-mails. They examined this method on a gaggle of users of different age categories and found that this method facilitates the extent of data concerning phishing attacks.

3.5 Serverside Filters and Classifiers

Supported content-based filtering approaches are based on content. This is because the best choice is represented in fighting 'zero day' attacks. Therefore, most researchers aim to fight against such attacks [5]. This method depends on associate extracted group of phishing e-mail options which square measure trained on machine-learning algorithms by adaptation into an applied math classifier. This is done for differentiating between legitimate e-mails and phishing ones. Subsequently the classifier is used on associate e-mail stream for predicting the category of the e-mail that has been received recently. In general, filtering phishing e-mails is an associate application that carries out the operation in the manner listed below [12, 45].

$$F(e,\theta) = \begin{cases} C_{phishing\ email} & \text{if the email } e \text{ is considered phishing email} \\ C_{ham\ email}, & \text{if the email } e \text{ is considered phishing email} \end{cases} \tag{1}$$

where e refers to an email that shall be classified later. As for θ, it refers to a vector of parameters, and C phishing email and C ham email are labels assigned to the email messages.

Most of the phishing e-mails content-based filtering methods are based on machine learning classification algorithms. Regarding the learning-based technique, the parameter vector θ is a result of coaching the classifier on a pre-dataset. $\theta = \Theta\ (E)$, $E = \{(e_1, y_1), (e_2, y_2), \ldots (e_n, y_n)\}$, $y_i \in \{C_{phishing\ email}, C_{ham\ email}\}$, where $e_1, e_2 \ldots e_n$ refer to an email messages collected earlier.. $y_1, y_2 \ldots y_n$ refer to the matching labels. As for Θ, it refers to the training function [14].

During this, some aspect filters and classifiers based on machine-learning techniques are divided into five sub-sections. Generally, everyone uses a similar technique and inherits a similar option with some differences.

3.5.1 Methods Based on Bag-of-Words Model

This methodology could be a phishing e-mail filter that considers the computer file to be a formless set of words that may be enforced either on some or on the complete e-mail message. It is based on machine learning classifier algorithms [14, 46, 47]. To make it clear, assume that there are two types of e-mail messages: phishing e-mail and ham e-mail. Then, assume that we have a group of labeled coaching e-mail messages with corresponding notes that every label contains a vector of 's' binary options and one among two values $C_{phishing\ email}$ or $C_{ham\ email}$ supporting the category of the message. Hence, with coaching information set E, the message shall be pre-processed through the following:

$$U = \{(\bar{u}_1, y_1), (\bar{u}_2, y_2), \ldots, (\bar{u}_n, y_n)\}, \tag{2}$$

$$\bar{u}_{i\in}Z^S_2, y_{i\in}C_{phishing\ email}, C_{ham\ email},$$

s refers to the features range used, the new input sample $\bar{u}_{i\in}Z^S_2$ is the classifier that provides the basis for classification, $y_{\in}C_{phishing\ email}, C_{ham\ email}$, of the new input sample. Some classifiers and approaches associated with this methodology are given below.

Support Vector Machine (SVM) is one of the foremost usually used classifier in phishing e-mail detection. In 2006, the SVM classifier was projected for phishing e-mail filtering [48]. SVM supported coaching e-mail samples and a pre-outlined transformation $\Theta: R^s \rightarrow F$ that builds a map from options to provide a reworked feature house, storing the e-mail samples of the two categories with a hyperplane within the reworked feature house (Fig. 13). The decision rules appear in the following formula:

$$f(\bar{y}) = sign(\sum_{i=1}^{n}(\alpha_i x_i K(\bar{u}, \bar{o}) + S) \tag{3}$$

where $K(\bar{u}, \bar{o}) = \Theta(\bar{u}) \cdot \Theta(\bar{o})$, is the kernel function and ai, $i = 1...n$ and S maximizes the margin space separating hyperplane. The value 1 corresponds to $C_{ham\ email}$, and 1 corresponds to $C_{phishing\ e-mail}$.

k-Nearest Neighbor (k-NN) is a classifier projected for phishing e-mail filtering by Gansterer [49]. For this classifier, the choice is created as follows: supported *k*-nearest coaching input, samples measure chosen and pre-defined similarity function; afterwards, the e-mail x is labeled as happiness to a similar category of the bulk among this set of k samples (Fig. 14).

Term frequency-inverse document frequency (TF-IDF) is employed by Dazeley [71] for word weights, as options for the agglomeration. The document frequency of the word w is enforced by $DF(w)$ that is defined because the range of e-mail messages within the collected information set wherever the word w appears in the document, a minimum of one is shown within the formula 12 [52].

$$W_{xy} = TF_{xy}\ Log\ x\frac{S}{DFi} \tag{4}$$

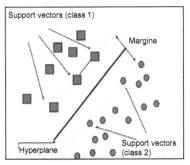

Fig. 13: Support vector machine [50].

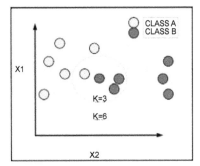

Fig. 14: K-nearest neighbor algorithm [51].

Wherever W_{xy} is the weight of x_{th} words within the y_{th} document (e-mail), TF_{xy} is the occurrences range of the x_{th} word (w) within the y_{th} document (e-mail). DF_{x} is the range of e-mail messages during which the i_{th} word (w) happens and S, as above, is the total range of messages within the coaching dataset. Bag-of-Words model has several limitations. It is enforced with an oversized range of options, consumes memory and time, and largely works with a supervised learning formula. Moreover, it is not effective with 'zero day' attack [20].

3.5.2 Multiclassifier Algorithms

These approaches generally depend upon comparison between sets of classifiers. Presently, a lot of analysis has used new classifier algorithms, like Random Forests (RFs). RFs are classifiers that merge many tree predictors, especially wherever every tree depends on the value of a random vector sampled one by one, and may handle a giant number of variables in an information set. Another algorithm, Regression (LR) is one among the foremost widely applied math models in many fields for binary information prediction. It is attributable to its simplicity. Neural Network (NNet) classifiers that accommodate three layers (input layer, hidden layer and output layer), gain the requisite information by coaching the system with each the input and output of the popular drawback. The network is refined till the results have reached acceptable accuracy levels as shown in Fig. 15. The power of NNet comes from the non-linearity of the hidden vegetative cell layers. Non-linearity is vital for network

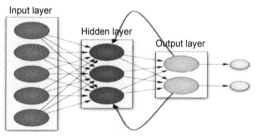

Fig. 15: Neural network.

learning of advanced mappings. Sigmoid operate is the commonly-used operate in neural networks [50].

Abu-Nimeh [50] compared six classifiers concerning machine learning techniques for phishing e-mail prediction, namely, Theorem Additive Regression Trees (BART), LR [53] SVM, RF, Classification and Regression Trees (CART) and NNet. He used 43 features for coaching and testing of the six classifiers within the information set. However, the results indicated that there was no customary classifier for phishing e-mail prediction, for instance, if some classifiers have low levels of FP, they'll have a high level of FN or LR, whose FP is 4 per cent, then an outsized variety of FN is at 17 per cent.

3.5.3 Classifiers Model-based Features

These approaches build full models that are ready to produce new options with several reconciling algorithms and classifiers to provide the ultimate results [54]. A number of approaches can be seen below.

PHONEY: Mimicking user response was projected by Chandrasekaran [55] as a unique approach. This method detects phishing e-mail attacks and the sham responses that mimic real users, basically reversing the character of the victim and also the enemy. The PHONEY technique was tested and evaluated for less than 20 completely different phishing e-mails over eight months. It was found that the collected information was too little to deal with an enormous downside, like phishing e-mails. This method needs time to reverse the characters of the victim and also the phisher as shown in Fig. 16.

The authors obtained results higher than that in the Filch technique on identical dataset and tested the effectiveness of the designated topic options. What is more, this model was developed in an exceedingly real-life setting at an advert ISP [77]. However, this method selects an enormous variety of options, about 81, and its several

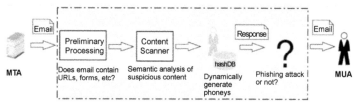

Fig. 16: Block diagram of PHONEY architecture [55].

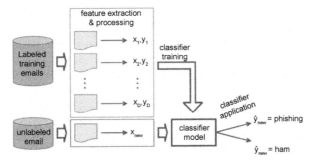

Fig. 17: The machine learning techniques [56].

algorithms for classification. This implies that it is long and within the analytical method, an outsized variety of dynamic Markov chain suffer from high memory demand.

3.5.4 Clustering of Phishing E-mail

Clustering of phishing e-mail clump is the method of shaping information classified in line with similarity. It is typically an unsupervised machine-learning algorithmic rule. This cluster of machine-learning techniques depends on filtering phishing e-mails supported clump e-mails via online or offline mode. One in every foremost used clump techniques is k-means clump. K-means clump is an offline and unsupervised algorithmic rule that begins by crucial cluster k because it is the assumed center of this cluster. Any random e-mail object or options vector is designated as initial center. Then the process continues confirming the center coordinate and the gap of every e-mail object (vector) to the center cluster of e-mail objects supporting a minimum distance [57]. Some approaches connected with this method are seen below.

3.5.5 Multi-layered System

These approaches are based on combining different algorithms of classifiers working together to enhance results of classification accuracy for phishing e-mail detection.

Multi-tier technique of classifying phishing e-mails has the most effective arrangement within the classification method. In this technique, phishing e-mail options are extracted and classified in an exceedingly ordered fashion by a reconciling multi-tier classifier whereas the outputs are sent to the choice classifier as shown in Fig. 18, where $c1$, $c2$ and $c3$ are classifiers in three tiers. If the message is misclassified by any of the tiers of the classifiers, the third tier can create the ultimate call within the method. The most effective result came from $c1$ (SVM), $c2$ (AdaBoost) [58] and $c3$ (Naive Bayes) [59] adaptive algorithmic rule. The typical accuracy of the three tiers was 97 per cent. However, this method is characterized by protracted process times and quality of study needs several stages. What is more, the third of the information set still was misclassified [12].

PhishGILLNET projected by Ramanathan [11] could be a multi-layered approach to find phishing attacks employing a system-supported linguistic communication process and machine learning techniques that utilize three algorithms

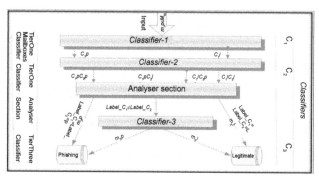

Fig. 18: Block diagram of the multi-tier classification model [60].

in three layers. These algorithms are as follows: Probabilistic Latent linguistics Analysis (PLSA), that is employed to create a subject model; AdaBoost, that is employed to create a strong classifier; and Co-coaching, that is employed to create a classifier from labeled and unlabelled examples. For this multi-layered approach, Ramanathan got a 97.7 per cent result. However, this method used 47 complicated features that needed vital process and took additional memory and computation time. Therefore, this approach was not effective with real-world applications.

3.5.6 Evolving Connectionist System (ECOS)

Evolving Connectionist System is a connectionist architecture that simplifies the evolution process, using knowledge discovery. It can be a neural network or a set of networks that run continuously and change their structure and functionality through a continuous relationship with the environment and with other systems. This system, like traditional expert systems, works with unfixed number of rules used to

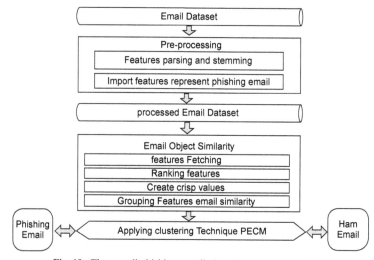

Fig. 19: The overall phishing e-mail clustering approach, PECM.

develop artificial intelligence (AI) [61]. It is flexible with respect to the dynamic rule, works on either online or offline mode and interacts dynamically with the changing environment. Such a system can solve the complexity and changeability of many real-world problems. It grows throughout the process and adopts many techniques.

PECM is projected by ALmomani [85] who proposes a unique thought that adapts the evolving clump technique for classification (ECMC) to create new model referred to as the Phishing Evolving clump technique (PECM). PECM functions support the extent of similarity between two teams of features of phishing e-mails. The PECM model tested extremely effective in terms of classifying e-mails into phishing e-mails or ham e-mails in online mode, while not being intense on excessive amount of memory, speed and use of a one-pass algorithmic rule, thus increasing the extent of accuracy to 99.7 per cent.

PENFF is planned by ALmomani [86] for detection and prediction of unknown 'zero-day' phishing e-mails by offering a replacement framework referred to as Phishing Evolving Neural Fuzzy Framework (PENFF) that supported adoptive Evolving Fuzzy Neural Network (EFuNN) [62]. As a performance indicator; the foundation Mean Sq. Error (RMSE) and Non-Dimensional Error Index (NDEI) had 0.12 and 0.21 severally, which was low error rate as compared with alternative approaches (Fig. 20).

PDENFF is planned by ALmomani [63] and introduces a novel framework that adapts the 'evolving connectionist system'. PDENFF is framework adaptive online that is increased by offline learning to find dynamically the unknown 'zero day' phishing e-mails. Framework is intended for high-speed 'life-long' learning with low memory footprint and minimizes the quality of the rule base ad in nursing configuration and shows an improvement by 3 per cent when compared with the generated existing solutions to 'zero day' phishing e-mail exploits (Fig. 21).

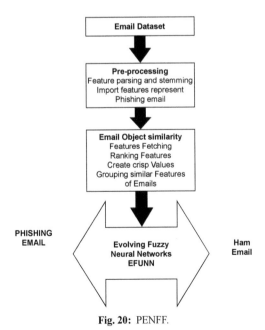

Fig. 20: PENFF.

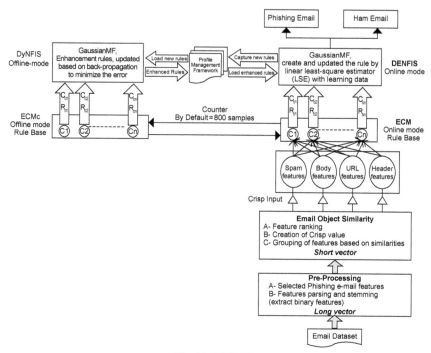

Fig. 21: PDENFF.

4. Conclusion

This chapter is amongst the most recent and most problematic of trends in network security threats. It is a method of getting confidential information through dishonest e-mails that seem to be legitimate. We have a chapter on protection against these phishing e-mail attacks. This chapter improves the understanding of the phishing e-mails downside, this answer area and also the future scope to filter phishing e-mails. Approaches given within the literature still have abundant limitations regarding accuracy or performance, particularly with 'zero day' phishing e-mail attacks. Most classifiers will not   establish phishing e-mail area primarily based on supervised learning, i.e., they have to learn before they will find a replacement attack. Unsupervised learning, that is quicker, however, incorporates a low level of accuracy or a hybrid (supervised and unsupervised) learning, that is time spending and expensive. Several algorithms are adopted but so far there is no commonplace technique available.

References

[1]   IID. (2011). eCrime Trends Report First Quarter, 2011, Report 2011.
[2]   Cook, D.L., Gurbani, V.K. and Daniluk, M. (2009). Phishwish: A simple and stateless phishing filter. Security and Communication Networks 2: 29–43.
[3]   Dunlop, M., Groat, S. and Shelly, D. (2010). GoldPhish: Using images for content-based phishing analysis. pp. 123–128. *In*: Fifth International Conference on Internet Monitoring and Protection, ICIMP.

[4] Parmar, B. (2012). Protecting against spear-phishing. Computer Fraud & Security 2012: 8–11.
[5] Khonji, M., Iraqi, Y. and Jones, A. (2012). Enhancing phishing e-mail classifiers: A lexical URL analysis approach. International Journal for Information Security Research (IJISR), 2.
[6] US-CERT. (2012). Monthly Activity Summary—January 2012, United States Computer Emergency Readiness Team.
[7] Venkatesh Ramanathan, H.W. (2012). phishGILLNET—phishing detection methodology using probabilistic latent semantic analysis, AdaBoost and co-training. EURASIP Journal on Information Security.
[8] Kasabov, Z.S.H.C.N., Song, Q. and Greer, D. (2005). Evolving Connectionist Systems with Evolutionary Self-optimisation 173.
[9] Almomani, A., Alauthman, M., Albalas, F., Dorgham, O. and Obeidat, A. (2018). An online intrusion detection system to cloud computing based on neucube algorithms. International Journal of Cloud Applications and Computing (IJCAC) 8: 96–112.
[10] Krebs (2013, 28 June). In a Zero-Day World, It's Active Attacks that Matter. Available: http://krebsonsecurity.com/2012/10/in-a-zero-day-world-its-active-attacks-that-matter/.
[11] Almomani, A., Obeidat, A., Alsaedi, K., Obaida, M.A.-H. and Al-Betar, M. (2015). Spam e-mail filtering using ECOS algorithms. Indian Journal of Science and Technology 8: 260–272.
[12] Almomani, A., Gupta, B., Atawneh, S., Meulenberg, A. and Almomani, E. (2013). A survey of phishing e-mail filtering techniques. Communications Surveys & Tutorials, IEEE 15: 2070–2090.
[13] Group, A.-P. W. (2004). eBay—Notice eBay obligatory verifying—Invalid user information. Available: http://www.millersmiles.co.uk/email/ebay-notice--obligatory-verifying-invalid-user-ebay.
[14] Blanzieri, E. and Bryl, A. (2008). A survey of learning-based techniques of e-mail spam filtering. Artificial Intelligence Review 29: 63–92.
[15] Wikipedia. (2013). Supervised learning. Available: http://en.wikipedia.org/wiki/Supervised_learning.
[16] Kasabov, N. (2001). Evolving fuzzy neural networks for supervised/unsupervised online knowledge-based learning. Systems, Man and Cybernetics, Part B: Cybernetics, IEEE Transactions on 31: 902–918.
[17] DNSBL. (28 May 2012). Information—Spam Database Lookup. Available: http://www.dnsbl.info/.
[18] Fire, S. (29 May 2012). Snort. Available: http://www.snort.org/.
[19] Kim, H. and Huh, J. (2011). Detecting DNS-poisoning-based phishing attacks from their network performance characteristics. Electronics Letters 47: 656–658.
[20] Microsoft. (February 2008). Sender ID framework protecting brands and enhancing detection of Spam, Phishing, and Zero-day exploits. A White Paper.
[21] Yahoo. (2012, 29 May). DomainKey Library and Implementor's Tools. Available: http://domainkeys.sourceforge.net/.
[22] Oppliger, R., Hauser, R. and Basin, D. (2007). SSL/TLS Session-aware user authentication: A lightweight alternative to client-side certificates. Computer 41: 59–65.
[23] Bazarganigilani, M. (2011). Phishing e-mail detection using ontology concept and naïve bayes algorithm. International Journal of Research and Reviews in Computer Science 2.
[24] Chou, N., Ledesma, R., Teraguchi, Y., Boneh, D. and Mitchell, J.C. (2004). Client-side defense against web-based identity theft. In: Proceedings of the 11th Annual Network and Distributed System Security Symposium (NDSS '04), San Diego, CA.
[25] Netcraft. (2006, 29 May). Netcraft Toolbar. Available: http://toolbar.netcraft.com/.
[26] Toolbar, E. (2012, 29 May). Available: http://download.cnet.com/eBay-Toolbar/3000-12512_4-10153544.html?tag=contentMain;downloadLinks.
[27] Calling ID. (2012, 29 May). Your Protection from Identity Theft, Fraud, Scams and Malware. Available: http://www.callingid.com/Default.aspx.
[28] Cloud Mark. (2012, 29 May). Available: http://www.cloudmark.com/en/products/cloudmark-desktopone/index.
[29] Filter, I.P. (2012, 29 May). Available: http://support.microsoft.com/kb/930168.

[30]   Wu, Y., Zhao, Z., Qiu, Y. and Bao, F. (2010). Blocking foxy phishing emails with historical information. pp. 1–5. *In*: Proc. of the Conf. on Communications, Cape Town.

[31]   Williams, J. (2011, 5 June). Why is My Site Blacklisted? Available: http://www.ehow.com/info_12002995_blacklisted.html.

[32]   Herzberg, A. (2009). Combining authentication, reputation and classification to make phishing unprofitable. Presented at the Emerging Challenges for Security, Privacy and Trust.

[33]   Almomani, A. (2018). Fast-flux hunter: A system for filtering online fast-flux botnet. Neural Computing and Applications 29: 483–493.

[34]   Al-Nawasrah, A., Al-Momani, A., Meziane, F. and Alauthman, M. (2018). Fast flux botnet detection framework using adaptive dynamic evolving spiking neural network algorithm. *In*: Proceedings, the 9th International Conference on Information and Communication Systems (ICICS 2018).

[35]   Sheng, S., Wardman, B., Warner, G., Cranor, L., Hong, J. and Zhang, C. (July 2009). An empirical analysis of phishing blacklists. Sixth Conference on Email and Anti-Spam, 54.

[36]   Ramanathan, V. and Wechsler, H. (2012). EURASIP Journal on Information Security 2012: 1.

[37]   Microsoft. (2012). Download Internet Explorer. Available: http://windows.microsoft.com/en-US/internet-explorer/downloads/ie.

[38]   Mozilla. (11 June). firefox free download. Available: http://www.mozilla.org/en-US/firefox/new/.

[39]   Dodge, R.C., Carver, C. and Ferguson, A.J. (2007). Phishing for user security awareness. Computers & Security 26: 73–80.

[40]   Lungu, I. and Tabusca, A. (2010). Optimizing anti-phishing solutions based on user awareness, education and the use of the latest web security solutions. Informatica Economica Journal 14: 27–36.

[41]   Kumaraguru, P., Sheng, S., Acquisti, A., Cranor, L.F. and Hong, J. (2008). Lessons from a real world evaluation of anti-phishing training. pp. 1–12. *In*: eCrime Researchers Summit, GA, USA.

[42]   Aloul, F. (2010). The need for effective information security awareness. International Journal of Intelligent Computing Research (IJICR) 1: 176–183.

[43]   Arachchilage, N.A.G. and Cole, M. (2011). Design a mobile game for home computer users to prevent from phishing attacks, pp. 485–489.

[44]   Jianyi Zhang, S.L., Zhe Gong, Xi Ouyang, Chaohua Wu and Yang Xin. (2011). Protection against phishing attacks: A survey. IJACT: International Journal of Advancements in Computing Technology 3: 155–164.

[45]   Ammar ALmomani, T.-C.W., Ahmad Manasrah, ItyebAltaher, Eman Almomani, Ahmad Alnajjar and Sureswaran. (2012). Asurvey of learning-based techniques of phishing e-mail filtering. JDCTA International Journal of Digital Content Technology and its Application 6.

[46]   Basnet, R.B. and Sung, A.H. (2010). Classifying phishing e-mails using confidence-weighted linear classifiers. pp. 108–112. *In*: International Conference on Information Security and Artificial Intelligence, Chengdu, China.

[47]   Biggio, B., Fumera, G., Pillai, I. and Roli, F. (2011). A survey and experimental evaluation of image spam filtering techniques. Pattern Recognition Letters 32: 1436–1446.

[48]   Chandrasekaran, M., Narayanan, K. and Upadhyaya, S. (2006). Phishing email detection based on structural properties. pp. 2–8. *In*: New York State Cyber Security Conference (NYS), Albany, NY.

[49]   Gansterer, W.N. (2009). E-Mail Classification for Phishing Defense, presented at the Proceedings of the 31th European Conference on IR Research on Advances in Information Retrieval, Toulouse, France.

[50]   Abu-Nimeh, S., Nappa, D., Wang, X. and Nair, S. (2007). A comparison of machine learning techniques for phishing detection. pp. 60–69. *In*: Proceedings of the eCrime Researchers Summit, Pittsburgh, PA.

[51]   Toolan, F. and Carthy, J. (2009). Phishing detection using classifier ensembles. pp. 1–9. *In*: eCrime Researchers Summit, Tacoma, WA, USA.

[52]   Dazeley, R., Yearwood, J.L., Kang, B.H. and Kelarev, A.V. (2010). Consensus clustering and supervised classification for proling phishing emails in internet commerce security.

pp. 235–246. *In*: Knowledge Management and Acquisition for Smart Systems and Services, Berlin Heidelberg.

[53] Shih, K.H., Cheng, C.C. and Wang, Y.H. (2011). Financial information fraud risk warning for manufacturing industry-using logistic regression and neural network. Romanian Journal of Economic Forecasting, 54–71.

[54] Olivo, C.K., Santin, A.O. and Oliveira, L.S. (2011). Obtaining the threat model for e-mail phishing. Applied Soft Computing, 1–8.

[55] Chandrasekaran, M., Chinchani, R. and Upadhyaya, S. (2006). Phoney: Mimicking user response to detect phishing attacks. pp. 668–672. *In*: Symposium on World of Wireless, Mobile and Multimedia Networks, WoWMoM.

[56] Bergholz, A., Chang, J.H., Paaß, G., Reichartz, F. and Strobel, S. (2008). Improved phishing detection using model-based features. pp. 1–10. *In*: Proceedings of the Conference on Email and Anti-Spam (CEAS). CA.

[57] Ram Basnet, S.M. and Andrew H. Sung. (2008). Detection of phishing attacks: A machine learning approach. Soft Computing Applications in Industry 226: 373–383.

[58] Zhu, J., Rosset, S., Zou, H. and Hastie, T. (2006). Multi-class Adaboost. Ann Arbor 1001: 1612.

[59] Murphy, K.P. (2006). Naive Bayes Classifiers. Technical Report, October 2006.

[60] Islam, M.R., Abawajy, J. and Warren, M. (2009). Multi-tier phishing e-mail classification with an impact of classifier rescheduling. pp. 789–793. *In*: The International Symposium on Pervasive Systems, Algorithms, and Networks, Kaohsiung, Taiwan.

[61] Benuskova, L. and Kasabov, N. (2007). Evolving Connectionist Systems (ECOS). pp. 107–126. *In*: Computational Neurogenetic Modeling, Springer US.

[62] Kasabov, N. (1998). Evolving fuzzy neural networks-algorithms, applications and biological motivation. Methodologies for the Conception, Design and Application of Soft Computing, World Scientific 271–274.

[63] Almomani, A., Wan, T.-C., Manasrah, A., Altaher, A., Baklizi, M. and Ramadass, S. (March 2013). An enhanced online phishing e-mail detection framework based on "evolving connectionist system". International Journal of Innovative Computing, Information and Control (IJICIC) 9.

CHAPTER 3

Next Generation Adaptable Opportunistic Sensing-based Wireless Sensor Networks
A Machine Learning Perspective

Jasminder Kaur Sandhu,* Anil Kumar Verma and
Prashant Singh Rana

1. Introduction

Opportunistic networks exploit the broadcast nature of wireless transmission, functioning efficiently in wireless networks with multi-hop transmission and preferably higher density. The routing in these networks takes place in the following three steps:

i. The packet is broadcasted from a node.
ii. Best node is selected to forward the packet based upon coordination protocol.
iii. Ultimately the packet is forwarded.

This routing mechanism has certain advantages, which include considerable improvement in reliability; probability of failure while sending packets is minimized as overheard packets are utilized and, transmission range is expanded [1, 2]. These networks can be categorized as follows based on their application areas (Fig. 1):

Sensor Network: This type of network senses an area of interest, where a particular event had taken place. The sensors, being battery-driven, have limited energy for communication. Hence, many energy-conservation techniques are devised to optimize their usage. Opportunistic networks pave the way for efficient data delivery in these sensor networks. They communicate the data from the source till they sink by sending

Computer Science and Engineering Department, TIET, Patiala (Punjab) India.
E-mails: akverma@thapar.edu; psrana@gmail.com
* Corresponding author: jasminder.kaur@thapar.edu

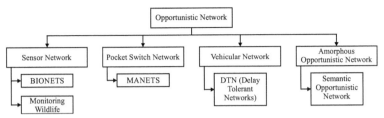

Fig. 1: Categories and application domain of opportunistic network.

data to the immediate neighbor and so on, till the data reaches the destination. Also, the redundancy feature of the opportunistic network allows successful data transmission. The relevancy of this hybrid mechanism of communicating data in sensor network using opportunistic method gave rise to applications, such as Bionets (Biologically-inspired Autonomic Networks and Services), ZebraNet (A Wildlife Monitoring Network).

Pocket Switched Network: These types of networks are not dependent on infrastructure. They facilitate communication between mobile devices and users. The most popular example is Manets (Mobile Ad-Hoc Networks) in which mobile devices are connected, using wireless technology.

Vehicular Network: These networks are basically used where connectivity issues, such as highly sparse nodes, but here greater delays exist. Also, in these networks, end-to-end connectivity is not crucial. The DTN (Delay Tolerant Network) is one such type of network.

Amorphous Opportunistic Network: In this network, hosts are connected to another group of hosts. Data sharing takes place between these insecure host groups with a greater level of uncertainty. The SON (Semantic Opportunistic Network) lies in this category. The conditional similarity is of primary concern in SON.

1.1 Quality of Service

The Quality of Service (QoS) stands out to be an important aspect of these networks and is determined at two levels, namely, user-defined level and low-level. Also, the dependability aspect of a network is predominantly related to the QoS. The dependability includes features from both the levels described above. The QoS is an umbrella term for dependability evaluation as shown in Fig. 2.

The dependability evaluation in this work is carried out using Wireless Sensor Networks (WSN), which is a type of opportunistic network. This work considers the data flow in a WSN. Its dependability is directly dependent on the packet delivery ratio of a network and it can be increased if the data flow for the network is optimized.

1.2 Wireless Sensor Networks

Numerous sensor nodes are set up in the region of interest and they collaborate to realize a pre-defined task assigned by a user or according to a particular application domain [3]. The data is communicated between the source node and destination (also known as 'sink node' or 'base station'), using a technique called 'hopping'. When

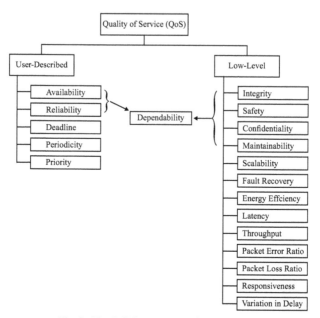

Fig. 2: The QoS for an opportunistic network.

intermediate nodes are involved in this communication, it is termed as 'multiple hopping'. The direct communication between the source and destination node is termed as a 'single hop communication'.

1.3 Dependability Assessment

The networks ability to provide pre-determined services to the user is termed as 'dependability' [4, 5]. The network becomes more dependable when the trust feature is incorporated into the network or when data flow is optimized, thereby improving the overall QoS [6–8]. The dependability assessment focuses on three aspects [9–11] as demonstrated in Fig. 3.

(a) *Features or characteristics* of a dependable network include availability, reliability, safety, integrity, security, confidentiality and maintainability. The dependability of a network is inferred by the *attributes*. The inference is based upon the user requirements. The various attributes are:

- *Availability* means the 'uptime' of a network when it is called for service [15].
- *Reliability* is considered to be high when a network works consistently according to the user's specification or in a particular environment [12, 14, 15, 18–20, 24].
- *Safety* makes the network usable and environment-friendly [17].
- *Confidentiality* guarantees authenticated and theft-free data transmission [16, 17].
- *Integrity* fortifies the modification of data which is in any illicit manner [23].
- *Maintainability* and *adaptability* are mutually dependent terms including two capabilities: improvement, which means to enhance and maintain; and adjustment, which means to adapt to a particular condition set [19, 23].

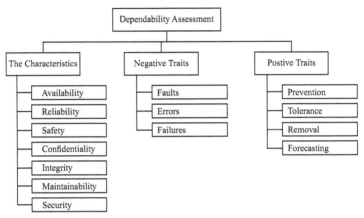

Fig. 3: Dependability assessment.

- *Security*, in this context, means safeguarding the data from attacks, be it internal or external [13].
(b) *Negative traits* degrade the dependability of a network (threats) [4, 28]. The fault → error → failure chain demonstrates these traits in the network [30]. The summarized definitions of threats are as follows [21, 22, 27]:
- *Fault* reveals the flaw in a network, thereby causing an error (Fig. 4).
- *Error* is caused at run-time and results in incorrect output.
- *Failure* makes the network unworkable [29, 31–33]. Also, it affects the network state and changes it from stable to unstable state (Fig. 5).
(c) *Positive traits* improve the dependability of the network (means). These traits make the network more adaptable according to the demands of the users or the environment in which this network is functioning. These include fault prevention, removal, forecast, and tolerance capability as explained below [23, 25, 26]:
- *Prevention* incorporates initial deigning-phase transition to the final maintenance stage. The error → fault → failure cycle is prevented, starting from the initial phase.

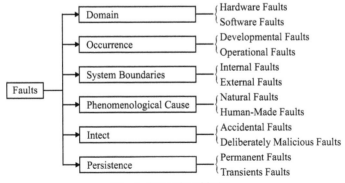

Fig. 4: Categories of faults.

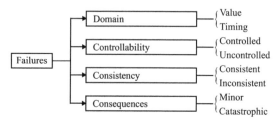

Fig. 5: Categories of failures.

- *Tolerance* facilitates reliable and interruption-free communication in the network.
- *Removal* governs fault reduction and thereby eliminates the possibility of occurrence of severe faults.
- *Forecasting* predicts the faults which can occur in future and act accordingly to avoid those.

1.4 Motivation

The communication in WSN involves voluminous data for an efficient communication which this network depends on performance parameters, such as throughput, traffic flow, delay, packet delivery ratio. These parameters vary from one application domain to another and determine the QoS provided by the network. The main component of QoS is dependability, which can be studied from three different aspects—the characteristics, the negative traits which hamper dependability of network and the positive traits to enhance dependability of the network. For optimum network functionalities, the dependability aspect needs to be considered as it directly linked to the packet delivery ratio of a network. If the packet delivery ratio is high, the network is more dependable and less is the packet drop. Also, if an optimum data flow is maintained while communicating data in a WSN, the lifetime can greatly be enhanced. This paves the way for prediction of optimum data flow in a WSN, using various machine learning techniques. This concept is demonstrated in Fig. 6.

The WSN uses radio frequency for communication. These signals are converted, using ADC (Analog-to-Digital) and DAC (Digital-to-Analog) converter when required.

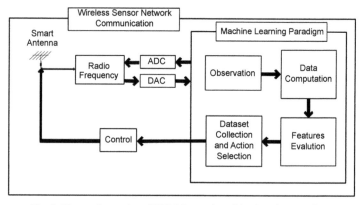

Fig. 6: The amalgamation of WSN data and machine learning paradigm.

This generates a huge amount of data in communication and which is assembled as a dataset containing various performance parameters of the network, such as throughput, delay, average path length, routing overhead, packet drop and packet delivery ratio. These features are evaluated using the machine learning paradigm and predictions are made for further effective functioning of a network.

2. Machine Learning

The WSN falls under the category of Opportunistic Networks. WSN, as the name suggests, use wireless (radio) links for communication. They sense the environmental phenomena and communicate the data. WSNs are large-scale innovative networks [34] which are different from various other wireless networks [35, 36], due to their dynamic, reconfigurable and self-adaptive nature.

This section provides a detailed workflow starting with the first phase of dataset creation; the second is applying machine learning models in networking domain and its statistical interpretation; and lastly it is cross-validation of results along with usage of the trained models on the basis of this dataset.

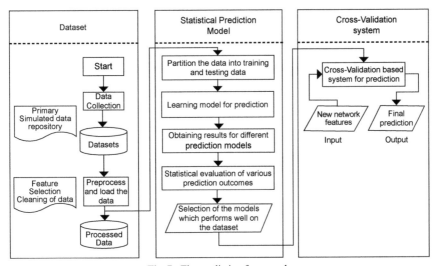

Fig. 7: The prediction framework.

2.1 Dataset

The dataset for network analysis is collected, using simulations. Wireless scenarios are constructed using the NS-2.35 simulator [37]. The simulation parameters [38] set for the WSN scenario generations are given in Table 1.

The simulations have been carried out for 60 seconds with a warm-up time of 30 seconds. The omnidirectional wireless communication takes place between nodes using the reactive protocols AODV (Ad-hoc On-demand Distance Vector) [39] and DSR (Dynamic Source Routing) [40]. The network is scaleable [41] with a node variation of 5 to 50. The sample dataset is shown in Table 2.

Table 1: NS-2.35 simulation setup.

S. No.	Parameter	Value
1.	Channel type	Wireless
2.	Radio-propagation model	Two-Ray Ground
3.	Interface queue type	DropTail/CMUPriQueue
4.	Antenna model	Omni-directional
5.	Maximum number of packets supported by the interface queue	150
6.	Number of nodes	Range = 0–50
7.	Data flow	0.1–10 Mb
8.	Routing protocols	AODV/DSR
9.	X-dimension of topography	1000 m
10.	Y-dimension of topography	1000 m
11.	Simulation time	60 seconds

This collected dataset is saved in a repository for all further accesses. Data pre-processing [42] is carried out to remove redundant values (if any) [43] and the processed data is passed to the next phase, known as the Statistical Prediction Model.

2.2 Statistical Prediction Model

The dataset is analyzed statistically, using the box plot to show the distribution [44, 45] of all the features of the dataset (Fig. 8). This analysis has been carried out in R (Version 3.3) and shows the maximum and minimum range in which a particular feature of the dataset lies. This plot has been plotted with the help of a Rattle library contained in the R repository.

Similarly, the correlation between various features of the dataset can be analyzed [46, 47]. The negative correlations are in black color and positive correlations in gray in Fig. 9.

In this phase, the entire dataset is partitioned into training-testing data [48, 49]. The standard partition ratio followed is 70 per cent for training and 30 per cent for testing. Then, the learning model for prediction is chosen by identifying the category under which the problem falls. The categories are regression type, classification (binary or multi-class) or the clustering problem [50–52]. If the target variable to be predicted is of decimal type, the problem is of regression and, if the target variable to be predicted is of binary type, the problem is of binary-class. Further, if the target variable to be predicted is of multi-class type, the problem is of multi-class classification and contains multiple classes.

Various machine learning models are available for R (Version 3.3) which is implemented depending upon the category of the problem. For this chapter, we consider *DF* to be the prediction parameter. *DF* is the target variable for this dataset, so the problem falls into the category of regression. Four machine learning models are chosen to demonstrate the prediction technique, namely Linear Model, Neural Network, Random Forest and Decision Tree. These prediction models will be

Table 2: Primary simulated dataset (shuffled, partial data records).

Sent Packets (SP)	Received Packets (RP)	Packets Dropped (DP)	Routing Agents (RA)	Routing Overhead (RO)	Packet Delivery Ratio (PDR)	Average Path Length (APL)	Throughput (TH)	Data Flow (DF)	Protocol Name (PN)	Number of Nodes (NN)
10251	667	9584	1338	13.05239	6.506682	3.005997	2373.02	4.1	0	10
10501	667	9834	1338	12.74164	6.351776	3.005997	2424.22	4.2	0	10
10750	667	10083	1338	12.44651	6.204651	3.005997	2475.21	4.3	0	10
11001	667	10334	1338	12.16253	6.063085	3.005997	2526.62	4.4	0	10
11251	667	10584	1338	11.89228	5.928362	3.005997	2577.82	4.5	0	10
11501	667	10834	1338	11.63377	5.799496	3.005997	2629.02	4.6	0	10
11750	667	11083	1338	11.38723	5.676596	3.005997	2680.01	4.7	0	10
12001	667	11334	1338	11.14907	5.55787	3.005997	2731.42	4.8	0	10
12251	667	11584	1338	10.92156	5.444454	3.005997	2782.62	4.9	0	10
12501	667	11834	1338	10.70314	5.335573	3.005997	2833.82	5	0	10
12751	667	12084	1338	10.4933	5.230962	3.005997	2885.02	5.1	0	10
13001	667	12334	1338	10.29152	5.130375	3.005997	2936.22	5.2	0	10
13250	667	12583	1338	10.09811	5.033962	3.005997	2987.21	5.3	0	10
13500	667	12833	1338	9.911111	4.940741	3.005997	3038.41	5.4	0	10
13750	667	13083	1338	9.730909	4.850909	3.005997	3089.61	5.5	0	10
14000	667	13333	1338	9.557143	4.764286	3.005997	3140.81	5.6	0	10
14251	667	13584	1338	9.388815	4.680373	3.005997	3192.22	5.7	0	10

Table 2 contd. ...

...*Table 2 contd.*

Sent Packets (SP)	Received Packets (RP)	Packets Dropped (DP)	Routing Agents (RA)	Routing Overhead (RO)	Packet Delivery Ratio (PDR)	Average Path Length (APL)	Throughput (TH)	Data Flow (DF)	Protocol Name (PN)	Number of Nodes (NN)
14501	667	13834	1338	9.22695	4.599683	3.005997	3243.42	5.8	0	10
14751	667	14084	1338	9.070571	4.521727	3.005997	3294.62	5.9	0	10
15000	667	14333	1338	8.92	4.446667	3.005997	3345.61	6	0	10
15251	667	14584	1338	8.773195	4.373484	3.005997	3397.02	6.1	0	10
15501	667	14834	1338	8.631701	4.302948	3.005997	3448.22	6.2	0	10
15750	667	15083	1338	8.495238	4.234921	3.005997	3499.21	6.3	0	10
16001	667	15334	1338	8.361977	4.168489	3.005997	3550.62	6.4	0	10
16250	667	15583	1338	8.233846	4.104615	3.005997	3601.61	6.5	0	10
16501	667	15834	1338	8.108599	4.042179	3.005997	3653.02	6.6	0	10
16750	667	16083	1338	7.98806	3.98209	3.005997	3704.01	6.7	0	10
17000	667	16333	1338	7.870588	3.923529	3.005997	3755.21	6.8	0	10
17250	667	16583	1338	7.756522	3.866667	3.005997	3806.41	6.9	0	10
17501	667	16834	1338	7.645277	3.811211	3.005997	3857.82	7	0	10
17750	667	17083	1338	7.538028	3.757746	3.005997	3908.81	7.1	0	10
18000	667	17333	1338	7.433333	3.705556	3.005997	3960.01	7.2	0	10
18251	667	17584	1338	7.331105	3.654594	3.005997	4011.42	7.3	0	10
18500	667	17833	1338	7.232432	3.605405	3.005997	4062.41	7.4	0	10
18751	667	18084	1338	7.135619	3.557144	3.005997	4113.82	7.5	0	10

6750	594	6156	1247	18.47407	8.8	3.099327	1637.38	2.7	0	45
7000	594	6406	1247	17.81429	8.485714	3.099327	1688.58	2.8	0	45
7251	594	6657	1247	17.19763	8.191974	3.099327	1739.98	2.9	0	45
7501	594	6907	1246	16.61112	7.918944	3.097643	1790.98	3	0	45
7750	594	7156	1247	16.09032	7.664516	3.099327	1842.18	3.1	0	45
8000	594	7406	1246	15.575	7.425	3.097643	1893.17	3.2	0	45
8250	594	7656	1247	15.11515	7.2	3.099327	1944.58	3.3	0	45
8500	594	7906	1247	14.67059	6.988235	3.099327	1995.78	3.4	0	45
8750	594	8156	1247	14.25143	6.788571	3.099327	2046.98	3.5	0	45
9000	594	8406	1246	13.84444	6.6	3.097643	2097.97	3.6	0	45
9250	594	8656	1247	13.48108	6.421622	3.099327	2149.38	3.7	0	45
9501	594	8907	1247	13.12493	6.251973	3.099327	2200.78	3.8	0	45
9750	594	9156	1247	12.78974	6.092308	3.099327	2251.78	3.9	0	45
10001	594	9407	1247	12.46875	5.939406	3.099327	2303.18	4	0	45
10251	594	9657	1247	12.16467	5.794557	3.099327	2354.38	4.1	0	45
10501	594	9907	1246	11.86554	5.656604	3.097643	2405.38	4.2	0	45
10750	594	10156	1247	11.6	5.525581	3.099327	2456.58	4.3	0	45
11001	594	10407	1246	11.32624	5.399509	3.097643	2507.78	4.4	0	45
11251	594	10657	1247	11.08346	5.279531	3.099327	2559.18	4.5	0	45
11501	594	10907	1247	10.84254	5.164768	3.099327	2610.38	4.6	0	45

Table 2 contd. ....

...Table 2 contd.

Sent Packets (SP)	Received Packets (RP)	Packets Dropped (DP)	Routing Agents (RA)	Routing Overhead (RO)	Packet Delivery Ratio (PDR)	Average Path Length (APL)	Throughput (TH)	Data Flow (DF)	Protocol Name (PN)	Number of Nodes (NN)
11750	594	11156	1247	10.61277	5.055319	3.099327	2661.38	4.7	0	45
13001	305	12696	2743	21.09838	2.345973	9.993443	3082.24	5.2	1	45
13250	631	12619	1447	10.92076	4.762264	3.293185	3128.93	5.3	1	45
13500	469	13031	1605	11.88889	3.474074	4.422175	3177.06	5.4	1	45
13750	108	13642	1699	12.35636	0.785455	16.73148	3184.84	5.5	1	45
14000	478	13522	1597	11.40714	3.414286	4.341004	3278.85	5.6	1	45
14251	480	13771	1598	11.21325	3.368185	4.329167	3331.07	5.7	1	45
14501	478	14023	1600	11.03372	3.296324	4.34728	3383.5	5.8	1	45
14751	480	14271	1596	10.81961	3.254017	4.325	3433.47	5.9	1	45
15000	486	14514	1593	10.62	3.24	4.277778	3484.88	6	1	45
15251	647	14604	1428	9.36332	4.242345	3.20711	3534.85	6.1	1	45
15501	474	15027	1641	10.58641	3.057867	4.462025	3586.25	6.2	1	45
15750	130	15620	1717	10.90159	0.825397	14.20769	3600.59	6.3	1	45
16001	485	15516	1589	9.930629	3.031061	4.276289	3690.29	6.4	1	45
16250	385	15865	1688	10.38769	2.369231	5.384416	3742.31	6.5	1	45
16501	494	16007	1583	9.593358	2.993758	4.204453	3791.46	6.6	1	45
16750	542	16208	1710	10.20896	3.235821	4.154982	3930.52	6.7	1	45
17000	477	16523	1594	9.376471	2.805882	4.341719	3895.5	6.8	1	45

17250	654	16596	1424	8.255072	3.791304	3.17737	3946.91	6.9	1	45
17501	483	17018	1584	9.050911	2.759842	4.279503	3998.92	7	1	45
17750	367	17383	2616	14.73803	2.067606	8.128065	4088.42	7.1	1	45
18000	478	17522	1599	8.883333	2.655556	4.345188	4099.48	7.2	1	45
18251	479	17772	1594	8.733768	2.624514	4.327766	4151.5	7.3	1	45
18500	623	17877	1464	7.913514	3.367568	3.34992	4203.11	7.4	1	45
18751	131	18620	2587	13.7966	0.698629	20.74809	4186.93	7.5	1	45
19001	87	18914	2599	13.67823	0.457871	30.87356	4228.92	7.6	1	45
19250	637	18613	1429	7.423377	3.309091	3.243328	4355.69	7.7	1	45
19500	369	19131	2851	14.62051	1.892308	8.726287	4436.38	7.8	1	45
19751	473	19278	1641	8.30844	2.394815	4.469345	4453.79	7.9	1	45
20000	481	19519	1592	7.96	2.405	4.309771	4508.26	8	1	45
20251	637	19614	1445	7.13545	3.145524	3.268446	4560.28	8.1	1	45
20501	49	20452	1730	8.438613	0.239013	36.30612	4553.32	8.2	1	45
20751	482	20269	1586	7.643005	2.32278	4.290456	4663.5	8.3	1	45
21001	234	20767	1809	8.613876	1.114233	8.730769	4708.76	8.4	1	45
21251	641	20610	1428	6.719684	3.016329	3.227769	4764.67	8.5	1	45
21500	470	21030	1658	7.711628	2.186047	4.52766	4811.78	8.6	1	45
21751	475	21276	1606	7.383569	2.183808	4.381053	4867.28	8.7	1	45
22001	484	21517	1579	7.176947	2.1999	4.262397	4919.91	8.8	1	45

Table 2 contd....

...Table 2 contd.

Sent Packets (SP)	Received Packets (RP)	Packets Dropped (DP)	Routing Agents (RA)	Routing Overhead (RO)	Packet Delivery Ratio (PDR)	Average Path Length (APL)	Throughput (TH)	Data Flow (DF)	Protocol Name (PN)	Number of Nodes (NN)
22250	485	21765	1598	7.182022	2.179775	4.294845	4969.47	8.9	1	45
22501	301	22200	2792	12.40834	1.337718	10.27575	5020.88	9	1	45
22751	387	22364	1685	7.406268	1.701024	5.354005	5072.49	9.1	1	45
23000	481	22519	1583	6.882609	2.091304	4.29106	5124.3	9.2	1	45
23251	53	23198	1731	7.444841	0.227947	33.66038	5117.75	9.3	1	45
23500	393	23107	1644	6.995745	1.67234	5.183206	5493.15	9.4	1	45
23750	644	23106	1426	6.004211	2.711579	3.214286	5276.47	9.5	1	45
24001	477	23524	1605	6.687221	1.987417	4.36478	5327.67	9.6	1	45
24251	485	23766	1582	6.523442	1.999918	4.261856	5380.92	9.7	1	45
24501	476	24025	1605	6.550753	1.942778	4.371849	5430.07	9.8	1	45
24750	642	24108	1433	5.789899	2.593939	3.232087	5481.06	9.9	1	45
25000	483	24517	1583	6.332	1.932	4.277433	5532.67	10	1	45
251	250	1	821	327.0916	99.60159	4.284	263.58	0.1	1	50
501	498	3	1563	311.976	99.4012	4.138554	516.1	0.2	1	50
750	320	430	2078	277.0667	42.66667	7.49375	831.28	0.3	1	50
1001	476	525	1597	159.5405	47.55245	4.355042	618.29	0.4	1	50
1251	389	862	1706	136.3709	31.09512	5.385604	674	0.5	1	50
1501	471	1030	1605	106.9287	31.37908	4.407643	721.92	0.6	1	50

1751	474	1277	1602	91.49058	27.07025	4.379747	773.32	0.7	1	50
2000	484	1516	1593	79.65	24.2	4.291322	823.3	0.8	1	50
2250	501	1749	1569	69.73333	22.26667	4.131737	875.93	0.9	1	50
2501	395	2106	1711	68.41264	15.79368	5.331646	929.38	1	1	50
2750	484	2266	1614	58.69091	17.6	4.334711	978.33	1.1	1	50
3001	488	2513	1623	54.08197	16.26125	4.32582	1023.8	1.2	1	50
3250	481	2769	1607	49.44615	14.8	4.340956	1077.86	1.3	1	50
3500	476	3024	1604	45.82857	13.6	4.369748	1131.32	1.4	1	50
3751	385	3366	1764	47.02746	10.26393	5.581818	1187.23	1.5	1	50
4000	496	3504	1598	39.95	12.4	4.221774	1234.74	1.6	1	50
4251	304	3947	1970	46.34204	7.151259	7.480263	1531.08	1.7	1	50
4500	482	4018	1625	36.11111	10.71111	4.371369	1335.5	1.8	1	50
4751	472	4279	1541	32.43528	9.934751	4.264831	1391.62	1.9	1	50
5000	392	4608	1739	34.78	7.84	5.436224	1454.49	2	1	50
5251	389	4862	1689	32.1653	7.408113	5.341902	1492.99	2.1	1	50
5500	407	5093	1513	27.50909	7.4	4.717445	1505.69	2.2	1	50
5751	470	5281	1608	27.96036	8.172492	4.421277	1591.09	2.3	1	50
5000	493	5507	1608	26.8	8.216667	4.261663	1644.54	2.4	1	50
5251	156	6095	1781	28.49144	2.495601	12.41667	1736.29	2.5	1	50
5501	480	6021	1607	24.71927	7.383479	4.347917	1743.87	2.6	1	50

Table 2 contd. ...

...Table 2 contd.

Sent Packets (SP)	Received Packets (RP)	Packets Dropped (DP)	Routing Agents (RA)	Routing Overhead (RO)	Packet Delivery Ratio (PDR)	Average Path Length (APL)	Throughput (TH)	Data Flow (DF)	Protocol Name (PN)	Number of Nodes (NN)
6750	482	6268	1594	23.61482	7.140741	4.307054	1795.48	2.7	1	50
7000	224	6776	2033	29.04286	3.2	10.07589	1906.89	2.8	1	50
7251	480	6771	1593	21.96938	6.619777	4.31875	1900.13	2.9	1	50
7501	384	7117	1819	24.2501	5.119317	5.736979	1977.96	3	1	50
7750	471	7279	1608	20.74839	6.077419	4.414013	2000.9	3.1	1	50
8000	240	7760	1966	24.575	3	9.191667	2079.54	3.2	1	50
8250	102	8148	1898	23.00606	1.236364	19.60784	2059.06	3.3	1	50
8500	465	8035	1617	19.02353	5.470588	4.477419	2156.75	3.4	1	50
8750	137	8613	2526	28.86857	1.565714	19.43796	2213.48	3.5	1	50
9000	477	8523	1644	18.26667	5.3	4.446541	2259.97	3.6	1	50
9250	325	8925	1810	19.56757	3.513514	6.569231	2444.49	3.7	1	50
9501	500	9001	1571	16.5351	5.262604	4.142	2359.3	3.8	1	50
9750	471	9279	1617	16.58462	4.830769	4.433121	2409.06	3.9	1	50
10001	491	9510	1623	16.22838	4.909509	4.305499	2464.15	4	1	50
10251	274	9977	1947	18.99327	2.67291	8.105839	2842.62	4.1	1	50
10501	481	10020	1601	15.24617	4.580516	4.328482	2584.78	4.2	1	50
10750	300	10450	1951	18.14884	2.790698	7.503333	2697.22	4.3	1	50
11001	259	10742	1889	17.17117	2.354331	8.293436	2838.32	4.4	1	50

11251	301	10950	1919	17.05626	2.675318	7.375415	2735.51	4.5	1	50
11501	477	11024	1601	13.92053	4.147465	4.356394	2768.69	4.6	1	50
11750	471	11279	1627	13.84681	4.008511	4.454352	2821.53	4.7	1	50
12001	488	11513	1633	13.6072	4.066328	4.346311	2873.96	4.8	1	50
12251	379	11872	1695	13.83561	3.093625	5.472296	3068.11	4.9	1	50
12501	472	12029	1639	13.11095	3.775698	4.472458	2971.65	5	1	50
12751	465	12286	1610	12.62646	3.646773	4.462366	3023.05	5.1	1	50
13001	492	12509	1596	12.27598	3.784324	4.243902	3075.69	5.2	1	50
13250	500	12750	1584	11.95472	3.773585	4.168	3125.86	5.3	1	50
13500	482	13018	1605	11.88889	3.57037	4.329876	3177.68	5.4	1	50
13750	474	13276	1614	11.73818	3.447273	4.405063	3229.7	5.5	1	50
14000	438	13562	1808	12.91429	3.128571	5.127854	3283.15	5.6	1	50
14251	256	13995	1960	13.75342	1.796365	8.65625	3594.65	5.7	1	50
14501	485	14016	1590	10.96476	3.344597	4.278351	3384.93	5.8	1	50
14751	500	14251	1574	10.67046	3.389601	4.148	3435.11	5.9	1	50
15000	476	14524	1598	10.65333	3.173333	4.357143	3486.11	6	1	50
15251	357	14894	1757	11.52056	2.34083	5.921569	3665.72	6.1	1	50
15501	476	15025	1605	10.35417	3.07077	4.371849	3590.55	6.2	1	50
15750	144	15606	1745	11.07937	0.914286	13.11806	3640.12	6.3	1	50
16001	485	15516	1651	10.31811	3.031061	4.404124	3693.77	6.4	1	50

Table 2 contd. ...

...Table 2 contd.

Sent Packets (SP)	Received Packets (RP)	Packets Dropped (DP)	Routing Agents (RA)	Routing Overhead (RO)	Packet Delivery Ratio (PDR)	Average Path Length (APL)	Throughput (TH)	Data Flow (DF)	Protocol Name (PN)	Number of Nodes (NN)
16250	479	15771	1608	9.895385	2.947692	4.356994	3740.88	6.5	1	50
16501	500	16001	1569	9.508515	3.030119	4.138	3791.87	6.6	1	50
16750	479	16271	1592	9.504478	2.859701	4.323591	3845.32	6.7	1	50
17000	386	16614	1699	9.994118	2.270588	5.401554	3894.48	6.8	1	50
17250	501	16749	1572	9.113043	2.904348	4.137725	3945.47	6.9	1	50
17501	500	17001	1573	8.988058	2.85698	4.146	3998.52	7	1	50
17750	320	17430	2024	11.40282	1.802817	7.325	4163.79	7.1	1	50
18000	483	17517	1659	9.216667	2.683333	4.434783	4102.14	7.2	1	50
18251	364	17887	1700	9.314558	1.994411	5.67033	4236.9	7.3	1	50
18500	153	18347	1887	10.2	0.827027	13.33333	4433.92	7.4	1	50
18751	482	18269	1612	8.596875	2.57053	4.344398	4253.49	7.5	1	50
19001	389	18612	1684	8.862691	2.047261	5.329049	4305.92	7.6	1	50
19250	187	19063	1923	9.98961	0.971429	11.28342	4458.09	7.7	1	50
19500	485	19015	1602	8.215385	2.487179	4.303093	4405.66	7.8	1	50

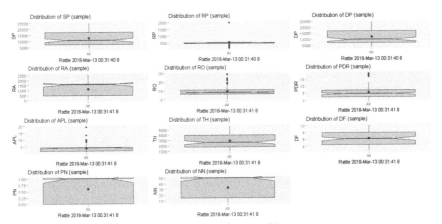

Fig. 8: Box plot for features of dataset.

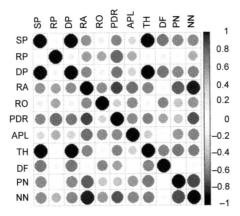

Fig. 9: Correlation between features of dataset.

evaluated on the basis of Correlation (denoted by 'r'), Coefficient of Determination (denoted by 'R^2'), Root Mean Square Error (denoted by '$RMSE$'), Accuracy (denoted by 'Acc') and Time Taken (denoted by 'TT'). Correlation provides information about how actual and predicted values are linked. It can be calculated by the Karl Pearson formula of correlation [53]. The value of correlation lies in [0,1]. The greater the value tends towards 1, the better is the correlation. The Coefficient of Determination ($R2$) is the primary outcome of regression evaluation and its value lies in the interval $0 < R2 < 1$. The greater the value tends towards 1, the better is the regression model [54]. If the value is zero, it means that the regression model is a failure. Mathematically, it is the square of correlation. Root Mean Square Error ($RMSE$) tells us how much error is present between the actual and predicted values [55]. The accuracy represents how close the predicted value is to the actual value and within acceptable error limits [56]. It predicts which model best suits a problem. Time taken tells the total time taken by a machine learning model for execution [56].

All the above models require a formula in *R* programming language for computation:

```
formula ← as. formula(paste(target,"~",paste(c(selected inputs),collapse = "+")))
```

Linear Model: This model is a linear amalgamation of dataset features or attributes [70]. It means that there is a linear relationship between the target and the other features of the dataset [71]. The code in *R* for linear model execution is:

```
                    Linear Model ← lm (formula, train Dataset)
```

Neural Network: These networks are constructed upon the concept of artificial neurons spread along multiple layers [72]. The non-linear patterns are easily recognized, using these models instead or linear one-to-one relationship [73]. The input layer deals with the neurons which only take input data. The input neurons are equal to the number of features/attributes of the dataset [74] and the output layer yields a single neuron to obtain the result in the case of a regression problem. The entire computation takes place in the hidden layer sandwiched in between the input and the output layers [75]. The code in *R* for neural network model execution is:

```
Neural Network Model ← nnet (formula, train Dataset, size=10, linout=TRUE,
        skip=TRUE, MaxNWts=10000, trace=FALSE, maxit=100)
```

The tuning parameters used in the neural network model are size, linout, skip, MaxNWts, trace and maxit.

Random Forest Model: It is an ensemble-based supervised learning model [76] whose accuracy is dependent on the total number of trees grown in the forest. Larger the number of trees the greater is the accuracy [77]. It yields better prediction accuracy in a majority number of regression scenarios because the variances of many input variables are captured, enabling a much higher number of features to cooperate in the prediction accuracy [78]. Feature importance is one such parameter which can be computed easily with the help of this algorithm [79, 80]. The code in *R* for random forest model execution is:

```
          Random Forest ← random Forest (formula,train Dataset,
                ntree=5,mtry=2,importance=TRUE)
```

The tuning parameters used in the random forest model are ntree, mtry and importance.

Decision Tree Model: Decision tree can be used for both the problems regression and classification [81]. The core concept of this model is a division of the dataset into smaller datasets [82, 83] which are conceptualized on the descriptive features. This division takes place until the smallest dataset is obtained containing the data points to be categorized under a single specific label. Every feature of the dataset is the root and the outcome is represented as a child [84, 85]. The splitting of a particular feature is dependent on two values—information gain [86] and the entropy [87] from the split. The continuous values in a regression problem are converted to categorical values before splitting them at the root node. In case of regression, the split is decided, based on the information gain parameter. The only disadvantage of the decision tree is that it over-fits noisy data. The code in *R* for decision tree model execution is:

```
Decision Tree ← rpart(formula, train Dataset, parms=list(split="information"),
                control=rpart.control (use surrogate=0, max surrogate=0))
```

The tuning parameters for decision tree are 'parms' and control.

The computed results in Table 3 can be statistically evaluated to obtain a final outcome. From this table, we can conclude that if we want a better result in terms of accuracy, we will opt for the linear model and neural network. Similarly, if we want better results in terms of error, we will prefer the neural network model. In a nutshell, for overall optimization, we need to decide which criteria best suit the application requirement.

Table 3: Model evaluation results.

Machine Learning Model	References	*r*	*R²*	*RMSE*	*Acc*	*TT*
Linear Model	[57, 63]	0.99	0.98	0.2	86.05	514.19
Neural Network	[58, 64, 65]	0.31	0.1	0.01	83.72	514.94
Random Forest	[59, 66, 67]	0.96	0.92	0.41	62.79	512.89
Decision Tree	[60, 68, 69]	0.95	0.9	0.47	48.84	513.48

2.3 Cross-validation System

For validating the chosen model, cross-validation is carried out. The most popular cross-validation technique is the K-fold cross-validation [61, 62]. We use 10-fold cross-validation and the machine learning models are repeatedly executed 10 times to validate the robustness of the results. The cross-validation results for correlation are shown in Table 4. Similar results can be achieved for rest of the performance evaluation parameters as described in Figs. 10, 11, 12, 13 and 14. These figures

Table 4: Ten fold cross-validation for correlation.

Runs	Linear Model	Neural Network	Random Forest	Decision Tree
1	0.99	0.31	0.96	0.95
2	1	0.31	0.95	0.95
3	0.98	0.3	0.95	0.94
4	0.99	0.3	0.96	0.9
5	0.99	0.32	0.96	0.94
6	0.99	0.31	0.95	0.95
7	0.99	0.31	0.97	0.9
8	0.98	0.31	0.96	0.93
9	1	0.3	0.96	0.95
10	0.99	0.31	0.96	0.95

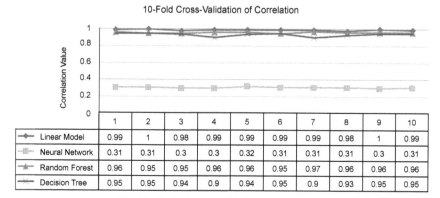

	1	2	3	4	5	6	7	8	9	10
Linear Model	0.99	1	0.98	0.99	0.99	0.99	0.99	0.98	1	0.99
Neural Network	0.31	0.31	0.3	0.3	0.32	0.31	0.31	0.31	0.3	0.31
Random Forest	0.96	0.95	0.95	0.96	0.96	0.95	0.97	0.96	0.96	0.96
Decision Tree	0.95	0.95	0.94	0.9	0.94	0.95	0.9	0.93	0.95	0.95

Fig. 10: Tenfold cross-validation for correlation.

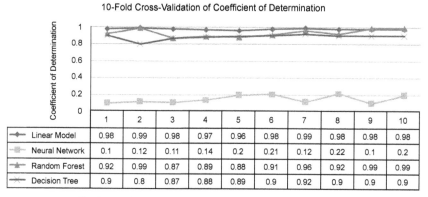

	1	2	3	4	5	6	7	8	9	10
Linear Model	0.98	0.99	0.98	0.97	0.96	0.98	0.99	0.98	0.98	0.98
Neural Network	0.1	0.12	0.11	0.14	0.2	0.21	0.12	0.22	0.1	0.2
Random Forest	0.92	0.99	0.87	0.89	0.88	0.91	0.96	0.92	0.99	0.99
Decision Tree	0.9	0.8	0.87	0.88	0.89	0.9	0.92	0.9	0.9	0.9

Fig. 11: Tenfold cross-validation for coefficient of determination.

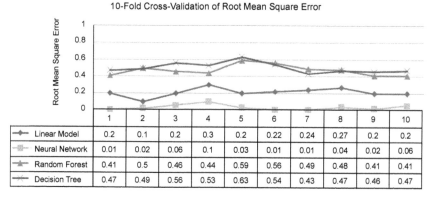

	1	2	3	4	5	6	7	8	9	10
Linear Model	0.2	0.1	0.2	0.3	0.2	0.22	0.24	0.27	0.2	0.2
Neural Network	0.01	0.02	0.06	0.1	0.03	0.01	0.01	0.04	0.02	0.06
Random Forest	0.41	0.5	0.46	0.44	0.59	0.56	0.49	0.48	0.41	0.41
Decision Tree	0.47	0.49	0.56	0.53	0.63	0.54	0.43	0.47	0.46	0.47

Fig. 12: Tenfold cross-validation for RMSE.

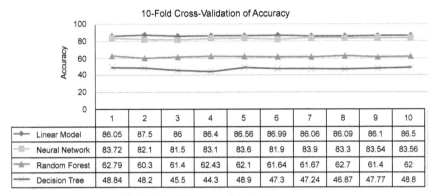

	1	2	3	4	5	6	7	8	9	10
Linear Model	86.05	87.5	86	86.4	86.56	86.99	86.06	86.09	86.1	86.5
Neural Network	83.72	82.1	81.5	83.1	83.6	81.9	83.9	83.3	83.54	83.56
Random Forest	62.79	60.3	61.4	62.43	62.1	61.64	61.67	62.7	61.4	62
Decision Tree	48.84	48.2	45.5	44.3	48.9	47.3	47.24	46.87	47.77	48.8

Fig. 13: Tenfold cross-validation for accuracy.

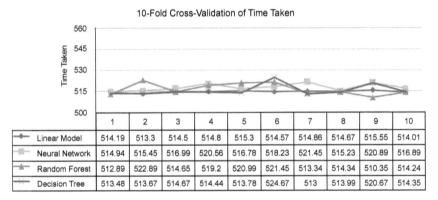

	1	2	3	4	5	6	7	8	9	10
Linear Model	514.19	513.3	514.5	514.8	515.3	514.57	514.86	514.67	515.55	514.01
Neural Network	514.94	515.45	516.99	520.56	516.78	518.23	521.45	515.23	520.89	516.89
Random Forest	512.89	522.89	514.65	519.2	520.99	521.45	513.34	514.34	510.35	514.24
Decision Tree	513.48	513.67	514.67	514.44	513.78	524.67	513	513.99	520.67	514.35

Fig. 14: Tenfold cross-validation for time taken.

show the results after performing 10-fold cross-validation for '*r*', '*R2*', '*RMSE*', '*Acc*' and '*TT*'.

Hence, a final conclusion can be drawn, based upon the application area requirement. This conclusion is termed as the Final Prediction (output). Also, after the entire process of prediction using machine learning models is completed, the models developed are saved. When a new network is introduced with some specific features, these models can be used to provide the desired output. In short, the past experience can be used to provide futuristic research directions.

3. Conclusion

Adaptability is the basic element of any wireless network deployed in the hostile environment. It governs the QS provided by the network and hence affects dependability. The fluctuating traffic rates need to be optimized for wireless networks. This paves the way for futuristic research in the domain of optimization and machine learning, which can provide better regulation of traffic flow. This chapter considers linear model, neural network, random forest and decision tree for predicting the data flow in a Wireless Sensor Network. The linear model outperforms the rest of the models in case of accuracy. The neural network model results in minimum value of root mean square error. The choice of a prediction model is based on the requirement of application domain, in which it is deployed.

References

[1]   Che-Jung Hsu, Huey-Ing Liu and Winston K.G. Seah. (2011). Opportunistic routing—A review and the challenges ahead. Computer Networks 55: 3592–3603.
[2]   Abdul Wahid, Gaurav Kumar and Khaleel Ahmad. (2014). Opportunistic networks: Opportunity versus challenges-survey. National Conference on Information Security Challenges 14–24.
[3]   Aqeel-ur-Rehman, Abu Zafar Abbasi, Norman Islam and Ahmed Shaikh. (2014). A review of wireless sensors and networks applications in agriculture. Computer Standards and Interfaces 36: 263–270.
[4]   Akyildiz, I.F., Su, W., Sankarasubramaniam, Y. and Cayirci, E. (2002). Wireless sensor networks: A survey. Computer Networks 38: 393–422.
[5]   Jun Zheng and Abbas Jamalipour. (2009). Wireless Sensor Networks—A Networking Perspective, Wiley & Sons Publication, IEEE Press.
[6]   Amir Ehsani Zonouz, Liudong Xing, Vinod M. Vokkarane and Yan Sun. (2016). Hybrid wireless sensor networks: A reliability, cost and energy-aware approach. IET Wireless Sensor Systems.
[7]   Elghazel, W., Bahi, J., Guyeux, C., Hakem, M., Medjaher, K. and Zerhouni, N. (2015). Dependability of wireless sensor networks for industrial prognostics and health management. Computers in Industry 68: 1–15.
[8]   Muhammad Adeel Mahmood, Winston K.G. Seah and Ian Welch. (2015). Reliability in wireless sensor networks: A survey and challenges ahead. Computer Networks 79: 166–187.
[9]   Arslan Munir, Joseph Antoon and Ann Gordon-Ross. (2019). Modeling and analysis of fault detection and fault tolerance in wireless sensor networks. ACM Transactions on Embedded Computing Systems 14(1): 3 3.1–3.43.
[10] Guangjie Han, Jinfang Jiang, Lei Shu, Jianwei Niu and Han-Chieh Chao. (2014). Management and applications of trust in wireless sensor networks: A survey. Journal of Computer and System Sciences 80: 602–617.
[11] Antonio Damaso, Nelson Rosa and Paulo Maciel. (2014). Reliability of wireless sensor networks. Sensors 14(9): 15760–15785.
[12] Joseph E. Mbowe and George S. Oreku. (2014). Quality of service in wireless sensor networks. Scientific Research Wireless Sensor Network 6(2): 19–26.
[13] Silva, B., Callou, G., Tavares, E., Maciel, P., Figueiredo, J., Sousa, E. and Neves, F. (2013). Astro: An integrated environment for dependability and sustainability evaluation. Sustainable Computing: Informatics and Systems 3(1): 1–17.
[14] Simona Bernardi, Jose Merseguer and Dorina Corina Petriu. (2013). Model-driven Dependability Assessment of Software System, Springer-Verlag, Berlin, Heidelberg.
[15] Hegde, S.B., Babu, B.S. and Venkataram, P. (2017). A cognitive theory-based opportunistic resource-pooling scheme for Ad hoc networks. Journal of Intelligent Systems 26(1): 47–68.
[16] Sazia Parvin, Farookh Khadeer Hussain, Jong Sou Park and Dong Seong Kim. (2012). A survivability model in wireless sensor networks. Computers and Mathematics with Applications 64: 3666–3682.
[17] Yongxian Song, Ting Chen, Juanli Ma, Yuan Feng and Xianjin Zhang. (2012). Design and analysis for reliability of wireless sensor network. Journal of Networks 7(12): 2003–2010.
[18] Ivanovitch Silva, Luiz Affonso Guedes, Paulo Portugal and Francisco Vasques. (2012). Reliability and availability evaluation of wireless sensor networks for industrial applications. Sensors 12(1): 806–838.

[19] Jaime Chen, Manuel Diaz, Luis Llopis, Bartolome Rubio and Jose M. Troya. (2011). A survey on quality of service support in wireless sensor and actor networks: Requirements and challenges in the context of critical infrastructure protection 34: 1225–1239.

[20] Yves Langeron, Anne Barros, Antoine Grall and Christophe Berenguer. (2011). Dependability assessment of network-based safety-related system. Journal of Loss Prevention in the Process Industries 24: 622–631.

[21] Chang, Y.R., Huang, C.Y. and Kuo, S.Y. (2010). Performance assessment and reliability analysis of dependable and distributed computing systems based on BDD and recursive merge. Applied Mathematics and Computation 217(1): 403–413.

[22] James, P.G. Sterbenz, David Hutchison, Egemen K. Cetinkaya, Abdul Jabbar, Justin P. Rohrer, Marcus Scholler and Paul Smith. (2010). Resilience and survivability in communication networks: Strategies, principles, and survey of disciplines. Computer Networks 54: 1245–1265.

[23] Zhengyi Le, Eric Becker, Dimitrios G. Konstantinides, Chris Ding and Fillia Makedon. (2010). Modeling reliability for wireless sensor node coverage in assistive testbeds. p. 6. *In*: Proceedings of the 3rd International Conference on Pervasive Technologies Related to Assistive Environments (PETRA '10), ACM, New York, Article 46.

[24] Dario Bruneo, Antonio Puliafito and Marco Scarpa. (2010). Dependability analysis of wireless sensor networks with active-sleep cycles and redundant nodes. pp. 25–30. *In*: Proceedings of the First Workshop on Dynamic Aspects in Dependability Models for Fault-Tolerant Systems (DYADEM-FTS '10), ACM, New York.

[25] Cardellini, Valeria, Emiliano Casalicchio, Vincenzo Grassi, Francesco Lo Presti and Raffaela Mirandola. (2009). Towards self-adaptation for dependable service-oriented systems. Architecting Dependable Systems-VI, Springer, Berlin, Heidelberg, pp. 24–48.

[26] Lihua Xu, Hadar Ziv, Thomas A. Alspaugh and Debra J. Richardson. (2006). An architectural pattern for non-functional dependability requirements. The Journal of Systems and Software 79: 1370–1378.

[27] Lance Doherty and Dana A. Teasdale. (2006). Towards 100% reliability in wireless monitoring networks. pp. 132–135. *In*: Proceedings of the 3rd ACM International Workshop on Performance Evaluation of Wireless Adhoc, Sensor and Ubiquitous Networks (PE-WASUN '06), ACM, New York.

[28] Amirhosein Taherkordi, Majid Alkaee Taleghan and Mohsen Sharifi. (2006). Dependability considerations in wireless sensor networks applications. Journal of Networks 1(6): 28–35.

[29] Michael G. Hinchey, James L. Rash, Christopher A. Rouff and Denis Gracanin. (2006). Achieving dependability in sensor networks through automated requirements-based programming. Computer Communications 29: 246–256.

[30] Claudia Betous-Almeida and Karama Kanoun. (2004). Construction and stepwise refinement of dependability models. Performance Evaluation 56: 277–306.

[31] Kaaniche, M., Laprie, J.C. and Blanquart, J.P. (2002). A framework for dependability engineering of critical computing systems. Safety Science 40(9): 731–752.

[32] Walter J. Gutjahr. (2000). Software dependability evaluation based on Markov usage models. Performance Evaluation 40: 199–222.

[33] Platis, A., Limnios, N. and Le Du, M. (1998). Dependability analysis of systems modeled by non-homogeneous Markov chains. Reliability Engineering & System Safety 61(3): 235–249.

[34] Shelke, Maya, Akshay Malhotra and Parikshit N. Mahalle. (2018). Congestion-aware opportunistic routing protocol in wireless sensor networks. pp. 63–72. *In*: Smart Computing and Informatics, Springer, Singapore.

[35] Mittal, Vikas, Sunil Gupta and Tanupriya Choudhury. (2018). Comparative analysis of authentication and access control protocols against malicious attacks in wireless sensor networks. pp. 255–262. *In*: Smart Computing and Informatics, Springer, Singapore.

[36] Tavakoli Najafabadi, R., Nabi Najafabadi, M., Basten, T. and Goossens, K.G.W. (2018). Dependable Interference-aware Time-slotted Channel Hopping for Wireless Sensor Networks.

[37] Malhotra, Sachin and Munesh C. Trivedi. (2018). Symmetric key based authentication mechanism for secure communication in MANETs. pp. 171–180. *In*: Intelligent Communication and Computational Technologies, Springer, Singapore.

[38] Sharma, Vivek, Bashir Alam and Doja, M.N. (2018). Performance enhancement of AODV routing protocol using ANFIS technique. pp. 307–312. *In*: Quality, IT and Business Operations, Springer, Singapore.

[39] Thirukrishna, J.T., Karthik, S. and Arunachalam, V.P. (2018). Revamp energy efficiency in homogeneous wireless sensor networks using Optimized Radio Energy Algorithm (OREA) and power-aware distance source routing protocol. Future Generation Computer Systems 81: 331–339.

[40] Kumar, M. Jothish and Baskaran Ramachandran. (2018). Predict-act congestion control (pacc) in wireless sensor networks for routing. Journal of Computational and Theoretical Nanoscience 15(1): 133–140.

[41] Intanagonwiwat, Chalermek, Ramesh Govindan and Deborah Estrin. (2000). Directed diffusion: A scalable and robust communication paradigm for sensor networks. pp. 56–67. *In*: Proceedings of the 6th Annual International Conference on Mobile Computing and Networking, ACM.

[42] Suresh, Rengan, Feng Tao, Johnathan Votion and Yongcan Cao. (2018). Machine learning approaches for multi-sensor data pattern recognition: K-means, deep neural networks, and multi-layer K-means. p. 1639. *In*: 2018 AIAA Information Systems-AIAA Infotech@ Aerospace.

[43] Zhou, Lina, Shimei Pan, Jianwu Wang and Athanasios V. Vasilakos. (2017). Machine learning on big data: Opportunities and challenges. Neurocomputing 237: 350–361.

[44] Lantz, Brett. (2013). Machine Learning with R., Packt Publishing Ltd.

[45] Pedregosa, Fabian, Gaël Varoquaux, Alexandre Gramfort, Vincent Michel, Bertrand Thirion, Olivier Grisel, Mathieu Blondel et al. (2011). Scikit-learn: Machine learning in Python. Journal of Machine Learning Research 12: 2825–2830.

[46] Li, Hongzhi, Wenze Li, Xuefeng Pan, Jiaqi Huang, Ting Gao, LiHong Hu, Hui Li and Yinghua Lu. (2018). Correlation and redundancy on machine learning performance for chemical databases. Journal of Chemometrics.

[47] Hall, Mark Andrew. (1999). Correlation-based Feature Selection for Machine Learning.

[48] Huang, Guang-Bin, Qin-Yu Zhu and Chee-Kheong Siew. (2006). Extreme learning machine: Theory and applications. Neurocomputing 70(1-3): 489–501.

[49] Akay, Mehmet Fatih. (2009). Support vector machines combined with feature selection for breast cancer diagnosis. Expert Systems with Applications 36(2): 3240–3247.

[50] Huang, Guang-Bin, Hongming Zhou, Xiaojian Ding and Rui Zhang. (2012). Extreme learning machine for regression and multiclass classification. IEEE Transactions on Systems, Man, and Cybernetics, Part B (Cybernetics) 42(2): 513–529.

[51] Nasrabadi, Nasser M. (2007). Pattern recognition and machine learning. Journal of Electronic Imaging 16(4): 049901.

[52] Witten, Ian H., Eibe Frank, Mark A. Hall and Christopher J. Pal. (2016). Data Mining: Practical Machine Learning Tools and Techniques, Morgan Kaufmann.

[53] Dormann, Carsten F., Jane Elith, Sven Bacher, Carsten Buchmann, Gudrun Carl, Gabriel Carré, Jaime R. García Marquéz et al. (2013). Collinearity: A review of methods to deal with it and a simulation study evaluating their performance. Ecography 36(1): 27–46.

[54] Guyon, Isabelle and André Elisseeff. (March 2003). An introduction to variable and feature selection. Journal of Machine Learning Research 3: 1157–1182.

[55] Salakhutdinov, Ruslan, Andriy Mnih and Geoffrey Hinton. (2007). Restricted Boltzmann machines for collaborative filtering. pp. 791–798. *In*: Proceedings of the 24th International Conference on Machine Learning, ACM.

[56] Brazdil, Pavel B., Carlos Soares and Joaquim Pinto Da Costa. (2003). Ranking learning algorithms: Using IBL and meta-learning on accuracy and time results. Machine Learning 50(3): 251–277.

[57] Robert, Christian. (2014). Machine Learning, A Probabilistic Perspective, pp. 62–63.

[58] Adeli, Hojjat and Shih-Lin Hung. (1994). Machine Learning: Neural Networks, Genetic Algorithms and Fuzzy Systems, John Wiley & Sons, Inc.

[59] Pal, Mahesh. (2005). Random forest classifier for remote sensing classification. International Journal of Remote Sensing 26(1): 217–222.

[60] Quinlan, J. Ross. (1986). Induction of decision trees. Machine Learning 1(1): 81–106.
[61] Bergmeir, Christoph, Rob J. Hyndman and Bonsoo Koo. (2018). A note on the validity of cross-validation for evaluating autoregressive time series prediction. Computational Statistics & Data Analysis 120: 70–83.
[62] Refaeilzadeh, Payam, Lei Tang, and Huan Liu. (2009). Cross-validation. pp. 532–538. *In*: Encyclopedia of Database Systems, Springer, US.
[63] Takigawa, Ichigaku, Ken-ichi Shimizu, Koji Tsuda and Satoru Takakusagi. (2018). Machine learning predictions of factors affecting the activity of heterogeneous metal catalysts. pp. 45–64. *In*: Nanoinformatics, Springer, Singapore.
[64] Zhang, Pengfei, Huitao Shen and Hui Zhai. (2018). Machine learning topological invariants with neural networks. Physical Review Letters 120(6): 066401.
[65] Pham, Hieu, Melody Y. Guan, Barret Zoph, Quoc V. Le and Jeff Dean. (2018). Faster Discovery of Neural Architectures by Searching for Paths in a Large Model.
[66] Blanco, Carlos M. Guio, Victor M. Brito Gomez, Patricio Crespo and Mareike Lieβ. (2018). Spatial prediction of soil water retention in a Páramo landscape: Methodological insight into machine learning using random forest. Geoderma 316: 100–114.
[67] Zimmerman, Naomi, Albert A. Presto, Sriniwasa P.N. Kumar, Jason Gu, Aliaksei Hauryliuk, Ellis S. Robinson, Allen L. Robinson and Subramanian, R. (2018). A machine learning calibration model using random forests to improve sensor performance for lower-cost air quality monitoring. Atmospheric Measurement Techniques 11(1): 291.
[68] Dumitrescu, Elena, Sullivan Hue, Christophe Hurlin and Sessi Tokpavi. (2018). Machine Learning for Credit Scoring: Improving Logistic Regression with Non Linear Decision Tree Effects.
[69] Sivakumar, Subitha and Rajalakshmi Selvaraj. (2018). Predictive modeling of students performance through the enhanced decision tree. pp. 21–36. *In*: Advances in Electronics, Communication and Computing, Springer, Singapore.
[70] Preacher, Kristopher J., Patrick J. Curran and Daniel J. Bauer. (2006). Computational tools for probing interactions in multiple linear regression, multilevel modeling, and latent curve analysis. Journal of Educational and Behavioral Statistics 31(4): 437–448.
[71] Guisan, Antoine, Thomas C. Edwards Jr, and Trevor Hastie. (2002). Generalized linear and generalized additive models in studies of species distributions: Setting the scene. Ecological Modeling 157(2-3): 89–100.
[72] Kawato, Mitsuo, Kazunori Furukawa and Suzuki, R. (1987). A hierarchical neural-network model for control and learning of voluntary movement. Biological Cybernetics 57(3): 169–185.
[73] Kosko, Bart. (1992). Neural Networks and Fuzzy Systems: A Dynamical Systems Approach to Machine Intelligence/Book and Disk, Vol. 1, Prentice Hall.
[74] Bose, Nirmal K. and P. Liang. (1996). Neural Network Fundamentals with Graphs, Algorithms and Applications, McGraw-Hill Series in Electrical and Computer Engineering.
[75] Schalkoff, Robert J. (1997). Artificial Neural Networks, Vol. 1. New York: McGraw-Hill.
[76] Svetnik, Vladimir, Andy Liaw, Christopher Tong, J. Christopher Culberson, Robert P. Sheridan and Bradley P. Feuston. (2003). Random forest: A classification and regression tool for compound classification and QSAR modeling. Journal of Chemical Information and Computer Sciences 43(6): 1947–1958.
[77] Strobl, Carolin, James Malley and Gerhard Tutz. (2009). An introduction to recursive partitioning: rationale, application, and characteristics of classification and regression trees, bagging, and random forests. Psychological Methods 14(4): 323.
[78] Guo, Lan, Yan Ma, Bojan Cukic and Harshinder Singh. (2004). Robust prediction of fault-proneness by random forests. pp. 417–428. *In*: Software Reliability Engineering, ISSRE 2004. 15th International Symposium on, IEEE.
[79] Grömping, Ulrike. (2009). Variable importance assessment in regression: linear regression versus random forest. The American Statistician 63(4): 308–319.
[80] Strobl, Carolin, Anne-Laure Boulesteix, Thomas Kneib, Thomas Augustin and Achim Zeileis. (2008). Conditional variable importance for random forests. BMC Bioinformatics 9(1): 307.

[81] Chen, Mu-Yen. (2011). Predicting corporate financial distress based on integration of decision tree classification and logistic regression. Expert Systems with Applications 38(9): 11261–11272.

[82] Fan, Chin-Yuan, Pei-Chann Chang, Jyun-Jie Lin and Hsieh, J.C. (2011). A hybrid model combining case-based reasoning and fuzzy decision tree for medical data classification. Applied Soft Computing 11(1): 632–644.

[83] Markey, Mia K., Georgia D. Tourassi and Carey E. Floyd. (2003). Decision tree classification of proteins identified by mass spectrometry of blood serum samples from people with and without lung cancer. Proteomics 3(9): 1678–1679.

[84] Im, Jungho and John R. Jensen. (2005). A change detection model based on neighborhood correlation image analysis and decision tree classification. Remote Sensing of Environment 3: 326–340.

[85] Cho, Yoon Ho, Jae Kyeong Kim and Soung Hie Kim. (2002). A personalized recommender system based on web usage mining and decision tree induction. Expert Systems with Applications 23(3): 329–342.

[86] Rokach, Lior and Oded Maimon. (2005). Top-down induction of decision trees classifiers—a survey. IEEE Transactions on Systems, Man, and Cybernetics, Part C (Applications and Reviews) 35(4): 476–487.

[87] Felicísimo, Ángel M., Aurora Cuartero, Juan Remondo and Elia Quirós. (2013). Mapping landslide susceptibility with logistic regression, multiple adaptive regression splines, classification and regression trees, and maximum entropy methods: A comparative study. Landslides 10(2): 175–189.

CHAPTER 4

A Bio-inspired Approach to Cyber Security

Siyakha N. Mthunzi,[1] Elhadj Benkhelifa,[1,]* Tomasz Bosakowski[2] and Salim Hariri[3]

1. Introduction

With little consensus on definitions to concepts, such as cyber security, cyberspace and most 'things' cyber [1], addressing cyber security is often inadequate due to its misinterpretations and mistranslations. Across literature, there is convergence in the view that cyber security, unlike traditional computer security, lacks the defining clarity of what the 'cyber' prefix contributes to the general security concept [1] and is perhaps, a source of confusion and misunderstanding across various perspectives [2]. In recognition of this lack of consensus, the authors of this chapter find it prudent for a general outline to the current cyber security definition landscape to be surmised by academia, industry, government and across professionals in general. A conception of cyber security-encompassing network and communication infrastructures and the human actor/user suggests cyber security in the context of the security of anything (including physical artefacts) that interacts with computer networks and communication infrastructures. The UK government, for instance, defines cyber security as "an interactive domain for digital networks across the world" [3]. Clearly, this view pivots the general citizenry at the crux of the security objective. The significance of the human actor/factor is evidenced in the European Commission's prediction of a major global breakdown in electronic communication services and networks (costing around €193 billion) due to malicious human action and terrorist attacks. An 'inclusive'

---

[1] Cloud Computing and Applications Research Lab, School of Computing and Digital Technologies, Staffordshire University, Mellor Building, College Road. Stoke-on-Trent, ST4 2DE.
  E-mail: siyakha.mthunzi@research.staffs.ac.uk
[2] School of Computing and Digital Technologies, Staffordshire University, Mellor Building, College Road. Stoke-on-Trent, ST4 2DE; T.Bosakowski@staffs.ac.uk
[3] University of Arizona.
* Corresponding author: E.Benkhelifa@staffs.ac.uk

definition of cyber security is posited by [2] as "the approach and actions associated with security risk management processes followed by organizations and states to protect confidentiality, integrity and availability of data assets used in cyberspace." This context of the cyber security concept includes guidelines, policies and collection of safeguards, technologies, tools and training to provide best protection for cyber environments and their users.

This chapter considers cyber security as a continuum of technologies and innovations for ensuring and assuring the security of data and networking technologies. It identifies the complexity and dynamic context of cyberspace as central to mitigating catastrophic cyber threats and attacks by drawing inspiration from Nature's complex and dynamic systems. This chapter explores how natural phenomenon in complex systems (including animalia and plants) that have survived through evolution, could be exploited as mechanisms for adaptive mitigation in complex cyber environments. Drawing from predation avoidance and anti-predation techniques employed by non-extinct prey animals and plants, this chapter hypothesizes how prey-inspired survivability could be adopted in cyber systems design and implementation.

2. Background

Recent years have witnessed an exponential increase in cyber crime, arguably exacerbated by the adoption of emerging technologies, such as IoT and cloud computing. In 2015 alone, most of the breaches considered as hacking incidents targeted customers' bank details, addresses and other personal information. Over the years, hacking incidents have grown to encompass all aspects of a modern economy, including transport, energy, banking, healthcare, telecommunications and in some instances, government organizations. Cyber-hacking incidents, including German and UK telecommunications giants Vodafone [4] and TalkTalk [5], respectively, act as perfect examples of the scale, frequency and implication of such attacks. Most recently in the UK, Dixons Carphone suffered a huge data breach involving 5.9 million payment cards and 1.2 million personal data records [6]. These breaches raise concerns about the readiness of security solutions in an environment of highly sophisticated, persistent and motivated adversaries, particularly if considered against their implications on utilities, such as power, transport, etc. With new inventions in smart health and healthcare being a sensitive area, security in such sectors is critical [7].

Traditional computer network environments allow for highly manageable security processes that restrain user permissions, restrict software, roll out updates across campuses, centrally manage intrusion detection and control network traffic through firewall policies. Routing and other reactive approaches ensure efficient and effective control of the security of networks and devices. Thus, the degree of control in meeting security goals varies considerably between user-case and environment. Regardless, this fine-tuned control allows at the minimum, adequate logging and understanding of large networked environments. However, with the advent and wide use of cyberspace, security is arguably more complex and requires different processes to maintain data confidentiality, integrity, availability (CIA) and systems stability. Technologies, such as IoT and Big Data, among others, make traditional firewall deployment a challenge, not least because of bandwidth and power wastage on devices in a multi-hopping routing environment under a Denial of Service attacks (DoS), but enforcing static security policies in highly mobile environment posing additional challenges. This is nearly impossible in cyber environments with a range of mobile and non-mobile devices and various communication interfaces; and networks are formed in a variety of manners, forms and structures. Furthermore, it can be argued that the CIA triage, an industry standard addressing the security domain [8], is deficient in the cyber domain. As [9] notes, other information characteristics attributed to cyber environments ought to be added to the CIA model [9].

Cyberspace's vision of complex and highly dynamic network and communication environments can be summarized as isolated nodes which are at risk from a variety of security threats, including forced networking for malicious purposes. Thus, the potential for cyber security requires a novel type of security system to help defend it. As the security crisis in cyberspace escalates and the frequency and complexity of the latest vulnerabilities and cyber-attacks increases, the mandate to adopt more effective solutions will grow in importance to implement simplified, animated and cost-effective cyber solutions. This should contribute to the productivity of existing solutions, including the human information security professional. Moreover, the existing traditional security technologies, such as firewalls, anti-virus scanners and user access control aid in restricting known threats. However, without additional intelligent components to oversee and integrate with the effectiveness of these security controls (as is the case in enterprise networks), if any of these are subverted, the results, as has been shown, are catastrophic.

3. Cyberspace

To examine the cyber landscape, it is important to understand the complexity that characterizes cyberspace. Complexity is a phenomenon that can be observed in a variety of systems, including physical and living systems. While a complete and unanimous definition of complexity is somewhat contentious across domains [10], the discussions in this chapter revolve around the scientific definition posited by [10]. Complex systems describe "phenomena, structure, aggregates, organisms, or problems that share the following common themes: they are inherently complicated or intricate, they are rarely completely deterministic, mathematical models of the system are usually complex and involve non-linear, ill-posed, or chaotic behavior, and the systems are predisposed to unexpected outcomes (emergent behavior)" [10].

While large-scale networks are inherently complex and true for cyberspace and cyber security [11], complexity itself is a useful concept in designing robust systems [12]. In this chapter, complexity in cyberspace describes the general sense of a system whose components are by design or function, or both, challenging to verify. In fact, the current chapter postulates complexity in cyberspace in relation to interactions within networks and communication systems, including system users and misusers and the unpredictability of emerging behaviors. As noted by [12], complex systems are such that interactions among systems components could be unintended, for instance, unintended interaction with system data which result in unpredictable system outputs and emergent behaviors not intended for that system. And as literature will demonstrate in the sections below, unpredictability is perhaps an inevitable attribute of cyberspace and its related technologies. Figure 1 illustrates the scale of cyberspace, technologies,

Fig. 1: Illustration of cyberspace.

network and communication systems and other paradigms. This figure is not intended to represent an exhaustive view of the entire cyber domain but to provide illustrative examples. These will be briefly described in bullet points below.

- Cloud computing is the de-facto computing platform and enabler of emerging technologies, such as Bitcoin.
- The Internet of Things (IoT) [13–16] heralds a vision of internet connected things: physical, and via a network to be able to exchange local and global information (themselves and their environments, respectively). IoT technology thus enables further innovations and services based upon to immediate access to the said generated information.
- Infrastructure, like Health, water and transport among many others are critical.
- Contested environments and cyber warfare have become critical in this new era of computer and communication networks.

Complex systems develop on a micro scale through evolution. In evolution, selection pertains to those attributes that foster organisms' survival benefits or disadvantages [17]. For instance, herbivorous mammals whose habitat has tall trees with the most nutrition and fruits atop, are likely to survive and reproduce if they are tall with long necks or can climb up trees. Similarly, elements within complex systems are generally subject to selection, whereupon those best suited for the environment are chosen. For example, products in a free-market economy are selected through market forces, politicians in a democracy through elections/voting and animals through natural pressures, such as predation and competition. Complex natural systems are plentiful with complex patterns of behavior (e.g., adaptation and learning) emanating from interactions among autonomous entities [18]. An example is the adaptation of memory and the self-learning mechanism employed by B-cells in identifying and destroying pathogens in the natural immune system [19]. Thus, adaptation facilitates changes in response to changes within an environment. Using feedback loops, small changes in input information often can trigger large-scale outputs.

Figure 2 illustrates the transformation of a set of components to a network of connected and interdependent components. The graphic design on the left

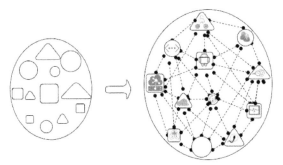

Fig. 2: The formation of complex systems: non-linear, connected, interdependent and adaptive.

shows the building components of a system, a set of autonomous components. Its transformation—graphic on the right, represents a complex system in which autonomous components (numerous) which grow exponentially are connected (dotted lines) and interdependent (black note at edge of dotted line). The nature of their connectivity defines the complexity of the system (in the global sense) rather than its characteristics. Autonomy enables components to adapt through local instructions and collectively synchronize (through cooperation and coordination) individual statuses, resulting in a bottom-up form of order.

With current innovation trends in cyberspace and the proliferation of new devices and platforms with multiparty collaborations, particularly involving third parties, coordinating and controlling interactions among these parties is often error prone. The unpredictable and dynamic nature of complex environments, such as cloud computing (itself a subset of cyberspace) requires intelligent systems control. While traditional computing systems generally maintained consistent control over inherent processes, control theory's [20] classical methodologies and assumptions provide better insight into handling control. The basic premise of control theory is motivated by enhanced adaptation in the presence of extreme unpredictable and dynamic changes [21]. Nonetheless, in non-linear and dynamic cyber environments, the control paradigm ought to be adaptable and dynamically configurable and/or re-configurable. Among a huge state-of-the art technology, classification by the authors in [22] identifies characteristics for such self-adaptions and these are outlined below as:

- Goals: The objectives a system should achieve, e.g., evolution, flexibility, multiplicity, duration, etc.
- Change: The cause of adaptation, e.g., source, type, frequency, anticipation, etc.
- Mechanism: The system's response towards autonomy, scope, duration, timeliness, etc.
- Effect: The impact of adaptations upon the system, e.g., critical, predictability, resilience, etc.

If it is acceptable that cyberspace is complex and therefore falls within the realm of complex systems described above, cyber security research should perhaps seek inspiration from well-established complex adaptive systems, such as those in Nature. Along this thrust, it is important to distinguish the cyber domain according to the its component sub-domains and investigate cyber security challenges.

3.1 Cyber Security Challenges

Cyber security, unlike traditional information security, in not only a process, but also a product and/or technology. As [9] demonstrates, the cyber domain encompasses characteristics beyond those commonly described by a traditional CIA triage, e.g., warfare, bullying, terrorism, etc. A home automation system, for instance, can be compromised without affecting the victim's information (this falls outside the CIA triage and common security attributes), but by targeting the victim's other assets (i.e., cyber crime). Thus, cyber security pertains to the protection of assets beyond those commonly referred to as information, including humans, the interests of nations, states and societies, to household appliances, etc. One may logically conclude perhaps, that cyber security extends to ethical issues related to its assets just as much as the legal. A range of other such dimensions are presented in NATO's national cyber security framework manual [3] and demonstrate the complexity of cyber security.

On the other hand, traditional computing infrastructures mean that security controls were managed within a contained systems [23] and static environments. In this sense, protections against threats was designed and planned based on the assumption that outcomes of security solutions were linearly related to threat. For instance, [24]'s game theoretic approach to protect critical infrastructure against terrorist threats assumes an initial threat score for a particular target according to original and inherent counter measures relevant to that threat. Based on this assumption, they suggest that choices of subsequent solutions will decrease the overall threat. Whilst in a general theoretic sense, functions that convert inputs into required system outputs can be designed and controlled given that all inputs are provided [25], literature shows that the complexity of cyberspace limits the amount of initial threat knowledge cyber security solutions have. It has been demonstrated that sophisticated and persistent adversaries and zero-day attacks are able to systematically plan their attacks and persist within the compromised networks [26]. Cyberspace enables adversaries to increase their attack surface, thus complicating vulnerability management and elevating the attack complexity. Cases in point include Stuxnet, Flame and Duqu, which obfuscate network traffic to evade detection [27]. Based upon the foregoing, this section identifies complexity as central to future cyber security solutions research.

Thus the thrust of future cyber security should aim to reduce complexity to the human cyber security solution (professionals) and build integrated solution capable of mitigating rapidly evolving cyber threats. Current approaches (generally top-down) to cyber security, where extensive efforts focus upon cyber security policy and regulations are inherently inadequate [28] as high-level failure or inadequacy is passed down to low-level elements in cyberspace (including the user and society). On the other hand, further extensive work in academia and industry has been devoted to designing attack and defense tools to efficiently counter cyber security threats, for instance, counter measures integrated into network protocols to ensure reliable information exchange [29]. While these approaches are sufficient for mitigating cyber security threats, it also means that the governance of cyber security risk is harder to implement. Moreover, as literature suggests, counter measures remain inadequate [30].

Figure 3 illustrates the foregoing and highlights the existing research challenges for cyber security. In this graphic design, cyberspace complexity and dynamism introduce data control failure issues and adaption failure issues to cyber security (top of the graphic design). Furthermore, service and resource failure remain a major area

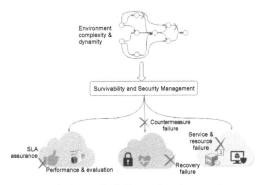

Fig. 3: Research challenges for cyber security.

of concern [31, 32]. In addition, assurance, performance and evaluation remain a constant challenge [33].

Now classified as a 'tier one' threat to national security by British government [34, 35], this shift in strategy at a national level suggests the significance of cyber security. Cyber attacks, such as those in Iraq [36], Iran [37] and during the 'Arab Spring' [38] undoubtedly demonstrate possible implications in future cyber attacks. On the international stage, the United Nations Group of Governmental Experts [39, 40] formally recognized the applicability of international law in relation to cyber activities and cyber issues in the context of international security. Nonetheless, despite a clear consensus as to the applicability of international law, adequately established state practices presents a significant challenge to application of cyber security laws [41].

Existing security frameworks, including Cloud Security Alliance (CSA),[1] National Institute of Standards and Technology (NIST),[2] the European Network and Information Security Agency (ENISA),[3] the UK's Centre of the Protection of National Infrastructure (CPNI),[4] etc., focus upon methods for cyber security aimed to, among other things, consolidate security risks and vulnerabilities. These frameworks aim to provide best practice guidelines for mitigating cyber threats. Table 1 shows example core standardization areas in key cyberspace applications. This is not intended as an exhaustive table either for the standardization areas or for key cyber application areas. However, of interest to the authors of this chapter is a clear demonstration to the urgent need for adequate cyber security (cyber incident management) standardization.

The US DoD [42] defines cyberspace intelligence, surveillance and reconnaissance as activities in cyberspace which result in the gathering of intelligence to support current and future operations [43] while subsequently drawing parallels between this definition and that of espionage, which is defined as being "the unauthorized intentional collection of information by states." Whilst the impact of espionage in cyberspace is

---

Table 1: Example core standardization areas in key cyberspace applications.

Core Areas of Cyber Security Standardization	Examples of Relevant SDOs	Examples of Some Key Applications				
		Cloud Computing	**Emergency Management**	**Industrial Control**	**Health IT**	**Smart Grid**
Cryptographic Techniques	IEEE; ISO TC 68 ISO/IEC JTC 1 W3C	Standards Mostly Available	Some Standards Available	Some Standards Available	Some Standards Available	Some Standards Available
Cyber Incident Management	ISO/IEC JTC 1 ITU-T PCI	Some Standards Available	New Standards Needed	Some Standards Available	Some Standards Available	Some Standards Available
Identity and Access Management	FIDO Alliance; IETF; OASIS OIDF ISO/IEC JTC 1 ITU-T; W3C	Standards Mostly Available	Standards Being Developed	New Standards Needed	Standards Being Developed	New Standards Needed
Information Security Management Systems	ATIS; IEC; ISA ISO/IEC JTC 1 ISO TC 223 OASIS; The Open Group	Some Standards Available	New Standards Needed	Some Standards Available	Some Standards Available	New Standards Needed
IT System Security Evaluation	ISO/IEC JTC 1; The Open Group	Some Standards Available	Standards Mostly Available	Some Standards Available	Some Standards Available	Some Standards Available
Network Security	3GPP; IETF; IEEE; ISO/IEC JTC 1 ITU-T; The Open Group; WiMAX Forum	Some Standards Available	Some Standards Available	Some Standards Available	Some Standards Available	Some Standards Available
Security Automation & Continuous	IEEE; IETF ISO/IEC JTC 1 TCG; The Open Group	Some Standards Available	Some Standards Available	New Standards Needed	Some Standards Available	New Standards Needed
Monitoring Software Assurance	IEEE; ISO/IEC JTC 1 OMG TCG; The Open Group	Some Standards Available	Some Standards Available	Some Standards Available	Some Standards Available	Some Standards Available
Supply Chain Risk Management	IEEE ISO/IEC JTC 1 IEC TC 65 The Open Group	Some Standards Available	Some Standards Available	Some Standards Available	Some Standards Available	Some Standards Available
System Security Engineering	IEC; IEEE; ISA ISO/IEC JTC 1 SAE International The Open Group	Some Standards Available	Standards Mostly Available	Some Standards Available	Some Standards Available	Some Standards Available

typically perceived to be less severe than that of offensive cyber operations, the hacking of the US Democratic National Committee's e-mails in 2016 demonstrates that such activities are still capable of causing significant damage [44]. Furthermore, evidence provided by [45], highlights the pervasiveness of economic and commercial espionage in cyberspace, with likely state-sponsored actors referred to as advanced persistent threats identified as having been conducting systematic hacking on a global scale to access intellectual property and sensitive data since 2014. With a huge explosion of multimedia big data including image, video, 3D, etc., the literature suggests the rise of multi-modal challenges in relation to privacy [46]. Zhang et al. (2017) in fact propose an anti-piracy framework they term as CyVOD to secure multimedia content copyrights against attacks [47].

Whilst historically espionage was limited to small-scale operations with specific targets, [48] argues that the potential scale of espionage operations in cyberspace and their ability to impact the civilian population means that this level of ambiguity and uncertainty in international law can no longer be tolerated. [45] suggests the need for legal reforms adequately reflecting the capabilities of modern technology. Furthermore, theory of expressive law [49] posits that these reforms influence state behavior and identify underlying intolerance in society. In addition, one can conclude such reforms as critical for establishing new domestic laws to reduce the overall pervasiveness of espionage activities in cyberspace. With an overview of security challenges in cyberspace and the subsequent gaps and limitations, the complexity of cyberspace imposes on a range of cyber strategies, the following subsection briefly explores how such gaps and limitations can be counteracted to enhance future cyber security operations.

3.1.1 Enhancing Cyber Security using Artificial Intelligence

To date, ongoing efforts focus towards soft computing and machine learning approaches to enhance computational intelligence [50]. In biology, several authors [51–53] and [54] to name a few, demonstrate the overwhelming use of machine-learning approaches for predicting the survivability of cancer patients. While machine-learning applications have been successfully applied in areas of science and computing security, remarkable growth of cyberspace, i.e., cloud computing, Internet of Things (IoT), Web technologies, mobile computing, digital economy, etc., machine learning approaches have not been consistently applied. A few of the machine learning's core attributes like scalability, adaptability and the ability to adjust to new and unknown changes suggest them as suitable for application in cyberspace. For instance, handling and processing of BigData or the capacity to perform computationally high calculations, which were perhaps challenging in yesteryears, are now significant opportunities that machine learning presents. For cyber security, two main areas are a good fit for machine learning: 1. data processing and 2. expert systems.

The field of artificial intelligence (AI) has advanced faster than anticipated over the past five years [55]. Projects such as DeepMind's AlphaGo [56] is an example of funding commitments towards AI implementations. The application of AI to cyber operations is seen by [55] as a key milestone that has transformative implications for cyber security, enabling States to do more with fewer resources in a manner similar to the impact cyber operations had on the potential scale of intelligence operations. [57–59] in fact identify five sub-fields of AI as possible areas to enhance both

Table 2: Examples of AI application as defensive and offensive countermeasures.

1 AI Subfield	2 Possible Cyber Utility (Defensive)	3 Possible Cyber Utility (Offensive)
4 Expert systems	5 Identifying deceptive media	6 Producing deceptive media
7 Machine learning	8 Threat detection and future forecasting of attacks	9 Detection of open-source vulnerabilities; countermeasure evasion
Deep learning	Attack detection and malware identification	Brute forcing existing countermeasure, e.g., password
Computer vision	Predicting cyber-attacks, detecting and identifying vulnerabilities and generating patching solutions	Discover new sophisticated and zero-day attacks
Natural language processing	Anomaly detection and Botnet identification	Social engineering

offensive and defensive cyber operations. Nonetheless, emerging consensus amongst researchers [57], [58] and [59] highlights the scale of implications introduced by AI. Thus, Allen and Chan [55] urge responsible, sustainable and effective use of AI, including legal, ethical and economic concerns. The table below presents subfields of AI, their possible application as offensive and defensive security countermeasures.

3.2 Need for Unconventional Approaches to Cyber Security

The motivation for seeking inspiration from other systems is necessary due to the observation that cyberspace and inherently cyber security environments are complex. In other fields, for instance robotics, analogies to the immune system are exploited to design self-organization mechanisms [12]. Biological concepts are central and contribute to robust implementations in a range of domains, including computing, financial modeling [60] and robotics [61] to name a few. The underlying strengths of biological systems lies in their distributed architecture, where autonomous entities make local decisions with global implications [62], for instance, the immune system is able adapt and self-protect by dynamically creating and destroying mutated or infected body cells, as it learns new threats and protects itself and its protective components [62]. As virtualization is commonplace in cyberspace, software defined platforms and networks rely heavily on it; however, security monitoring becomes harder as the attack surface is both new and wider [63]. Thus, the robustness of cyber technologies and infrastructures is determined by the quality of underlying virtualization, while sophistication of technological resources influence the level of implementations [64].

Nature effectively demonstrates self-organization, adaptability, resilience and other successful phenomena [65]. The strengths in natural systems reside in the ability of autonomous entities to make local decisions, continuously coordinate and share information, to maintaining a global form of order [18]. The predator-prey dynamics, for instance, highlights the importance and consequences of interactions between two species and how the functions of a community depend on the characteristics of that community.

Biological systems have been a subject of research across the computing continuum stretching back to the 1980s [66]. In recent years, several surveys [67, 65, 68, 69,

19], have dedicated their efforts towards evaluating biologically-inspired algorithms in computing-related applications. With the growth in demand on networked systems and reliance on internet connectivity provided through an assortment of devices and infrastructures [19], it is imperative that computing systems are adaptive, resilient, scalable and robust enough to withstand failure, and dynamic enough to cope with changes. Bio-inspired approaches are argued to provide consistency in performance over a long period of time [61] and indeed have in common, the complexity attributes and the relative success of inter-networked environments [67]. For instance, the self-organizing (SO) attribute of bio-systems employed in wireless ad hoc networks means that clustering of routing nodes enhances the scalability of data-forwarding protocols [19]. As such, the network is rendered robust and can adapt to frequent topological changes.

Other works elucidate the genius of Nature by forwarding the argument that systems inspired by biology deliver significant results to enable exploration [70] and unique advances beyond the imagination of yesteryears [71]. This postulation is evident in performance optimization and enhancement of highly distributed and heterogeneous environments, such as data centers, grids and clouds [72]. Nonetheless, as has been argued in literature, not all assumptions on biological algorithms are without limitations [73]. Despite the mentioned limitations, there remains undoubtedly a huge potential in the use of bio-inspired methods as unconventional solutions to problems in the computing continuum. It is logical to be relatively optimistic considering how understanding physiological and ecological factors in biology has enabled medical innovation to cure and in some cases, even eradicate diseases [74].

Figure 4 is a taxonomic example of bio-inspired approaches distinguished according to their physiological and ecological attributes. Phenomena such as the evolutionary implications of predation on adaption and counter-adaption, determine behaviors in species. Behaviors, including predation risk and cost assessment in foraging species [75], change in response to outcomes of interactions between entities and their environment [76].

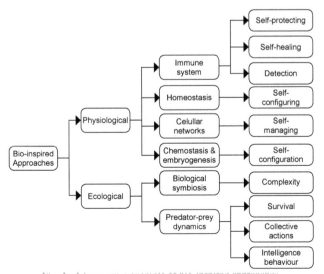

Fig. 4: A taxonomic example of bio-inspired approaches.

Based upon interactions and behaviors that exist between a predator and its prey, a range of innovations have been developed. To this end, [77] applies the predator-prey dynamic: the principles of the cost of predation in particular, as a new approach for malware prevention. Similarly [78] posits the predator-prey paradigm as the beneficial worm (predator) that terminates an intruding worm, and the malicious worm (prey) being undesirable in a system. [79] work is grounded on the self-management attributes of a zebra herd against predators. [80, 81] aimed to resolve virus and worm challenges in distributed systems is based on attributes of successful predators found in predating communities. The authors in [82] explain that conventional security approaches focus on performing complex analysis on restricted datasets, whereas unconventional mechanisms process small amounts of data in a 'simple distributed fashion' over a wide array of inputs. Many forms of biologically-inspired artificial intelligence systems are shown to be successful when applied to an IDS, with Genetic Algorithms (GA) being particularly successful [83]. Despite their success, these approaches tend to improve on the efficiency of older techniques and many issues, such as single-entry point analysis remain. When considering the defense of cloud systems against network aware malware, it is important to note that such systems now encompass multiple nodes based within complex network structures. Whilst conventional security defenses (single entry point analysis) are suitable for single machines, new unconventional defense systems are required to secure these complex networks. As such, the application of distributed biological defense systems to computer networks would be more suitable to solve many problems, be it malicious software or others.

Bio-inspired approaches applied in traditional distributed computing systems demonstrate useful characteristics derived from their biological parentage—attributes which define origin, development and progression, or ecological; interactional factors between organs and organisms in their natural environments, for instance, the design of IDSs based upon negative selection of T-cells that bind and kill infected or harmful cells [19]. Alternatively, the adaptation of the memory and self-learning mechanism employed by B-cells in identifying and destroying pathogens for designing IDSs [19] applies to physiological metaphors of the immune system. Given the significant success of biological systems, it seems logical to investigate theoretical underpinnings that describe the core elements of this chapter, their application as plausible approaches in the cyber security continuum. To counteract insecurity, unconventional mechanisms and strategies of proactive defense, synonymous with those found in Nature, including deceptive strategies, such as honeypots, can be implemented as passive decoys against adversaries, or as active lures to distract adversaries [84]. Countermeasures have been suggested, including aggressive approaches, such as 'white worms' [85], which actively pursue malicious software with the intent to destroy it. Deceptive techniques, such as address hopping [86] protect data in transit by masking the actual visibility of a transmitting device from a possible attacker.

The following section investigates the biological systems further, considering complexity and self-organization in natural systems as possibly fit for cyber security. Foremost, this section will present a brief overview of common biological systems and their application areas. This will be followed by exploring existing applications on computing security in general, followed by an evaluation of concepts' applicability in cyber security. In the subsequence subsections, the predator-prey dynamic, the prey's predations avoidance and anti-predation mechanisms are speculatively applied as a cyber security case scenario. Predation avoidance is speculatively viewed as an exploitable mechanism to enable survivability in cyber space. Previous works by [87–89] provide comprehensive reviews that contribute towards the unconventional context of the current section.

4. Review of Bio-inspired Methods

Among the many works, the authors in [65] present a comprehensive survey of bio-inspired networking protocols, citing a substantial number of sources alluding to the fact that immune-inspired algorithms form the basis for network security, anomaly and misbehavior detection [65]. The authors associate epidemiology to content distribution in computer networks, including the analysis of worms and virus spreading on the internet. In [67] they concur by associating intrusion detection and malware propagation with AIS and epidemiology, respectively, as complimentary bio-inspired domains. In other works, the authors in [90] proposed a trust and reputation model (BTRM-WSN) inspired by the ant colony, as a strategy to leverage trust selection according to the most reputable path [92]. Although their model is designed for wireless sensor networks, it is reasonable to assume that, the underlying trust model can be extended to cloud computing environments by adapting the ant colony system [92] in which the paths to fulfill defined conditions are built by leaving pheromone residues so that the trailing ants can follow a trusted route. Other models, including the Trust Ant Colony System (TACS) [93], AntRep algorithm based on swam intelligence [94], Time Based Dynamic Trust Model (TBDTM) [95] to name a few, have been proposed for distributed systems. Nevertheless, it is imperative to emphasize the need for comprehensive testing and evaluation before their use in cloud environments [96].

Inspired by the reliability of gene identification and assignment inherent in biological systems, Wang et al. propose the Family-gene-based model for Cloud Trust (FBCT) to address existing limitations inherent in PKI-based systems, which include challenges in identifying nodes within cloud environments, access control and third party authentication system [97]. According to the authors in [99], by adopting biological principles in family genes, their model provides solutions for trust in the cloud computing domain. Works by [79] explored the use of biological metaphors as a basis for designing, modeling and implementation of a cloud-based web service, which is able to deal with counter stability issues that arise from long-running processes and security attacks [79]. According to the authors, their proposed zebra herd-inspired approach not only simplified complex technical challenges, but also enhanced new designs for automated self-management processes for system administrators. Table 3 summarizes some inspirational bio-systems, categorizing them according to their application area, and the strengths and weaknesses of each system.

Faced with a combination of persistent and sophisticated adversaries, it is important that cyber security countermeasures are developed, based on the foundations harvested from Nature. Existing solutions simply fail as they do not adapt and escalate their security strategies to counteract the intensity and shear aggressiveness of an adversary [98]. [26] suggest security countermeasures as only successful in traditional networks. Thus, these authors postulate the rise in popularity of Adaptive Cyber Defense (ACD) approaches, such as bio-inspired systems, based upon their ability to optimize unpredictability and maximize the adaptive configurations in attack surface,

Table 3: Summary of bio-inspired approaches in cyber environments.

Algorithm	Description	Application	Author
Multiple Sequence Alignment (MSA) algorithm	Protein structure	Web traffic classification and sequence alignment	[101]
IDS detector optimization algorithm with co-evolution	Co-evolution in populations	Optimizing IDS intrusion detection	[102]
Data Security strategy for	Immune systems	Stored data security	[103]
Secure Data Storage	Physiological & behavioral patterns	Biometric authentication cloud stored data	[104]
AIS for phishing Detection	T-lymphocytes life cycle	Phishing detection for e-mails	[105]
Integrated Circuit Metrics (ICMetrics)	Human properties and features	High entropy public/private key generation scheme	[106]
Biologically-inspired Resilience	Cells and organisms (sea chameleon)	Manages cloud security & leverage resilience	[98, 107]
Data Hiding for Resource Sharing Based on DNA	DNA sequences	Data hiding for confidentiality and integrity of cloud data	[108]
Organic Resilience Approach for assuring resiliency against attacks and failure	Immunology (inflammation & immunization)	Threat detection, automated re-organization for assurance	[109]
Security based on face recognition	Facial features	Authentication and authorization	[110]
Family-gene-based model for Cloud Trust (FBCT)	Gene identification and assignment	User authentication, access control and authorization	[111]
Agent of network danger evaluation	Immune system		[112]
Supervised learning classifier with real-time extraction (UCSSE)	Genetic-based machine learning	Adaptive and automatic signature discovery in IDS	[113]
Fraud detection and improper use of both computer system and mobile telecommunication operations	Immune system	Monitoring and detection of fraudulent intruders	[114]
Extension of predator prey model	Predator-prey communities	Tackles automated mobile malware in networks	[80]
Computer immune system	Innate immune phase	IDS mimics immune antigens to create signature strings	[115]
AntNet	Ant colony optimization algorithm & OS theory	Agents concurrently traverse a network and exchange information synonymous with stigmergy in insects	[116]

thereby raising the cost of an attack for the adversary [26]. Complexity, large-scale virtualization and the extremely distributed nature of the cyberspace means that accountability, auditability and trust in such ubiquitous environments becomes pivotal [99, 100].

Current examples of classic bio-inspired approaches in the computing continuum include theories and algorithms. According to [66], algorithms are useful for describing systems with discrete state space, i.e., how and why systems transition occurs [66]. For instance, algorithms based on mechanisms governing the behaviors of ant colonies, human immune system, bees swarming, fish, predator and prey interactions and communities, etc., have been modeled to produce highly efficient, complex and distributed systems [71]. Prominent areas in computing where bio-inspired algorithms have been applied include, but is not limited to, autonomic computing [117], artificial life [118], biomimetic [119], organic computing [120], and genetic algorithms [121]. In addition, theories such as the Self and Non-self, and Danger Theory [62], have been coined with their primary premise on inspirations from biology. Further developments in bio-inspired approaches also necessitated the formalization of the Concurrency Theory as a formal framework for modeling parallel, distributed and mobile systems [66]. In the ensuing subsection, we will highlight bio-inspired approaches applied in solving security issues in distributed and cloud computing systems. Bio-inspired approaches in this context imply mechanisms employed to facilitate and/or enhance the protection of data in distributed systems; as related to networked workstations and servers, the network itself including communication devices, etc., and cloud computing.

4.1 Bio-inspired Approaches

Artificial natural immune systems are applied in a variety of areas, and particularly lauded for its success in IDS [102]. Immune detectors determine the performance of the detection component of the immune system, a core component of the immune system [122]. Several works, including [112, 115, 123, 124] to name a few, employ immune-inspired approaches for developing computer security mechanism-based on the self-adaptive, self-learning, self-organizing, parallel processing, and distributed coordination attributes of AISs. In addition, the authors in [105] propose AIS phishing detection is inspired by part of the immune system's response mechanism to pathogens; immature T-lymphocytes life cycle. By generating memory detectors from a static training data set and immature detectors through mutation, the proposed system detects incoming phishing e-mails through memory detectors, while the immature detectors detect phishing e-mails with unknown signatures. Fang et al. posit the notion that their proposed systems performed well in phishing detection. Nonetheless, the authors contend to the fact that using a static instead of dynamic fire-threshold value on their detectors, their system suffers from deficiencies [105]. Similarly, [125] explore the use of immune-inspired concept of apoptosis for in computer network security, which essentially describes the immune system's programmed action of destroying infected or mutated cells [125]. A comprehensive review of phishing e-mail-filtering techniques is presented by [126], while works by [127] review current literature and present a range of solutions proposed against identified attacks.

Genetic algorithms (GA) are stochastic search methods inspired from principles in biological systems where problem solving is indirect through an evolution of solutions, with subsequent generations of solutions in turn yielding the best solution to a problem [128]. Along similar lines, [129] proposed GTAP gene-inspired algorithm for user authentication where users from a 'family' are identified by a unique gene certificate (synonymous with unique signatures) and users are authenticated upon a positive analysis of their gene code [129]. According to the authors, simulation results for GTAP demonstrated superiority in safety and security by countering the deficiencies in safety passwords and ambiguity of subject information in certificates presented in traditional mechanisms [129]. In other works [130], genetic algorithms are implemented in cryptography to evaluate and enhance the complexities of the encryption systems. An interested reader is referred to a complete guide for cryptographic solutions for computer and cyber security presented by [131]. In cryptanalysis, where an attack mechanism in implemented to assess the strength of an encryption system, GA are argued to be highly successful in substitution ciphers [132] and transposition ciphers [133]. Although neural networks are generally popular in pattern recognition and classification and noise filtering, they are useful in other areas, including the use of biometrics in security [128]. The key to their success is their accuracy in feature extraction and efficiency in classification, i.e., low rejection rate and high positive classification [128]. Along these lines, [134] proposed the Network Nervous System as a mechanism for effective protection against distributed denial of service (DDoS) attacks, based on the biological metaphor of the human central nervous system; distributed information gathering and processing, coordination and identification activities. Their work rests on the basis that traditional security tools fail to cope with the escalating power of attacks on computing infrastructures [134].

Ant colonies have been applied for routing traffic optimization; for instance, in works by [116] who evaluate an optimization algorithm, AntNet, in which agents concurrently traverse a network and exchange information synonymous with stigmergy in insects. According to the authors, this algorithm exhibited superior performance in contrast to its competitors [135]. [136] proposed FBeeAd-Hoc as a security framework for routing problems in mobile ad hoc networks (MANET) using fuzzy set theory and digital signature [136]. Works by [80] extend on previous work on the predator model, to propose countermeasures against automated mobile malware in networks. The author proposes additional entities, including immunization, persistent and seeking predators, modeled from the self-propagating, self-defending and mobility attributes found in predating animals as solutions to challenges mentioned above. Their works premise on the notion that traditional countermeasures, which are generally centralized, fail adequately to solve security challenges existing in distributed systems [80]. According to the authors, their model does not only counter effect malware attacks on a computer, but effectively distributes updates and patches to the infected computer, which in essence immunizes it from future attacks [80]. [137] suggest the use of predator models as inspirational solutions against viruses and worms.

4.2 Considerations Before Applying Bio-inspired Approaches

Before biological systems are applied, there are several problems that should be considered. A panel discussing issues concerning biological security systems [82] describes a number of these. The first and perhaps most important is that biological systems and computer systems do not share an end goal. Whereas biological systems aspire to survive, the goal of many computer systems is to accomplish a computational task. This would serve to create doubt as to whether the model would remain effective when its goals are not the same. The authors [82] suggest that the survival of biological systems is an inherent issue within computer systems for the reason that biological systems will perform sacrifices to achieve the goal of survival, while in computer networks, downtime from a system is not something that is permitted, particularly if there is sensitive data that is to be protected. It is then argued that to achieve the characteristics wanted from biological systems (self-repair, organization, defense), then the whole system must be implemented; not just small sub-systems. This may be regarded as difficult if the systems may not be implemented properly due to contrasting goals. Nonetheless, it may be argued that software defined systems (SDS) maybe a useful link for integrating unconventional applications, platforms and infrastructures and managing them via APIs. Hence, developing cyber-security solutions should consider the diversity of entities in cyberspace and the dynamic changes diversity brings, i.e., requirement, goal, security, survivability, etc. Integrating self-awareness, self-adaptation, self-organization, to name a few, into cyberspace enables adaption to change at run-time.

5. Prey-inspired Survivability

Survivability in cyber environments, as in Nature, is affected by a range of factors, including interactive high-order behaviors of both human factors and actors. Asymmetric warfare theory [138] makes the human factor/actor more apparent by alluding to time and stealth advantage that attackers possess. The human factor/actor is demonstrated in the cyber context by the possibility for side-channel attackers to implant arbitrary code into a neighbor's VM environment with little to no chances of detection [139, 140]. Considering the above, it is necessary to pose the question of how to best evaluate cyber infrastructure's survivability. This is a common challenge in complex and time-varying systems and requires methodological-evaluation approaches that accommodate intermediary states that cyberspace resembles. As [141] suggest, the traditional binary view of survivability as ineffective in complex environments.

Clearly, the complexity of evaluating and later assuring survivability in cyber environments is currently a challenging issue and requires a composite approach. There is an urgent need to combine traditional and complex formalisms to enhance secure deployment, provision and access to cyberspace systems. This chapter suggests drawing parallels between predation avoidance in animal communities and capabilities of security systems to survive compromise in cyberspace environments, in which the goal is to protect the assets by hiding its visibility and increasing the complexity of being observed. By increasing the complexity of an asset, this adeptly increases the cost of an attack, increases the complexity of executing an exploit and gives an advantage to the defender [142]. In [143]'s model, for instance, deceptive measures are employed to enhance intelligence for the defender, while thwarting an adversary's capabilities to observe, investigate and learn about a target. Predator-Prey Systems (PPS) demonstrate complex relationships through interacting entities in which one depends on the other for food and survival [144]. Nevertheless, evidence in literature suggests that predator-prey models have found limited application in core cyber security domains, including cloud computing security systems. Thus, predator-prey analogies can be developed to capture unique diversification mechanisms which ensure survivability in both homogeneous and heterogeneous prey species.

5.1 *Anti-predation and Predation Avoidance Mechanisms for Cyber Security*

The presence of strong predation-avoidance responses in Nature's prey species demonstrates that past species interactions affect present distributions and may play an important role in the ongoing assembly of contemporary communities. Such avoidance behaviors in a growing number of species fundamentally alters our view of the processes affecting species distributions and the process of community assembly [145]. In vervet monkey (*Cercopithecus aethiops*), vigilance is an anti-predator behavior shared between males and females; however much, with higher levels of vigilance is performed by males who spend more time on treetops [146]. The importance of vigilance in vervet groups provides the best chances of survival by reducing the risk for individuals. On the other hand, consideration in group size effect highlights an overall increase in vigilance in larger group sizes and improved reaction to approaching predators. Furthermore, flight, alarm calls and response to alarm calls in vervet are adapted as responses to alarm calls associated with specific predator species [147]. As in fishes, alarm signals in vervet monkeys perform multiple-duties, ranging from predator deterrents, or distress signals to call in mobbers [148].

Social behaviors in Thomson's gazelles, such as their alert posture, galloping, stotting, and soft alarm calls are argued to release alertness and flight information to avoid predation [149]. In some rare cases, adult Thomson gazelles (*Eudorcas Thomsonii*) will attempt to evade predators by lying down, yet in some instances, mothers are known to adopt aggressive defense strategies to divert predators from hunting their fawns [150, 149]. According to the authors, Thomson gazelles generally do not fight back predators when hunted [149]. As noted by [151], predation avoidance in Thompson gazelles is also associated with their grouping behaviors, i.e., larger groups have improved predator detection capabilitie, and their vulnerability factor against their greatest predator; cheetah (Acinonyx jubatus), significantly reduces in larger groups. Evidence in literature supports the claim that totting in Thompson gazelles is a vital tool for avoiding predation.

Evidence in literature supports the claim that stotting in Thompson gazelles is a vital tool for avoiding predation as evidenced from the hypothesis of using stotting to startle or confuse a predator and as an anti-ambush evasion technique (Caro, 1986). Evidence in from Heinrich's (1979) works suggests at least five predation-avoidance strategies employed by caterpillars (*Pyrrharctia Isabella*) against predating birds—restrict their feeding to underside of leaves, forage at night, use leaves for movement while foraging, distance themselves from an unfinished leaf, or snip it off altogether [152]. Like vervet and Thomson's gazelle communities discussed above, group living is argued to positively enhance an individual's protection, as warning signals, defensive movement and regurgitating noxious chemicals may increase survivability [153]. Indeed, this is true considering the activities of males in vervet communities, who take up high positions on treetops to scan their surroundings and raise alarms when they detect predation threats. Thus, males are functionally associated with observation, vigilance and are perceived as most active against predators [146].

The choice of predation avoidance or anti-predation mechanism is hugely important in Meerkat (*Suricata suricatta*) communities as they live under high predation pressures, while occupying challenging foraging niche [154]. As such, social learning (developed and molded by experience) and effective co-operation initiate key survival behaviors, including fleeing non-specific predators, mobbing against predating snakes, functional referential alarm calls, or running to bolt holes in response to aerial predators [154]. In addition, Meerkat depend hugely on group living through communal vigilance [155], which, unlike response to alarm calls, avoids imminent predation; vigilance occurs in the absence or presence of a predator or danger [156]. Zebras' fleeing responses to predating lions are described according to their proactive responses to a prior assessed risk level and reactive responses when predation is imminent [157]. According to the authors, responses against predation also extend to elusive behaviors, where zebras remove themselves as far away from an encounter habitat (usually waterholes) as possible. In contrast to animal prey, plant prey significantly their predation cost to potential predators as the handling time and processing of plant tissue is more taxing. In the following, the current section will focus upon five successful prey species to explore survival mechanisms. Vigilance, alarm calls, mobbing and group living are anti-predator behaviors shared among vervet monkeys (*Cercopithecus aethiops*). Within this community, vervet males are associated with higher levels of vigilance [146]. Vigilance in larger groups increases, which improves reaction to approaching predators. Furthermore, flight and response in vervets is adapted to alarm calls associated with specific predator species [147]. Survival techniques employed by prey entities include changes in functioning, behaviors and structure, enabling them to avoid detection and hence, predation. The foregoing is summarized in Table 4 where survival mechanisms for each natural community is distinguished as either a predation-avoidance behaviour or an anti-predator technique. Predation avoidance and anti-predation mechanisms describe the main objectives of diversification, which in turn define how prey species behave to improve selection and survivability [158]. Anti-predation mechanisms describe prey techniques, which reduce the probability of predation, while predation avoidance describes the mechanism that the prey uses to remove itself from the same habitat as the predator.

Based upon the above, it is possible to develop analogies to capture unique diversification mechanisms that ensure survival in both homogeneous and heterogeneous prey species. Both mechanisms (predation avoidance and anti-predation) describe the main objectives of diversification, which define how prey species behave in order to improve selection and survivability [158]. As killing of prey by predators is a focus of mathematical modeling as it is easily observable, [159] suggests that anti-predation behaviors is most critical to prey survival.

Mechanisms for cyber security and cyber environments would thus consider both, the subjective and objective selection of anti-predator and predation-avoidance mechanisms (techniques and behaviors). Mechanisms may exist as specific (where mechanisms are effective against a specific predator), or non-specific (where strategies are effective against all predators) [160]. Exploiting prey-survival attributes as a blueprint for designing processes and mechanisms for cyberspace offers the following benefits:

- Survivable preys possess unique attributes that are well adapted to their environments. One may conceptualize the design of cyber agents capable of escaping and/or counter-attacking a 'predator'. [161]'s mathematical formulations illustrate this efficacy.

Table 4: Examples of prey survival mechanisms.

Anti-predation	Survival mechanism/ behavior	Plants	Flat-tail horned lizard	Vervets	Thomson gazelles	Moth caterpillar	Meerkat	Equus quagga (Zebra)
	Alarm calling	x	x	✓	✓	x	✓	✓
	Chemical-def	✓	x	x	x	x	x	x
	Fight-back	✓	✓	✓	x	✓	✓	✓
	Stotting	x	x	x	✓	x	x	x
	Group living	✓	x	✓	✓	✓	✓	✓
	Mobbing	x	x	✓	x	x	✓	x
	Aposematic	✓	x	x	x	✓	x	✓
	Mimicry	✓	✓	x	x	✓	x	x
Predation avoidance	Camouflaging	✓	✓	x	✓	✓	x	✓
	Masquerade	✓	✓	x	x	✓	✓	✓

- Survivable prey species possess strong and successful mechanisms that demonstrate the far-reaching implications that historical interactions have on future species [145]. By understanding such mechanisms, it is possible to adopt/ adapt such processes for future cyber security.
- Prey analogies that characterize non-extinct prey can be developed. [79] explored the use to biological metaphors for designing, modeling and implementing a web services capable of counteracting stability issues that arise from long-running processes and security attacks. Moreover, cloud computing, itself a core element of cyberspace, is a metaphor for the internet [162] where services are provided as metered resources in electricity-like manner [163].
- Developing analogies from Nature requires methodological approaches to translate apt prey functions for cyber security environments [89]. This requires an understanding of relationships between the complexity of prey systems and their local stability; Theoretical ecology [164] provides an in-depth knowledge in this domain. Moreover, the complexity theory [165] proves the basis to deconstruct the complexity of cyberspace and Nature systems.

6. Research Directions in Survivability Assurance in Cyberspace

Recent trends towards bio-inspired designs have ushered in the development of methods for creating analogies to combine attributes or objectives in multi-domain systems in a formal manner [166]. BioTRIZ [167, 168] and other analogical reasoning tools summarized in the table below are some common examples. While conceptual design (CD) provides insights into the functions, working principles and a general

Table 5: Comparison of analogy development methods.

10 Design Method	11 Description	12 Is knowledge of bio-system required?
13 Functional Modeling	14 Well defined categories and scale to develop functional models	15 Yes
16 BioTRIZ	17 Allows invention to solve problems with contradiction	18 No
19 BID Lab Search Tool	20 Natural language processing tool to search bio-words using engineering words	21 No
22 Biology Thesaurus	23 Extensible to a range of tools other than functional methods	24 Undefined
25 IDEA-INSPIRE & DANE	26 Provides organized database for bio-design	27 Undefined

layout of a system's structure [169] are deficient since a one-to-one transfer of Nature concepts to cloud computing requires lateral thinking.

On the other hand, TRIZ (the theory of inventive problem solving) is a useful systematic methodology that provides logical approaches for innovative and inventive creations [170]. TRIZ has been adapted to suit other environments, such as information technology [171]. To capture key characteristics from Nature, this paper follows Beckmann's approach [172], but specifically focuses upon cloud computing rather than information technology in general. Since the original TRIZ principles provide abstract solution models [172], any new labels (we term prey-centric and cloud-centric) are also abstract. Hence, any further developments generate further minute abstract solutions.

As postulated by [172], abstraction means that TRIZ principles are applicable in a wide range of fields. TRIZ provides the added capacity to identify a solution [172], whereupon conceptual solutions can further be developed into specific factual solutions. Indeed, conceptualising solutions are informed by identified specific problems, which in turn inform the choice of the problem-solving approach and tool.

Existing analogical design approaches rely largely on thematic mapping processes and subjective choices of the components of a biological system or sub-systems, as by focusing upon the impact of ecological diversification [173], population dynamics [174], or simple constructs of an arms race [175]. This suggests the wrong notion that analogical design requires comprehensive understanding of cyber security technologies, but just the basics of the arms race [176]. In fact, as [177] argues, the designer's knowledge of both domains helps to infer information that facilitates bio-inspired designs from a problem or solution-driven perspective. Hence, the concept of analogies mentioned in this chapter proffer that, given an old problem (P_{old}), i.e., predate-survive dynamic, with an old solution (S_{old}), i.e., predation-avoidance and anti-predation, a new problem (P_{new}), i.e., cyberspace survivability can be conceptualized with new partial and perhaps null solutions (S_{new}), i.e., prey-inspired predation avoidance analogies in the solutions space sold to S_{new}.

Designing cyber-security solutions should consider the diversity of entities and the dynamic changes diversity brings, i.e., requirement, goal, security, survivability, etc. Integrating self-awareness, self-adaptation, self-organization, to name a few, into cloud environment design enables service composition to adapt to changes at run-time. On the one hand, underlying designs cannot be static and inadequate for synthesizing dynamic and distributed service compositions. On the contrary, designs should enable distributed coordination of entities necessary to achieve agreeable levels of survivability. In addition to integrating the three selves (self-aware, self-configure, self-organize), automation supports necessary adaptation; for instance, at design time, holistic synthesis of the cloud logic entails automated and fully distributed coordination of the involved entities and services. During execution, automation facilitates adaptation through 'self-attributes' synthesizing dynamically-evolving entities. Multi-agent-based systems are lauded for complex behaviors among interacting autonomous agents. By extending multi-agent capabilities into cyber environments, challenges, including security, survivability and availability, can be better managed. As suggested by [178], integrating multi-agent technologies can unlock even higher performing, complex, autonomous and intelligent applications and scalable yet reliable infrastructures.

Key Terminology and Definitions

Bio-inspired: Inspired by methods and mechanisms in biological systems, this is a short version for biologically inspired, a cross-domain field of study that aims to bring together cross-domain concepts, methods, techniques, etc. Common areas of study include evolution (genetic algorithm), ants and termites (emergent systems), life (artificial life and cellular automata), immune system (artificial immune system), etc.

Artificial Life: This bio-inspired domain pertains to study of systems that have relation to natural life. Also, commonly referred to as Alife, this domain encompasses experimentation; simulation and modeling, of natural life processes (e.g., adaptive behaviors) and its evolution. For instance, one would study in biological life in order to construct a system that behaves like a living organism.

Cyber Security: Cyber security is a continuum of technologies and innovations for ensuring and assuring the security of data and networking technologies, the approach and actions associated with security risk management processes followed by organizations and states to protect confidentiality, integrity and availability of data assets used in cyberspace. The concept includes guidelines, policies and collection of safeguards, technologies, tools and training to provide best protection for state of cyber environment and its users.

Cyber Defense: This refers to mechanisms (tools, techniques and strategies) implemented to defend cyberspace, especially critical infrastructure against malicious and potentially catastrophic attacks. Proactive cyber defense includes implied mechanisms implemented in anticipation of an attack. Aggressive cyber defense is a form of proactive defense whereupon ethical hackback aims to fight back attackers with an aim to frustrate and increase cost of attacks, track source of attack, or even destroy attack capabilities.

Survivability: Survivability has traditionally been described as a mission, i.e., a capability of a system to provide services in a timely manner, bearing in mind that precautionary countermeasures will fail. Dependability is a property for a computing system to rely upon for delivery of a service placed upon it.

Machine Learning: 'Machine learning is the science of getting computers to learn and act like humans do and improve their learning over time in autonomous fashion, by feeding them data and information in the form of observations and real-world interactions. It is part of research on artificial intelligence, seeking to provide knowledge to computers through data, observations and interacting with the world. That acquired knowledge allows computers to correctly generalize to new settings' [179].

Predator-prey: 'An interaction between two organisms of unlike species in which one of them acts as predator that captures and feeds on the other organism that serves as the prey. In ecology, predation is a mechanism of population control. Thus, when the number of predators is scarce the number of preys should rise. When this happens, the predators would be able to reproduce more and possibly change their hunting habits. As the number of predators rises, the number of prey decline' [180].

References

[1]   Betz, D.J. and Stevens, T. (2013). Analogical reasoning and cyber security. Secur. Dialogue 44(2): 147–164.

[2]   Schatz, D., Bashroush, R. and Wall, J. (2017). Towards a more representative definition of cyber security. Journal of Digital Forensics, Security and Law 12(2): 8.

[3]   Klimburg, A. (2012). National Cyber Security Framework Manual.

[4]   BBC online UK, 'Vodafone Germany hack hits two million customers', BBC. [online]. Available: https://www.bbc.co.uk/news/technology-24063621. [Accessed: 10-Jun-2018].

[5]   BBC online UK, 'TalkTalk hack' 'affected 157,000 customers', BBC, 2015. [online]. Available: https://www.bbc.co.uk/news/business-34743185. [Accessed: 10-Jun-2018].

[6]   BBC online UK, 'Dixons Carphone admits huge data breach', BBC, 2018. [online]. Available: https://www.bbc.co.uk/news/business-44465331. [Accessed: 13-Jun-2018].

[7]   Ghoneim, A., Muhammad, G., Amin, S.U. and Gupta, B. (2018). Medical image forgery detection for smart healthcare. IEEE Commun. Mag. 56(4): 33–37.

[8]   Iso Iec. (2005). BS ISO/IEC 27002:2005 Information technology—Security techniques—Code of practice for information security management. ISO, p. 130.

[9]   Von Solms, R. and Van Niekerk, J. (2013). From information security to cyber security. Comput. Secur. 38: 97–102.

[10] Foote, R. (2007). Mathematics and complex systems. Science 318(5849): 410–412.

[11] Wen, G., Yu, W., Yu, X. and Lü, J. (2017). Complex cyber-physical networks: From cybersecurity to security control. J. Syst. Sci. Complex 30(1): 46–67.

[12] Polack, F.A.C. (2010). Self-organisation for survival in complex computer architectures. Lect. Notes Comput. Sci. (including Subser. Lect. Notes Artif. Intell. Lect. Notes Bioinformatics) 6090 LNCS: 66–83.

[13] Vermesan, O. and Friess, P. (2014). Internet of Things Applications—From Research and Innovation to Market Deployment.

[14] Alam, S., Sogukpinar, I., Traore, I. and Coady, Y. (2014). In-cloud malware analysis and detection: State of the art. pp. 473–478. *In*: Proc. 7th Int. Conf. Secur. Inf. Networks—SIN '14.

[15] Miorandi, D., Sicari, S., De Pellegrini, F. and Chlamtac, I. (2012). Internet of things: Vision, applications and research challenges. Ad Hoc Networks 10(7): 1497–1516.

[16] Tewari, A. and Gupta, B.B. (2017). A lightweight mutual authentication protocol based on elliptic curve cryptography for IoT devices. Int. J. Adv. Intell. Paradig. 9(2-3): 111–121.

[17] Hoverman, J.T. and Relyea, R.A. (2009). Survival trade-offs associated with inducible defences in snails: The roles of multiple predators and developmental plasticity. Funct. Ecol. 23(6): 1179–1188.

[18] Sayed, A.H. (2014). Adaptive networks. Proc. IEEE 102(4): 460–497.

[19] Zheng, C. and Sicker, D.C. (2013). A survey on biologically inspired algorithms for computer networking. Commun. Surv. Tutorials, IEEE 15(3): 1160–1191.

[20] Smith, C.L. (1979). Fundamentals of control theory. Chemical Engineering (New York) 86(22): 11–39.

[21] Landau, I.D. (1999). From robust control to adaptive control. Control Eng. Pract. 7: 1113–1124. .

[22] Andersson, J., De Lemos, R., Malek, S. and Weyns, D. (2009). Modeling dimensions of self-adaptive software systems. pp. 27–47. *In*: Lecture Notes in Computer Science (including subseries Lecture Notes in Artificial Intelligence and Lecture Notes in Bioinformatics), Vol. 5525 LNCS.

[23] Zissis, D. and Lekkas, D. (2012). Addressing cloud computing security issues. Futur. Gener. Comput. Syst., Vl 28(3): 583–592.

[24] Hausken, K. and He, F. (2016). On the effectiveness of security countermeasures for critical infrastructures. Risk Anal. 36(4): 711–726.

[25] Checkland, P. (1981). Systems Thinking, System Practice.

[26] Cybenko, G., Jajodia, S., Wellman, M.P. and Liu, P. (2014). Adversarial and Uncertain Reasoning for Adaptive Cyber Defense: Building the Scientific Foundation, Springer, pp. 1–8.

[27] Virvilis, N. and Gritzalis, D. (2013). The big four—what we did wrong in advanced persistent threat detection? pp. 248–254. *In*: Availability, Reliability and Security (ARES), 2013 Eighth International Conference on.

[28] Harknett, R.J. and Stever, J.A. (2011). The new policy World of Cybersecurity. Public Adm. Rev. 71(3): 455–460.

[29] Wang, W. and Lu, Z. (2013). Cyber security in the Smart Grid: Survey and challenges. Comput. Networks 57(5): 1344–1371.

[30] C.A. (2016). The Eight Providers that Matter Most and How They Stack Up. The Forrester WaveTM.

[31] Shahriar, N., Ahmed, R., Chowdhury, S.R., Khan, A., Boutaba, R. and Mitra, J. (2017). Generalized recovery from node failure in virtual network embedding. IEEE Trans. Netw. Serv. Manag. 14(2): 261–274.

[32] Ren, S. et al. (2013). A coordination model for improving software system attack-tolerance and survivability in open hostile environments. Int. J. Adapt. Resilient Auton. Syst. 3(2): 175–199.

[33] Baheti, H. and Gill, R. (2011). Cyber—Physical Systems. The Impact of Control Technology, pp. 161–66.

[34] HM Government. (2010). A Strong Britain in an Age of Uncertainty: The National Security Strategy, Vol. 16, No. Supplement.

[35] Joint Committee on the National Security Strategy. (2016). National security strategy and strategic defence and security review. First Rep. Sess. 2016–17, Vol. HL, Paper 1, No. HC 153: 1–96.

[36] Mount, M. and Quijano, E. (2009). Iraqi insurgents hacked predator drone feeds, U.S. official indicates. CNN. [online]. Available: http://www.cnn.com/2009/US/12/17/drone.video. hacked/.

[37] Langner, R. (2013). To Kill a Centrifuge: A Technical Analysis of What Stuxnet's Creators Tried to Achieve, Arlington, VA Langner Gr., pp. 1–36.

[38] Ryan, Y. (2011). Anonymous and the Arab uprisings. Aljazeera.com, pp. 1–5.

[39] Segal, A. (2015). Net {Politics} » {The} {UN}'s {Group} of {Governmental} {Experts} on {Cybersecurity}. Council on Foreign Relations—Net Politics.

[40] Hurwitz, R. (2012). Depleted trust in the cyber commons. Strateg. Stud. Q. VO 6(3): 20.

[41] Fidler, D.P. (2016). Cyberspace, terrorism and international law. J. Confl. Secur. Law 21(3): 475–493.

[42] United States Defense Force. (February 2013). Joint Publication 3–12 Cyberspace Operations. United States Def. Force 12: 62.

[43] Pun, D. (2017). Rethinking espionage in the modern era. Chic. J. Int. Law 18(1).

[44] Osnos, E., Remnick, D. and Yaffa, J. (2017). Trump, putin and the new cold war. New Yorker 1: 1–77.

[45] PwC and BAE. (2017). Operation Cloud Hopper. PwC Web Site.

[46] Gupta, B.B., Yamaguchi, S. and Agrawal, D.P. (2017). Advances in security and privacy of multimedia big data in mobile and cloud computing. Multimed. Tools Appl.

[47] Zhang, B.B., Sun, Z., Zhao, R., Wang, C., Chang, J. and Gupta, C.K. (21017). CyVOD: A novel trinity multimedia social network scheme. pp. 18513–18529. *In*: Multimedia Tools and Applications.

[48] Deeks, A. (2016). Confronting and adapting: Intelligence agencies and international law. SSRN Electron. J.

[49] Cooter, R. (1998). Expressive law and economics. J. Legal Stud. 27(S2): 585–607.

[50] Greibach, S. (2010). Lecture Notes in Computer Science.

[51] Magoulas, G.D. and Prentza, A. (2001). Machine learning in medical applications. pp. 300–307. *In*: Machine Learning and Its Applications: Advanced Lectures.

[52] Delen, D., Walker, G. and Kadam A.a. (2005). Predicting breast cancer survivability: A comparison of three data mining methods. Artif. Intell. Med. 34(2): 113–127.

[53] Kourou, K., Exarchos, T., Exarchos, K., Karamouzis, M. and Fotiadis, D. (2015). Machine learning applications in cancer prognosis and prediction. Computational and Structural Biotechnology Journal 13: 8–17.

[54] Shukla, N., Hagenbuchner, M., Win, K.T. and Yang, J. (2018). Breast cancer data analysis for survivability studies and prediction. Comput. Methods Programs Biomed. 155: 199–208.

[55] Allen, G. and Chan, T. (2017). Artificial intelligence and national security. Belfer Cent. Sci. Int. Aff.

[56] Silver, D. et al. (2017). Mastering the game of Go without human knowledge. Nature 550(7676): 354–359.

[57] Tyugu, E. (2011). Artificial intelligence in cyber defense. 2011 3rd Int. Conf. Cyber Confl., pp. 1–11.

[58] Patil, P. (2016). Artificial intelligence in cyber security. Int. J. Res. Comput. Appl. Robot. 4(5): 1–5.

[59] Hallaq, B., Somer, T., Osula, A.-M., Ngo, K. and Mitchener-Nissen, T. (2017). Artificial intelligence within the military domain and cyber warfare. Eur. Conf. Inf. Warf. Secur. ECCWS, pp. 153–157.

[60] Brabazon, A. and O'Neill, M. (2006). Biologically inspired algorithms for financial modelling. Springer Science & Business Media.

[61] Oates, R., Milford, M., Wyeth, G., Kendall, G. and Garibaldi, J.M. (2009). The implementation of a novel, bio-inspired, robotic security system. pp. 1875–1880. *In*: Robotics and Automation, 2009. ICRA'09. IEEE International Conference on.

[62] Sobh, T.S. and Mostafa, W.M. (2011). A co-operative immunological approach for detecting network anomaly. Appl. Soft Comput. 11(1): 1275–1283.

[63] Ziring, N. (2015). The future of cyber operations and defense. Warfare 14: 1–7.

[64] Low, C., Chen, Y. and Wu, M. (2011). Understanding the determinants of cloud computing adoption. Ind. Manag. Data Syst. 111(7): 1006–1023.

[65] Dressler, F. and Akan, O.B. (2010). A survey on bio-inspired networking. Comput. Networks 54(6): 881–900.

[66] Priami, C. (2009). Algorithmic systems biology. Commun. ACM 52(5): 80–88.

[67] Meisel, M., Pappas, V. and Zhang, L. (2010). A taxonomy of biologically inspired research in computer networking. Comput. Networks 54(6): 901–916.

[68] Nakano, T. (2011). Biologically-inspired network systems: A review and future prospects. Syst. Man, Cybern. Part C Appl. Rev. IEEE Trans. 41(5): 630–643.

[69] Ribeiro, C.C. and Hansen, P. (2012). Essays and surveys in metaheuristics, Vol. 15, Springer Science & Business Media.

[70] Thakoor, S. (2000). Bio-inspired engineering of exploration systems. J. Sp. Mission Archit. 2(1): 49–79.

[71] Hinchey, M. and Sterritt, R. (2012). 99% (biological) inspiration. pp. 177–190. *In*: Conquering Complexity.

[72] Wakamiya, N. and Murata, M. (2007). Bio-inspired analysis of symbiotic networks. Springer, pp. 204–213.

[73] Breza, M. and McCann, J. (2008), Lessons in implementing bio-inspired algorithms on wireless sensor networks. pp. 271–276. *In*: Adaptive Hardware and Systems, 2008. AHS'08. NASA/ESA Conference on.

[74] Levy, S.B. and Marshall, B. (2004). Anti-bacterial resistance worldwide: causes, challenges and responses. Nat. Med. 10: S122–S129.

[75] Higginson, A.D. and Ruxton, G.D. (2015). Foraging mode switching: The importance of prey distribution and foraging currency. Anim. Behav. 105: 121–137.

[76] DiRienzo, N., Pruitt, J.N. and Hedrick, A.V. (2013). The combined behavioural tendencies of predator and prey mediate the outcome of their interaction. Anim. Behav. 86(2): 317–322.

[77] Ford, R., Bush, M. and Bulatov, A. (2006). Predation and the cost of replication: New approaches to malware prevention. Comput. Secur. 25(4): 257–264.

[78] Tanachaiwiwat, S. and Helmy, A. (2009). Encounter-based worms: Analysis and defense. Ad Hoc Networks 7(7): 1414–1430.

[79] Finstadsveen, J. and Begnum, K. (2011). What a webserver can learn from a zebra and what we learned in the process. p. 5. *In*: Proceedings of the 5th ACM Symposium on Computer Human Interaction for Management of Information Technology.

[80] Gupta, A. and DuVarney, D.C. (2004). Using predators to combat worms and viruses: A simulation-based study. pp. 116–125. *In*: Computer Security Applications Conference, 2004. 20th Annual.

[81] Toyoizumi, H. and Kara, A. (2002). Predators: Goodwill mobile codes combat against computer viruses. pp. 11–17. *In*: Proceedings of the 2002 Workshop on New Security Paradigms.

[82] Somayaji, A., Locasto, M. and Feyereisl, J. (2007). Panel: The future of biologically-inspired security: Is there anything left to learn? pp. 49–54. *In*: NSPW'07, September 18–21, 2007, North Conway, NH, USA.

[83] Sakellari, G. and Loukas, G. (2013). A survey of mathematical models, simulation approaches and testbeds used for research in cloud computing. S.I. Energy Effic. Grids Clouds 39(0): 92–103.

[84] López, C.F.L. and Reséndez, M.H. (2008). Honeypots: Basic concepts, classification and educational use as resources. *In*: Information Security Education and Courses.

[85] Lu, W., Xu, S. and Yi, X. (2013). Optimizing Active Cyber Defense, Springer, pp. 206–225.

[86] Shi, L., Jia, C., Lü, S. and Liu, Z. (2007, April). Port and address hopping for active cyber-defense. pp. 295–300. *In*: Pacific-Asia Workshop on Intelligence and Security Informatics, Springer, Berlin, Heidelberg.

[87] Mthunzi, S.N. and Benkhelifa, E. (2017). Trends towards bio-inspired security countermeasures for cloud environments. pp. 341–347. *In*: Proceedings—2017 IEEE 2nd International Workshops on Foundations and Applications of Self* Systems, FAS*W 2017.

[88] Welsh, T. and Benkhelifa, E. (2018). Perspectives on resilience in cloud computing: Review and trends. pp. 696–703. *In*: Proc. IEEE/ACS Int. Conf. Comput. Syst. Appl. AICCSA, Vol. 2017 October.

[89] Mthunzi, S. and Benkhelifa, E. (2018). Mimicking Prey's Escalation Predation-avoidance Techniques for Cloud Computing Survivability Using Fuzzy Cognitive Map, pp. 189–196.

[90] Mármol, F.G. and Pérez, G.M. (2009). Security threats scenarios in trust and reputation models for distributed systems. Comput. Secur. 28(7): 545–556.

[91] Dang, D.H. and Cabot, J. (2013). Automating inference of OCL business rules from user scenarios. *In*: Proceedings—Asia-Pacific Software Engineering Conference, APSEC 1(4): 156–163.

[92] Dorigo, M. and Gambardella, L.M. (1997). Ant colony system: A cooperative learning approach to the traveling salesman problem. Evol. Comput. IEEE Trans. 1(1): 53–66.

[93] Marmol, F.G., Perez, G.M. and Skarmeta, A.F.G. (2009). TACS, a trust model for P2P networks. Wirel. Pers. Commun. 51(1): 153–164.

[94] Wang, W., Zeng, G. and Yuan, L. (2018). Ant-based reputation evidence distribution in P2P networks. pp. 129–132. *In*: Grid and Cooperative Computing, 2006. GCC 2006. Fifth International Conference.

[95] Zhuo, T., Zhengding, L. and Kai, L. (2006). Time-based dynamic trust model using ant colony algorithm. Wuhan Univ. J. Nat. Sci. 11(6): 1462–1466.

[96]   Firdhous, M., Ghazali, O. and Hassan, S. (2012). Trust management in cloud computing: A critical review. Int. J. Adv. ICT Emerg. Reg. 4(2): 24–36.

[97]   Wang, T., Ye, B., Li, Y. and Yang, Y. (2010). Family gene-based Cloud Trust model. pp. 540–544. *In*: ICENT 2010 International Conference on Educational and Network Technology.

[98]   Hariri, S., Eltoweissy, M. and Al-Nashif, Y. (2011). Biorac: Biologically inspired resilient autonomic cloud. p. 80. *In*: Proceedings of the Seventh Annual Workshop on Cyber Security and Information Intelligence Research.

[99]   Takabi, H., Joshi, J.B.D. and Ahn, G.-J. (2010). Security and privacy challenges in cloud computing environments. IEEE Secur. Priv. 6: 24–31.

[100]   Ko, R.K.L. et al. (2011). Trust Cloud: A framework for accountability and trust in cloud computing. pp. 584–588. *In*: Services (SERVICES), 2011 IEEE World Congress on.

[101]   He, G., Yang, M., Luo, J. and Gu, X. (2015). A novel application classification attack against Tor. Concurr. Comput. Pract. Exp. 27(18): 5640–5661.

[102]   Liang, X. and Fengbin, Z. (2013). Detector optimization algorithm with co-evolution in immunity-based intrusion detection system. pp. 620–623. *In*: Measurement, Information and Control (ICMIC), 2013 International Conference on, Vol. 1.

[103]   Jinyin, C. and Dongyong, Y. (2013). Data security strategy based on artificial immune algorithm for cloud computing. Appl. Math 7(1L): 149–153.

[104]   Govinda, K. and Kumar, G. (2013). T Secure data storage in cloud environment using environment using fingerprint. Asian J. Comput. Sci. Inf. Technol. 2(5).

[105]   Fang, X., Koceja, N., Zhan, J., Dozier, G. and Dipankar, D. (2012). An artificial immune system for phishing detection. pp. 1–7. *In*: Evolutionary Computation (CEC), 2012 IEEE Congress on.

[106]   Tahir, R., Hu, H., Gu, D., McDonald-Maier, K. and Howells, G. (2012). A scheme for the generation of strong cryptographic key pairs based on ICMetrics. pp. 168–174. *In*: Internet Technology and Secured Transactions, 2012 International Conference for.

[107]   Welsh, T. and Benkhelifa, E. (2018). Embyronic Model for Highly Resilient PaaS, pp. 197–204.

[108]   Abbasy, M.R. and Shanmugam, B. (2011). Enabling data hiding for resource sharing in cloud computing environments based on DNA sequences. pp. 385–390. *In*: Services (SERVICES), 2011 IEEE World Congress on.

[109]   Carvalho, M., Dasgupta, D., Grimaila, M. and Perez, C. (2011). Mission resilience in cloud computing: A biologically inspired approach. pp. 42–52. *In*: 6th International Conference on Information Warfare and Security.

[110]   Wang, C. and Yan, H. (2010). Study of cloud computing security based on private face recognition. pp. 1–5. *In*: Computational Intelligence and Software Engineering (CiSE), 2010 International Conference on.

[111]   Wang, T., Ye, B., Li, Y. and Yang, Y. (2010). Family gene-based cloud trust model. pp. 540–544. *In*: Educational and Network Technology (ICENT), 2010 International Conference on.

[112]   Yang, J., Liu, X., Li, T., Liang, G. and Liu, S. (2009). Distributed agents model for intrusion detection based on AIS. Knowledge-based Syst. 22(2): 115–119.

[113]   Shafi, K. and Abbass, H.A. (2009). An adaptive genetic-based signature learning system for intrusion detection. Expert Syst. Appl. 36(10): 12036–12043.

[114]   Boukerche, A., Juca, K.R.L., Sobral, J.B. and Notare, M.S.M.A. (2004). An artificial immune based intrusion detection model for computer and telecommunication systems. Parallel Comput. 30(5): 629–646.

[115]   Kephart, J.O. (1994). A biologically-inspired immune system for computers. pp. 130–139. *In*: Artificial Life IV: Proceedings of the Fourth International Workshop on the Synthesis and Simulation of Living Systems.

[116]   Baran, B. and Sosa, R. (2000). A new approach for AntNet routing. pp. 303–308. *In*: Computer Communications and Networks, 2000. Proceedings. Ninth International Conference on.

[117]   Kephart, J.O. and Chess, D.M. (2003). The vision of autonomic computing. Computer (Long. Beach, Calif) 36(1): 41250.

[118] Harmer, P.K., Williams, P.D., Gunsch, G.H. and Lamont, G.B. (2002). An artificial immune system architecture for computer security applications. Evol. Comput. IEEE Trans. 6(3): 252–280.

[119] Oraph, B.T. and Morgens, Y.R. (2008). Cloud computing. Commun. ACM 51(7).

[120] Schmeck, H. (2005). Organic computing-a new vision for distributed embedded systems. pp. 201–203. *In*: Object-oriented Real-time Distributed Computing, 2005. ISORC 2005. Eighth IEEE International Symposium on.

[121] Agostinho, L., Feliciano, G., Olivi, L., Cardozo, E. and Guimarães, E. (2011). A bio-inspired approach to provisioning of virtual resources in federated clouds. pp. 598–604. *In*: Proceedings—IEEE 9th International Conference on Dependable, Autonomic and Secure Computing, DASC 2011.

[122] Ji, Z. and Dasgupta, D. (2009). V-detector: An efficient negative selection algorithm with 'probably adequate' detector coverage. Inf. Sci. (Ny) 179(10): 1390–1406.

[123] Kim, J., Bentley, P.J., Aickelin, U., Greensmith, J., Tedesco, G. and Twycross, J. (2007). Immune system approaches to intrusion detection—A review. Natural Computing 6(4): 413–466.

[124] Gonzalez, F.A. and Dasgupta, D. (2003). Anomaly detection using real-valued negative selection. Genet. Program. Evolvable Mach. 4(4): 383–403.

[125] Saudi, M.M., Woodward, M., Cullen, A.J. and Noor, H.M. (2008). An overview of apoptosis for computer security. pp. 1–6. *In*: Information Technology, 2008. ITSim 2008. International Symposium on, Vol. 4.

[126] Almomani, A., Gupta, B.B., Atawneh, S., Meulenberg, A. and Almomani, E. (2013). A survey of phishing e-mail filtering techniques. IEEE Commun. Surv. Tutorials 15(4): 2070–2090.

[127] Gupta, B.B., Arachchilage, N.A.G. and Psannis, K.E. (2017). Defending against phishing attacks: Taxonomy of methods, current issues and future directions. Telecommun. Syst.

[128] Hordijk, W. (2005). An overview of biologically-inspired computing in information security. pp. 1–14. *In*: Proceedings of the National Conference on Information Security, Coimbatore, India.

[129] Sun, F. and Cheng, S. (2009). A gene technology-inspired paradigm for user authentication. pp. 1–3. *In*: Bioinformatics and Biomedical Engineering, 2009. ICBBE 2009. 3rd International Conference on.

[130] Isasi, P. and Hernandez, J.C. (2004). Introduction to the applications of evolutionary computation in computer security and cryptography. Comput. Intell. 20(3): 445–449.

[131] Gupta, S., Agrawal, B. and Yamaguchi, D.P. (2016). Handbook of Research on Modern Cryptographic Solutions for Computer and Cyber Security, IGI Global.

[132] Verma, A.K., Dave, M. and Joshi, R.C. (2007). Genetic algorithm and tabu search attack on the mono-alphabetic substitution cipher in adhoc networks. Journal of Computer Science.

[133] Toemeh, R. and Arumugam, S. (2015). Breaking transposition cipher with genetic algorithm. Elektron. ir Elektrotechnika 79(7): 75–78.

[134] Shorov, A. and Kotenko, I. (2014). The framework for simulation of bioinspired security mechanisms against network infrastructure attacks. Sci. World J. 2014.

[135] Di Caro, G. and Dorigo, M. (1998). AntNet: Distributed stigmergetic control for communications networks. J. Artif. Intell. Res. 317–365.

[136] Rafsanjani, M.K. and Fatemidokht, H. (2015). FBee AdHoc: A secure routing protocol for BeeAdHoc based on fuzzy logic in MANETs. AEU-International J. Electron. Commun. 69(11): 1613–1621.

[137] Grimes, R. (2001). Malicious mobile code: Virus protection for Windows. O'Reilly Media, Inc.

[138] Baskerville, R. (2004). Agile security for information warfare: A call for research. ECIS 2004 Proc., p. 10.

[139] Ristenpart, T., Tromer, E., Shacham, H. and Savage, S. (2009). Hey, you, get off of my cloud: Exploring information leakage in third-party compute clouds. pp. 1991–212. *In*: Proceedings of the 16th ACM Conference on Computer and Communications Security.

[140] Mthunzi, S.N., Benkhelifa, E., Alsmirat, M.A. and Jararweh, Y. (2018). Analysis of VM Communication for VM-based Cloud Security Systems, pp. 182–188.

[141] Liang, Q. and Modiano, E. (2017). Survivability in time-varying networks. IEEE Trans. Mob. Comput. 16(9): 2668–2681.

[142] McQueen, M.A. and Boyer, W.F. (2009). Deception used for cyber defense of control systems. *In*: Proceedings of the 2nd Conference on Human System Interactions, HSI, Vol. 9.

[143] Yuill, J., Denning, D.E. and Feer, F. (2006). No Title. Using Decept. to hide things from hackers Process. Princ. Tech.

[144] Colomer, M.A., Margalida, A., Sanuy, D. and Perez-Jimenez, M.J. (2011). A bio-inspired computing model as a new tool for modeling ecosystems: the avian scavengers as a case study. Ecol. Modell. 222(1): 33–47.

[145] Resetarits, W.J. (2001). Colonization under threat of predation: avoidance of fish by an aquatic beetle, Tropisternus lateralis (Coleoptera: Hydrophilidae). Oecologia 129(1): 155–160.

[146] Baldellou, M. and Henzi, S.P. (1992). Vigilance, predator detection and the presence of supernumerary males in vervet monkey troops. Anim. Behav. 43(3): 451–461.

[147] Isbell, L.A. (1994). Predation on primates: ecological patterns and evolutionary consequences. Evol. Anthropol. Issues, News, Rev. 3(2): 61–71.

[148] Smith, R.J.F. (1992). Alarm signals in fishes. Rev. Fish Biol. Fish. 2(1): 33–63.

[149] Walther, F.R. (1969). Flight behaviour and avoidance of predators in Thomson's gazelle (Gazella thomsoni Guenther 1884). Behaviour 34(3): 184–220.

[150] Fitzgibbon, C.D. (1990). Anti-predator strategies of immature Thomson's gazelles: Hiding and the prone response. Anim. Behav. 40(5): 846–855.

[151] Fitzgibbon, C.D. (1990). Why do hunting cheetahs prefer male gazelles? Anim. Behav. 40(5): 837–845.

[152] Heinrich, B. (1979). Foraging strategies of caterpillars. Oecologia 42(3): 325–337.

[153] Hunter, A.F. (2000). Gregariousness and repellent defences in the survival of phytophagous insects. Oikos 91(2): 213–224.

[154] Thornton, A. and Clutton-Brock, T. (Apr. 2011). Social learning and the development of individual and group behaviour in mammal societies. Philos. Trans. R. Soc. London, Series B, Biol. Sci. 366(1567): 978–987.

[155] Le Roux, A., Cherry, M.I., Gygax, L. and Manser, M.B. (2009). Vigilance behaviour and fitness consequences: Comparing a solitary foraging and an obligate group-foraging mammal. Behav. Ecol. Sociobiol. 63(8): 1097–1107.

[156] Voellmy, I.K., Goncalves, I.B., Barrette, M.-F., Monfort, S.L. and Manser, M.B. (2014). Mean fecal glucocorticoid metabolites are associated with vigilance, whereas immediate cortisol levels better reflect acute anti-predator responses in meerkats. Horm. Behav. 66(5): 759–765.

[157] Courbin, N. et al. (2015). Reactive responses of zebras to lion encounters shape their predator-prey space game at large scale. Oikos.

[158] Jr, E.D.B., Jr, D.R.F. and III, E.D.B. (1991). Predator avoidance and antipredator mechanisms: distinct pathways to survival. Ethol. Ecol. Evol. 3(1): 73–77.

[159] Wang, X. and Zou, X. (2017). Modeling the fear effect in predator–prey interactions with adaptive avoidance of predators. Bull. Math. Biol. 79(6): 1325–1359.

[160] Matsuda, H., Hori, M. and Abrams, P.A. (1996). Effects of predator-specific defence on biodiversity and community complexity in two-trophic-level communities. Evol. Ecol. 10(1): 13–28.

[161] Waltman, P., Braselton, J. and Braselton, L. (2002). A mathematical model of a biological arms race with a dangerous prey. J. Theor. Biol. 218(1): 55–70.

[162] Moothedan, M. and Joseph, C. (2016). Survey on SLA for PaaS Clouds 6(1): 57–61.

[163] Brynjolfsson, E., Hofmann, P. and Jordan, J. (2010). Cloud computing and electricity. Commun. ACM 53(5): 32.

[164] Allesina, S. and Pascual, M. (2008). Network structure, predator—Prey modules, and stability in large food webs. Theor. Ecol. 1(1): 55–64.

[165] Sadeghi, A.-R., Schneider, T. and Winandy, M. (2010). Token-based Cloud Computing, Springer, pp. 417–429.

[166] Glier, M. and McAdams, D. (2011). Concepts in biomimetic design: Methods and tools to incorporate into a biomimetic design course. ASME.

[167] Cheong, H. and Shu, L.H. (2013). Using templates and mapping strategies to support analogical transfer in biomimetic design. Des. Stud. 34(6): 706–728.

[168] Sullivan, T. and Regan, F. (2011). Biomimetic design of novel antifouling materials for application to environmental sensing technologies. J. Ocean Technol. 6(4): 42–54.

[169] Frillici, F.S., Fiorineschi, L. and Cascini, G. (2015). Linking TRIZ to conceptual design engineering approaches. Procedia Eng. 131: 1031–1040.

[170] Ilevbare, I.M., Probert, D. and Phaal, R. (2013). A review of TRIZ, and its benefits and challenges in practice. Technovation 33(2-3): 30–37.

[171] Beckmanna, H. (2015). Method for transferring the 40 inventive principles to information technology and software. In: Procedia Engineering 131: 993–1001.

[172] Russo, D. and Spreafico, C. (2015). TRIZ 40 Inventive principles classification through FBS ontology. Procedia Eng. 131: 737–746.

[173] Gorman, S.P., Kulkarnni, R.G., Schintler, L.A. and Stough, R.R. (2004). A predator prey approach to the network structure of cyberspace. pp. 1–6. *In*: Proceedings of the Winter International Synposium on Information and Communication Technologies.

[174] Kumar, M., Mishra, B.K. and Panda, T.C. (2016). Predator-prey models on interaction between Computer worms, Trojan horse and antivirus software inside a computer system. Int. J. Secur. its Appl. 10(1): 173–190.

[175] Director, T.T. et al. (2013). On the Cusp of Evolutionary Change.

[176] Mazurczyk, W. and Rzeszutko, E. (2015). Security–A Perpetual War: Lessons from Nature 17(1): 16–22.

[177] Nagel, J.K.S., Nagel, R.L., Stone, R.B. and McAdams, D.A. (2010). Function-based, biologically inspired concept generation. Artif. Intell. Eng. Des. Anal. Manuf. 24(4): 521–535.

[178] Talia, D. (2012). Clouds meet agents: Toward intelligent cloud services. IEEE Internet Computing 16(2): 78–81.

[179] Daniel Faggella. (2017). What is Machine Learning? [online]. Available: https://www.techemergence.com/what-is-machine-learning/. [Accessed: 30-Jun-2018].

[180] Biology and Dictionary, O. (2018). Predator-prey relationship. [online]. Available: https://www.biology-online.org/dictionary/Predator-prey_relationship. [Accessed: 30-Jun-2018].

CHAPTER 5

Applications of a Model to Evaluate and Utilize Users' Interactions in Online Social Networks

Izzat Alsmadi[a] and *Muhammad Al-Abdullah*[b,*]

1. Introduction

In the current internet world, Online Social Networks (OSNs) are ranked as the highest visited websites by users all over the world. According to Facebook,[1] "There are over 2.07 billion monthly users as of the third quarter of 2017. Among those users are 16 million local businesses which use the website to extend their market as of May 2013. There are 4.75 billion pieces of content shared daily as of May 2013."

OSN users consist of individual users, companies, or state departments who are required to maintain their presence in the different OSNs for varying reasons, such as maximizing social capital and building reputation, maintaining competitive advantage, or facilitating citizens' transactions respectively.

The high volume of data exchanged in OSNs make knowledge extraction a crucial goal for both practitioners and researchers. Additionally, recent business acquisitions, such as Nokia by Microsoft and WhatsApp by Facebook, show that Information Technology interests are shifting from hardware and software to information.

As such, data in social networks is a target for analysis for people and entities from a wide variety of domains, including but not limited to social science, security and intelligence, business analysis and marketing. However, data in social networks is collected and analyzed with or without users' explicit knowledge. There are many claims that indicate illegal privacy intrusions for OSNs to collect and analyze private

[a] Texas A&M, San Antonio, One University Way, San Antonio, TX 78224.
[b] University of San Francisco, 2130 Fulton St., San Francisco, CA 94117.
* Corresponding author: malabdullah@usfca.edu

[1] https://zephoria.com/top-15-valuable-facebook-statistics/

conversations and contents between the different users. According to [1], "Due to the proliferation of using various online social media, individual users put their privacy at risk by posting and exchanging enormous amounts of messages or activities with other users. This causes serious concern about leaking private information out to malevolent entities without the users' consent," (p. 265).

Furthermore, one kind of data of interest for analysis is opinion mining or sentiment analysis in which people's opinions about (or recommendations on) a particular product, news, event, or businesses are explored. Recommendation systems are concerned with utilizing data from OSNs to provide recommendations on subjects like a product, an expert, or a candidate employer or employee. However, one of the problems in those recommendation systems is that they are usually context-dependent. Each recommendation system should start by specifying (e.g., based on a user query), the subject of the recommendation. Also, users' data is analyzed in real time to evaluate users' opinions or recommendations on that particular query. These two factors and others contribute to the complexity of the process.

As a solution, reputation ranking models have been proposed, e.g. [1, 2]. They balance the protection of OSNs users' privacy while gathering information about their relations, opinions, or activities by collecting statistical data about the nodes and their activities without looking into the actual content of those activities. The models also provide static recommendation about the users themselves on OSNs, which ultimately promote information quality in OSNs where users will be more concerned with their friends' selection and the activities they post, as this may eventually impact their own reputation. Such referral systems can have several possible applications. For example, they can be used to rank search queries, credit reporting assessment, or referrals for professional, educational, etc., purposes.

This chapter extends the reputation-ranking models and offers the following explicit new contributions:

- It evaluates the reputation ranking model on an actual dataset collected from Facebook [3].
- It extends the equation of weighted edges that describe trust between friends in OSNs.
- It provides solutions for practical problems described in the earlier hypothesis model [1].
- It extends the use of 'classical, not weight-based cliques' and implements a model of weighted cliques that use normalized weights to evaluate the strength or bond in each clique.

Our proposed reputation model is inspired by two currently popular interaction-based ranking models:

(1) *Page Rank (PR)*: Page rank is one of the most significant algorithms used in search engines in particular and the internet in general to rank web pages retrieved, based on a search query. As most search queries retrieve a large number of matched or semi-matched results, it is important to develop an algorithm to rank the retrieved results or web pages. Based on search queries, most page-ranking algorithms retrieve web pages and rank them based on their popularity. A website or web page can get more popular when more people either visit this website or page or point to it (i.e., Inlinks). Inlinks is a metric used to indicate the number of web

pages pointing to the subject web page. In our OSN reputation model, Inlinks is represented by the number of friends. Page visits or page views are represented in our model by the number of interactions from friends with the subject or node activities.

(2) *H-index*: Despite its simplicity, H-index is still the most popular citation index to evaluate or compare researchers' contributions. For individual authors, H-index formula calculates the index based on two attributes—the number of publications of the author and the citations for each publication. Ultimately, the H-index value represents the number of publications of the author in which their citations exceed a certain number. For example, an author with an H-index of 10 indicates that this author has 10 publications in which each one of those 10 papers is cited by 10 or more papers. Similar to the earlier comparison with page ranks, H-index number of publications can be seen as number of friends in our reputation rank while citation number is represented by friends' interactions with subject node activities. In fact, friends' interactions with the source node can be seen as recommendations or referrals which are synonymous to H-index citations.

The rest of this chapter is organized as follows: Section 2 discusses how an entity's reputation is derived by describing each component of the reputation. Section 3 describes how the proposed reputation model is applied to privacy assessment in OSNs. Section 4 gives an overview of related work in terms of existing models of trust, reputation and privacy in OSNs. Section 5 provides an illustration of the expected applications of the proposed reputation model. Finally, Section 6 concludes this work and suggests future work directions.

2. Reputation Model

The focus of the proposed reputation system is on how to derive a reputation score for each node or user in OSNs. In this work, a graph (G) is considered where an entity is a vertex and the social relationship between two entities is represented by edges. That is, if two entities are connected with an edge, the two entities are friends with each other. Nonetheless, the degree of trust that one entity has towards the other entity can vary, implying that trust relationship is subjective and asymmetric. Thus, each node is assigned a weighted reputation score so that users with different volumes of interactions can be compared with each other. Two major inputs are considered in this calculation:

(1) *The interactions between the user as a node with other users or nodes.* In graph models, those interactions are modeled as edges. Unlike classical approaches, the proposed model differentiates the (thickness) of the edge or the relation between friends through the volume of interactions between the friends [1]. This weighted edge is bi-directional where the value in one direction is typically different from the other direction. One of the motivations in this scope is to be able to distinguish as a finer detail the weight of the relation between friends instead of solely the classical binary classification (i.e., as a friend or not).

(2) *User activities.* In integration with the first component, users' reputation can be calculated based on their activities and their friends' interactions with those activities. In OSNs, users can create posts, tweets, or upload pictures, among

others. Regardless of the nature of such activities based on the different OSNs, friends respond to those activities (e.g., by likes, comments, retweets). The model takes into account the volume of friends' interactions with those activities as well as a normalization process based on weights in order to be able to compare activities from different users regardless of their volume. The models provided in the literature cannot further distinguish whether friends' interactions with the particular activity was positive or negative. This is reasoned as the limitations of the data from which the models were built. The data does not incorporate nodes interactions' content to not violate the privacy of the sample participants. Therefore, our proposed model extends to include sentiment analysis to evaluate the polarity of users' interactions with each other as another dimension to weight relations between friends.

Alsmadi et al.'s model is extended by extracting cliques from OSNs. Unlike the classical use of maximal cliques, that is concerned with evaluating whether a relation exists or not between two particular nodes (i.e., binary zero or one), the proposed model calculates the weight of each edge and ultimately the weighted strength in the whole clique.

3. Friends' Weighted Trust

In a graph-based social networks, each user is represented by a node. Edges between nodes indicate a friendship relation between those two particular nodes. Typically, friendship link in OSNs is represented in a binary form (i.e., exists or not). However, our proposed model calculates weights for the different relations based on the interactions between the different friends or nodes. This makes the edge values between two particular friends different from one side to the other in most cases. As each OSN can have different types of activities between friends, a weight for each activity is calculated based on the volume of interactions between those friends in this activity divided by total activity interactions (i.e., from all friends). The same weight is calculated for each activity in the OSNs and an overall average is calculated based on all activities. As an initial estimator, all different types of activities (e.g., likes, photos, tags, tweets, retweets, favorites) were given the same weights. Eventually those different activities' weights will differ.

The proposed reputation model aims at providing each user or node in OSNs with a single value that best identifies the reputation score. Previously, we calculated weighted edges for each node friends based on their interaction with node activities. Nodes' weighted reputations can be simply the summation of weighted edges (i.e., from all friends) for the subject node.

4. Experiments and Analysis

Several OSN datasets are publicly available. However, there are certain constraints to exist in the selected dataset as it should include information about the different types of interactions between friends in the OSN. Accordingly, we used the dataset described in [3]. The dataset includes information collected from Facebook including 221 unique users and 2421 edge interactions between those users. The dataset also includes information about three types of interactions between friends in Facebook: number of common page likes, number of common photo tag, and number of wall posts.

We developed a program to calculate weighted edges, reputation and weighted cliques based on the equations described in [1].

5. Weighted Edges

Table 1 illustrates the top 50 edge relations out of the total of 2421. As the last one in the top 50 edge relations has the value of (0.431), this shows that the majority of social network relations can be classified as (weak), given that the number of 50 out of 2421 is 43 per cent. Classically, all 2421 will be given the value one (to represent a relation exists) in comparison with zero (to represent no relation). Based on our algorithm, the

Table 1: Top 50 weighted nodes relation.

Source Node	Destination Node	Weighed Edge	Source Node	Destination Node	Weighted Edge
32	33	1	141	145	0.640
54	110	1	100	102	0.614
65	67	1	209	216	0.604
69	79	1	201	215	0.592
110	115	1	139	142	0.583
117	118	1	109	115	0.574
123	124	1	72	108	0.564
128	129	1	211	212	0.553
131	133	1	184	185	0.519
140	142	1	127	128	0.519
142	145	1	146	149	0.518
149	150	1	88	106	0.514
183	205	1	219	218	0.513
212	213	1	157	166	0.493
120	121	0.805	219	155	0.486
115	118	0.707	218	183	0.482
116	119	0.706	127	129	0.480
129	131	0.693	104	123	0.465
130	131	0.681	138	142	0.464
185	186	0.674	162	171	0.454
113	114	0.666	63	75	0.444
148	149	0.666	181	215	0.444
199	215	0.666	72	109	0.435
103	106	0.654	52	53	0.435
216	4	0.640	200	18	0.431

value of one in weighted edges will always exist for those users in social networks who have only one friend. Weighted edge is one when each one of the three activities (likes, photos and walls) is given one-third of the weight. For one user, each one of those activities will be one (as numerator equals the denominator). We argue, however, that this rarely exists in reality (i.e., for a user to have only one friend) and it exists in the collected dataset largely due to limitations in the collection process. The second important information that weighted edges can provide us as an alternative to 'binary' edges where values can be either zero or one, is to know the difference in relation between two friends from the two directions. Originally, or in binary relations, when a relation exists, it will be one from both sides. In our case, weighted edges can be completely different and vary between zero and one. For example, none of those edges listed in the top 50 weighted edges exists also in the opposite direction. This means that those top 50 relations are strong from one side only. In OSNs, we can see patterns of such unbalanced edge-relations. For example, a celebrity or an ego user is expected to have many friends who are following her/him with significant weighted edges from their direction and zero or low weight from celebrity direction. There are examples of odd edges that this method can be used to automatically detect them. For example, Spam in social networks is a recently growing problem where many Spammers request an ad only to post their advertisements. Such users, who have very low level of activity interaction with friends, can be easily detected based on our weighted edge algorithm. Weighted edges can be also used as a simple method to classify friends based on their interactions. For example, friends with 70 per cent and above of weighted edges can be classified as 'close friends' and those between 70 per cent and 30 per cent can be classified as 'normal friends' and last level below 30 per cent as 'lightweight friends'.

6. Weighted Reputations

As described earlier, the goal of defining weighted reputations is to have a single and unique value for each user in OSNs to represent and compare this individual with other individuals in the same OSNs. This value should promote information quality, privacy and trust. This can be calculated from the number of friends for each user along with the volume of interactions from those friends with activities created or initiated by source user or node. In terms of information quality, users should be encouraged to create relevant posts that can trigger a lot of interactions. A user-trust to her/his different friends can be broadly classified based on how much such friends interact with the user activities. In practice, close friends interact with most of the activities created by a user. More interactions can be a simple indicator of more involvement or (trust) from friends. As for privacy, the aim is to enable a privacy system to automatically classify friends based on their interactions and eventually decide their privacy level on the source node or user activities. Similar to the values in weighted edges, values in weighted reputation vary between zero as minimum when a user has no friends, and one when the user has only one friend. All other cases are in between.

In the collected dataset, 42 nodes are shown to have a weighted reputation of one as they only have one friend in the collected dataset. Table 2 illustrates a sample of high-value weighted reputations excluding those with '1' values. The weighted

Table 2: A sample of top weighted reputations.

Node	Weighted Reputation	Node	Weighted Reputation
209	0.984559	33	0.533333
197	0.978355	61	0.528629
200	0.958937	175	0.526922
203	0.958559	154	0.508592
185	0.94143	36	0.498904
35	0.931713	151	0.497155
177	0.897661	152	0.480681
82	0.880556	56	0.470763
176	0.874618	13	0.458354
173	0.872204	98	0.453455
184	0.867593	37	0.430915
195	0.866925	167	0.419159
194	0.859571	60	0.411366
207	0.852864	62	0.395504
86	0.85277	38	0.38165
172	0.84245	31	0.376713
137	0.840903	46	0.37594
206	0.839206	42	0.366785
53	0.829154	47	0.359772
204	0.822254	7	0.346994
162	0.813433	3	0.343509
20	0.806756	10	0.338657
189	0.802676	198	0.333333
192	0.797044	215	0.333333
51	0.795301	217	0.333333
191	0.791895	34	0.328131
171	0.785188	6	0.305952

reputation is independent of the number of friends and, for example, the highest values are for those who have only one friend. This makes it a relative reputation from the friends' perspective rather than a general reputation score.

7. Trust in Cliques

The extraction of clique relations in social networks can show information related to groups and how those groups are formed or interact with each other. In social networks, a clique represents a group of people where every member in the clique is a friend with every other member in the clique. Those cliques can have containment relation with each other where large cliques can contain small size cliques. This is why most algorithms in this scope focus on finding the maximum clique. In our model, we evaluated adding the weight factor to clique algorithm to show more insights about the level of involvement of users in the clique. In this regard, clique activities are not like those in group discussions where an activity is posted by one user and is visible by all others. Our algorithm calculates the overall weight of the clique based on individual edges. Users post frequent activities. Their clique members, just like other friends, may or may not interact with all those activities.

Table 3 illustrates weighted cliques, the largest cliques in the dataset. The first column includes IDs of node members of the clique. The second column includes the number of members in each clique. The third column shows the clique strength as the summation of all weighted edges in the clique. The last column represents the weighted clique strength which divides the total clique by the number of edges in the clique. Normal clique strength value is constantly increasing with the number of nodes or edges in the clique. However, the weighted clique value eliminates the dependency on the number of nodes or edges. As such, we can see, for example, that the smallest clique in the table has the highest weighted clique. This is not a trend however, as the next highest weighted clique is the one with 20 nodes. This indicates that trust or interaction between members in the first clique is the highest. When we study social networks in a larger context, highest and lowest weighted cliques can be of special interests. We can also look at variations in clique strengths over a certain period of time. For example, increase of the weighted clique in a certain month over a period of time can trigger further investigations of what could cause such significant sudden increase (e.g., a clique collaborative activity or event).

Table 3: Total and weighted cliques.

Clique Members	No. of Members	Total Clique	Weighted Clique
221, 215, 190, 187, 177, 176, 171, 151	8	1.61	28.77
221, 215, 190, 187, 177, 175, 171, 170, 151	9	1.43	19.96
221, 215, 190, 181, 177, 176, 172, 171, 169, 165, 164, 162, 157, 156, 153, 152, 151	17	6.08	22.36
221, 215, 190, 181, 177, 175, 172, 171, 170, 169, 165, 164, 162, 157, 156, 153, 152, 151	18	9.75	22.08
221, 215, 181, 177, 175, 172, 171, 170, 169, 168, 166, 165, 164, 162, 157, 156, 153, 152, 151	19	8.52	24.94
216, 209, 206, 193, 192, 191, 180, 179, 175, 169, 167, 160, 158, 151, 98, 20, 18, 9, 4, 2	20	10.62	27.96
216, 209, 193, 192, 191, 180, 179, 175, 170, 169, 167, 163, 160, 158, 151, 98, 20, 18, 9, 4, 2	21	11.64	27.73

8. Related Work

This section discusses the research contributions in the areas of reputation, trust and privacy in OSNs. Specifically, the section focuses on the literature of the applications of those contributions in the scope of knowledge extraction and usage in OSNs.

[4] proposed a trust-based recommendation system which preserves privacy considerations. Friends of friends are recommended to a friend based on certain matching attributes. Typically, a recommendation system should be able to know personal or private information about an individual in OSNs in order to be able to recommend a person, profession or product based on those attributes. Such recommendation systems are similar in principle to match-maker programs that connect a couple based on common interests. Recommendation systems are also context-independent which means that all personal friends can be consulted in the recommendation system without the ability to distinguish that some of these friends may not be good recommenders given the particular context of the query. Guo et al.'s model tackled such issues with a model which extends recommendation beyond the first hop or degree friends (e.g., friends of friends and so on). The model defined a continuous trust level of values between zero (no trust) to one (a complete trust). In the same vein, [5] proposed allowing users in the social networks to interact beyond their direct friends while at the same time, preserve privacy and not exposing information to those strangers. The paper's case study focused on the issue of anonymizing the existence of connections between nodes in social networks. Usually this is not public information and can be only seen between friends in the first degree. The paper considered network anonymization specifically for the purpose of aggregating network queries [6] proposed a model to visualize trust relations in OSNs. The proposed authentication system recommends to users whether their newly requested invitations are trustable or not and this is one of the use cases of our proposed reputation model where, if a user in OSNs receives an invitation from a stranger, they can use the requester's reputation rank as a determinant for the decision whether to accept or reject the invitation. Our idea is to transfer trusts from the friends of the stranger to the invitee through this reputation rank formula. Our weighted edges or friendship formulas aim to show to the invitee not only the relation of the inviter, if it exists with the invitee friends, but also the weight of those relations since the bare existence of a relation or not can be confusing in some cases. In many cases, users can be tricked to add friends which means that such friends should not be allowed to interact with all regular user activities as soon as they are added to the friends' list. This approach can also be a method to counter spam spread in social networks where users can accept or deny messages from strangers or even friends automatically based on predefined metrics. Many celebrities or ego users allow adding new friends by default or without any processing. [7] discussed the association of friendship connections in OSNs and geographical locations and prediction of the location of future friends. Dey et al. (2012) evaluated the age factor of users in OSNs in their ability to protect their private information. They proposed a method to estimate age of Facebook-users based on interactions with friends. More recently, it is impossible to crawl Facebook users publicly due to the extra privacy requirements that Facebook enforced [8] proposed a mobile-based friend of friend's detection algorithm. That is similar in principle to matching algorithms that connect people based on their common attributes. In the friend-finding scope also, [9] proposed federated social networks. The approach is based on exposing user contact to find those contacts in other OSNs. As an alternative, the algorithm can look for candidate friends based on a similarity score with some common attributes. One problem with such approaches is that they need to expose the user's private information, such as user friends, their contacts, or some other

user-related attributes, like their work or geographic area. [10] conducted a survey to evaluate the challenge of balancing between privacy and information-sharing in OSNs. The study was conducted on users from Facebook and MySpace and showed that users in Facebook have more trust in their OSNs and are willing to share or expose more private information when they interact with their friends. Similarly, Zhang et al. (2010) included a general discussion on security concerns in OSNs. [11] evaluated the access control model in Facebook. Policies are implicitly used by OSN users to decide visibility on their profiles and activities. Users can specify static policies that are applied to all their accounts or they can decide fine-grained policies for their activities, case by case. Users sometimes may fall unintentionally into privacy problems due to their lack of knowledge of how policies in such OSNs work. In the subject of access control models also, [12] proposed a rule-based access control and sharing enforcement model in OSNs. The authors describe the depth of the relation and the length of the shortest path as two significant attributes to describe relations between users in OSNs.

[13] offered an automatic privacy assessment system in OSNs that handle cases, such as when a user in the social network may post an activity, image, or video in which other people or friends can be part of, without an explicit permission from their friends in their posted activities.

9. Applications of the Reputation Model

In this section, we describe examples of how our proposed reputation system can be used in selection of possible applications. In hybrid interaction and history-based privacy system in OSNs, our goal is to develop an automatic privacy assessment system for users which helps them to make reasonable privacy-assessment decisions for their activities and profiles with the following constraints and guidelines:

- Each node in the OSN will have relations with all other nodes in the social network with a weighted value that varies between zero (not friends) up to 1 (very close friends). The initial assumption is that only users who are friends (e.g., weighted edge > 0) can see other users' profiles and activities.

- The system can start with an initial setting of two options: (1) open-privacy, where all users with an edge-value > 0, can see users' profile and all posted activities. As an alternative, and for manual consideration, the users can temporarily block some of their friends from seeing their profiles and activities. This static decision will keep those friend values as zeros, which will make them temporary unfriendly, (2) closed-privacy in which no user can see any activity unless explicitly added by the subject user. This decision is memorized by the system for future activities and will only be nullified by a second explicit decision.

- As friends interact with user activities, their edge-trust values are gradually increased. This applies to closed-privacy models only, since the open-model assumes all friends can see all the others' activities initially. Eventually, the privacy level is decided by the level of interaction between these friends. This can keep increasing or decreasing with time based on their future interactions. This means that future privacy level is decided by the current interaction level with the user.

9.1 Friends' Recommendation System

One of the main functionalities in OSNs is to promote expansion of the usage of their social networks through the promotion of more interaction from users with network applications or with each other, using the network tools. Most of the algorithms to recommend new friends use similar attributes related to the person, country, profession, interests or common friends. Our goal is to expand the earlier proposed interactive-based reputation system and provide friends' recommendations based on individual strengths of relations with friends and the friends' interactions with the user activities. Our study of relations in social networks showed that the majority of friends in any user in OSNs are silent or low-profile friends who either rarely or occasionally interact with user activities. For example, in Facebook the three main friends' interactive actions are: likes, comments, or wall tags. Wall tags seem to be the highest indicator of relationship-strength, especially as it requires a especial permission by the node owner. Commenting on photos, videos, or posts is the middle-level activity where a number of friends will make comments periodically. The volume of such comments can be another indicator of the closeness or the weight of the relationship. The number of likes seems to be the largest and the lowest friendship-indicator. We surveyed a large number of users in Facebook and noticed that in many cases, there are several possible common features that can be extracted based on studying users' activities in OSNs in addition to their friends' interactions with those activities. These include, for example, how much those users are involved in the social network. Despite many common features, if two users are different in the volume of involvement in the social network, they may not be considered with similar profiles. Our model includes also the level of interactions with groups, or public pages, among others. Unlike classical friend matching, algorithms that may use similar groups or research interests, need to further evaluate the users' involvements in those groups or public pages. Users can register in many groups or public websites while they may rarely interact with their activities. Two users that heavily interact with a certain group or a public website should be seen as good candidate-friends even if they have many different attributes.

As a recommendation system for accepting invitations from strangers, our reputation system assists nodes in making decisions whether to accept or reject the invitation to connect that is sent by strangers. The model is interaction based, which means that the model will elaborate that this is a stranger to the invitee based on lack of interaction; it also shows who has this stranger interacted with from the invitee's friends so that they can be consulted about the stranger. Professional referrals in OSNs are seen by many businesses as necessary tools for marketing and the assessment of users' perceptions of commercial products. In addition to recommendations on products (e.g., books, electronics, etc.), recommendations can be used in seeking professional or expert referrals on services (e.g., a physician, consultant, educator, clinic, restaurant or brand). Users may trust their friends and their recommendations. Some professional related OSNs, such as LinkedIn allow users to seek recommendations from their friends. Interactions between friends can help to distinguish the nature of the relationship between these friends (i.e., personal or professional). Such classification should be also considered when considering referral recommendations from friends. For example, a professional recommendation from a personal friend can fall within the category of (conflict of interests). We assume here that OSN is capable of determining the nature of the relation that exists between those two friends (i.e., personal or professional). This requires modeling referral decisions based on four elements: (1) recommended, (2) recommender, (3) the nature of their relation, and (4) the context in which the recommended person is seeking the recommendation. As shown in Fig. 1, the recommender is in the middle since s/he needs to know all the components of the system. In addition, the recommendation should not only include information about the recommended person and the context, but also about the recommender and her/his relation to the recommender.

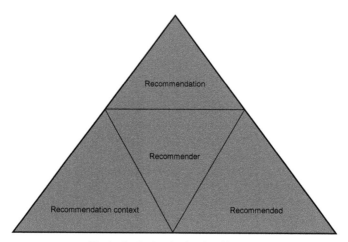

Fig. 1: Professional referral architecture.

9.2 Credit Reporting

Several references indicate that credit reporting agencies are already using or aggregating information from OSNs. We argue that our proposed reputation rank for OSN nodes can be used directly as part of credit or background reporting. It allows users to transparently monitor and be able to improve their credit assessment that will impact the decisions of whether to give these users credits, or any other possible decisions. In classical credit-and-background reporting, an individual score can be negatively impacted by incidents that users are not aware of or have no power to change. Reputation score is not an alternative to credit reporting that involves user financial decisions and activities; instead, it is a complement that shows a new or different dimension related to the user online presence or online behavior in particular. Users can improve their reputation score through their friends' selections and through the kind of activities they create.

Reports show that some reporting agencies use illegal tools to access private conversations between users in OSNs. All reporting organizations depend on the aggregation of a large amount of data about users from different sources (e.g., utility bills, cell phones, government agencies, OSNs, etc.). In many cases, they may pay to acquire such data or may acquire it illegally. The availability of such wealthy data publicly through OSNs encourages those data aggregations to use them as one of their major sources of information. Our reputation model proposes an alternative, where users and OSNs can control and manage the process. Our model proposes a statistical approach that does not require looking into the content of either public or private conversation. Based on the goal or usage of credit reporting, the nature of extracted information can vary. In general, the reputation rank itself can be used similar to the credit score where, for example, the user will qualify for a rent application if their credit score is more than a score A. OSNs should be able to generate these models. Although some information used in the processing maybe private, however, final report should only include general statistics that do not violate users' privacy.

9.3 Spam Detection in OSNs

Spam is very popular in e-mails where users frequently receive unsolicited e-mails sent for a large number of e-mail users, largely for marketing or financial purposes. Spam problem starts also growing rapidly in OSNs. In some cases, spammers register in publicly open pages to keep posting their spam messages. In other cases, they may seek friendships with many users to be able to spread their marketing messages in those users' activities. Many OSN users may accept friendship requests from anonymous users, intentionally or unintentionally. In addition, rarely do OSN users screen their friendship lists for friends to verify whether they want to keep or remove some names. Their accounts can possibly be exposed to security problems or at least spammers. Based on our model, we proposed to alert users periodically about users, who rarely interact with their activities. Users may screen those names and make decisions whether they want to connect or not with those names. This can be a convenient alternative to manual investigation specially for users, who have a large number of friends. This is also an alternative to content-based spam detection which may produce a possible number of false negative or positive elimination of friends.

9.4 Assessments of Networks Evolution

The formation of groups in OSNs is dynamic and can change from one network to another. Information or knowledge extraction related to groups' formations, interactions, or interests can be related to many research fields. However, there are several challenges that face such knowledge extraction, including the large amount of data and the need to have a focused context upon which search can be concentrated. Instrumentation or automation for such a process can then be possible. Through this paper, we show examples of how to measure dynamically the strength of relations among members in OSNs' agile groups. Those groups or cliques are not formally formed or defined.

9.5 Positive or Negative Interactions

Our original reputation model is based on statistical data that does not require analysis of the content of interactions. Such interactions are considered regardless of their nature of being a positive or negative. Nonetheless, semantic analysis can be conducted on public posts and interactions to see if the friends' comments with each other are positive or negative. Ultimately, friendship—weighted edges can vary from (-1) to (1) rather than from (0) to (1). For example, a friend can be heavily interacting with a user, but based on sentiment analysis, such interactions are always negative and hence the overall weight value of the relation should be negative.

9.6 Context-driven Privacy Assessment

We propose an automatic context-driven privacy system that can alert users, based on their posts or activities on who can or should see their activities. There are two modes in which this can operate: (1) a manual mode in which activity authors can decide the level of privacy about the activity. The privacy assessment system can then decide who can or can't view, comment on, or propagate the activity based on the users' friendship classification in author's profile; (2) an automated mode in which the privacy assessment system can decide the activity confidentiality classification. Sentiment analysis methods can be used to make a final judgment on the activity to be one class label from several alternatives (e.g., normal, classified or confidential). Based on this initial confidentiality classification, the second step will be similar to the manual mode where different friends will be assigned different visibility or access levels.

10. Twitter Dataset

The majority of publicly-existing datasets from social networks include attributes about the nodes without including attributes that describe the relations between the different nodes. Using Twitter data, we developed a tool[2] to extract the different types of interactions that may occur between users.

In comparison with the previous Facebook dataset that we used and described earlier, we found that interactions between friends in Twitter, in most cases, are not as explicit as in Facebook (e.g., comment on a friend post, or wall tag). Users in Twitter can like (favorite) their friends Tweets. They can also Retweet those Tweets. However, in Twitter, the high interactive comment-sessions in which all friends can respond to a user post or Tweet do not exist. In addition to Tweets, Retweets and Favorites, users in Twitter can indirectly interact with their friends through (mentions) or (hashtags). These (friends) can be websites or organizations rather than just Twitter users. Figure 2 shows our generated dataset files based on Twitter entities.

- Tweets: represent the main Twitter entity; all other entities can be extracted within the Tweet entity. For example, in our Twitter Parsing tool, the line ('tweets = twitter.GetTwitts(userName, 100). result;') generates the collection of Tweets that can be used to extract all other entities. The (100) in the code indicates Twitter rate limit.
- *Contributions* are the users who contribute to the authorship of the Tweet. This typically includes one ID, other than the original Tweet author.
- *Coordinates* represent the geographical location of the Tweet as longitude instead of latitude.
- *Followers* can be extracted from the (user) entity of the Tweet where these are other users following the Tweet author.

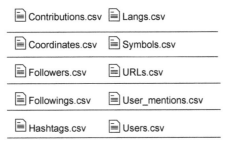

Fig. 2: Twitter parsed entities.

- *Followings* can be extracted from the (user) entity of the Tweet where these are other users that the Tweet author is following.
- *Hashtags*, if the Tweet includes a Hashtag, the Hashtag entity will include the Hashtag's text.
- *Languages* represent the machine-detected language(s) of the Tweet text.
- *URLs* are the collection of URLs or links that can be extracted from the Tweet text.
- *User Mentions* are the collection of screen names collected from the Tweet text.
- *Symbols* are the collection of financial symbols extracted from the Tweet text.

11. Conclusion

In an earlier work, [1] proposed a reputation score model in OSNs that aims to promote information accuracy and enhance knowledge extraction from OSNs and promote quality in users' activities and interactions. This chapter provides an overview of the extended reputation model as well as the applications of how to use this reputation model. The chapter also develops a public dataset from Twitter to be used as a prototype for the proposed weighted reputation model and its applications. With the huge amount of information exchanged through OSNs, it becomes necessary for these networks to promote information quality, specially as information in these networks is an integral part of the internet body of knowledge in general.

Key Terminology and Definitions

Online Social Network (OSN): An online platform or website which people, professionals, businesses, or even government agencies use to build social relations with other respective online users, who share a common interest being it career, ideology, business, or a service.

Weighted reputation model: A model that has been proposed in the field of OSNs and is used in preserving online relations between the different nodes (individuals, businesses, or government agencies). The model is not only studying if two nodes are connected, it also preserves the interaction level between the two nodes without analyzing the content of the interaction. This allows analyzing OSN nodes relations without breaching any security or privacy concerns.

References

[1] Alsmadi, I., Xu, D. and Cho, J.H. (2016). Interaction-based Reputation Model in Online Social Networks.
[2] Cho, J.H., Alsmadi, I. and Xu, D. (2016). Privacy and social capital in online social networks. pp. 1–7. The Proceedings of the Global Communications Conference (GLOBECOM), IEEE.
[3] Dev, H., Ali, M.E. and Hashem, T. (2014). User interaction-based community detection in online social networks. pp. 296–310. *In*: International Conference on Database Systems for Advanced Applications, Springer, Cham.
[4] Guo, L., Yang, S., Wang, J. and Zhou, J. (2005). Trust model based on similarity measure of vectors in P2P networks. pp. 836–847. *In*: Proceedings of Grid in Cooperative Computing, GCC.
[5] Zhou, B. and Pei, J. (2008). Preserving privacy in social networks against neighborhood attacks. pp. 506–515. *In*: Data Engineering, 2008, ICDE 2008. IEEE 24th International Conference. IEEE.
[6] Kim, T.H.J., Yamada, A., Gligor, V., Hong, J. and Perrig, A. (2013). Relationgram: Tie-strength visualization for user-controlled online identity authentication. pp. 69–77. *In*: International Conference on Financial Cryptography and Data Security, Springer, Berlin, Heidelberg.
[7] Backstrom, L., Sun, E. and Marlow, C. (2010). Find me if you can: Improving geographical prediction with social and spatial proximity. pp. 61–70. *In*: Proceedings of the 19th International Conference on Worldwide Web, ACM.
[8] Von Arb, M., Bader, M., Kuhn, M. and Wattenhofer, R. (2008). Veneta: Serverless friend-of-friend detection in mobile social networking. pp. 184–189. *In*: Networking and Communications, 2008, WIMOB'08. IEEE International Conference on Wireless and Mobile Computing, IEEE.
[9] Dhekane, R. and Vibber, B. (2011). Talash: Friend finding in federated social networks. *In*: Proceedings of LDOW.
[10] Dwyer, C., Hiltz, S. and Passerini, K. (2007). Trust and privacy concern within social networking sites: A comparison of Facebook and MySpace. AMCIS 2007 Proceedings, p. 339.
[11] Fong, P., Anwar, M. and Zhao, Z. (2009). A privacy preservation model for facebook-style social network systems. Computer Security–ESORICS 2009: 303–320.
[12] Carminati, B., Ferrari, E. and Perego, A. (2009). Enforcing access control in web-based social networks. ACM Transactions on Information and System Security (TISSEC) 13(1): 6.
[13] Squicciarini, A.C., Xu, H. and Zhang, X.L. (2011). CoPE: Enabling collaborative privacy management in online social networks. Journal of American Society for Information Science and Technology 62(3): 521–534.

Additional References

Acemoglu, D., Bimpikis, K. and Ozdaglar, A. (2014). Dynamics of information exchange in endogenous social networks. Theoretical Economics 9(1): 41–97.
Agarwal, V. and Bharadwaj, K.K. (2013). A collaborative filtering framework for friends recommendation in social networks based on interaction intensity and adaptive user similarity. Social Network Analysis and Mining 3(3): 359–379.
Basuchowdhuri, P., Anand, S., Srivastava, D.R., Mishra, K. and Saha, S.K. (2014). Detection of communities in social networks using spanning tree. pp. 589–597. *In*: Advanced Computing, Networking and Informatics-2, Springer, Cham.
Bejar, A. (2010). Balancing Social Media with Operations Security (OPSEC) in the 21st Century. Naval War Coll Newport RI Joint Military Operations Dept.
Caliskan Islam, A., Walsh, J. and Greenstadt, R. (2014). November. Privacy detective: Detecting private information and collective privacy behavior in a large social network. pp. 35–46. *In*: Proceedings of the 13th Workshop on Privacy in the Electronic Society, ACM.
Cascavilla, G., Conti, M., Schwartz, D.G. and Yahav, I. (2015, August). Revealing censored information through comments and commentaters in online social networks. pp. 675–680. *In*: Advances in

Social Networks Analysis and Mining (ASONAM), 2015 IEEE/ACM International Conference, IEEE.

Dey, R., Jelveh, Z. and Ross, K. (2012, March). Facebook users have become much more private: A large-scale study. pp. 346–352. *In*: Pervasive Computing and Communications Workshops (PERCOM Workshops), 2012 IEEE International Conference on, IEEE.

Dressler, J.C., Bronk, C. and Wallach, D.S. (2015, October). Exploiting military Op-Sec through open-source vulnerabilities. pp. 450–458. *In*: Military Communications Conference, MILCOM, 2015-2015, IEEE.

Engelstad, P.E., Hammer, H., Yazidi, A. and Bai, A. (2015, September). Advanced classification lists (dirty word lists) for automatic security classification. pp. 44–53. *In*: Cyber-Enabled Distributed Computing and Knowledge Discovery (CyberC), IEEE.

Guo, L., Zhang, C. and Fang, Y. (2015). A trust-based privacy-preserving friend recommendation scheme for online social networks. IEEE Transactions on Dependable and Secure Computing 12(4): 413–427.

Hunter, R.F., McAneney, H., Davis, M., Tully, M.A., Valente, T.W. and Kee, F. (2015). Hidden social networks in behavior change interventions. Journal Information 105(3).

Lawson, S. (2014). The US military's social media civil war: Technology as antagonism in discourses of information-age conflict. Cambridge Review of International Affairs 27(2): 226–245.

Lees, D.W. (2016). Understanding effects of Operations Security (OPSEC) awareness levels of military spouses through the lenses of training and program management: A qualitative study. Ph.D. dissertation, Creighton University, Omaha, Nebraska.

Moe, T.A. (2011). Social Media and the US Army: Maintaining a Balance. Army Command and General Staff Coll Fort Leavenworth KS School of Advanced Military Studies.

Phillips, K.N., Pickett, A. and Garfinkel, S.L. (2011). Embedded with Facebook: DoD Faces Risks from Social Media. Marine Corps Tactical Systems Support Activity Camp Pendleton CA.

Rea, J., Behnke, A., Huff, N. and Allen, K. (2015). The role of online communication in the lives of military spouses. Contemporary Family Therapy 37(3): 329–339.

Usbeck, F. (2015). The power of the story: Popular narratology in pentagon reports on social media use in the military. Poetics of Politics: Textuality and Social Relevance in Contemporary American Literature and Culture 258: 313.

Wang, Z., Liao, J., Cao, Q., Qi, H. and Wang, Z. (2015). Friendbook: A semantic-based friend recommendation system for social networks. IEEE Transactions on Mobile Computing 14(3): 538–551.

Yahav, I., Schwartz, D.G. and Silverman, G. (2014, August). Detecting unintentional information leakage in social media news comments. pp. 74–79. *In*: Information Reuse and Integration (IRI), 2014 IEEE 15th International Conference, IEEE.

Yang, X., Guo, Y. and Liu, Y. (2013). Bayesian-inference-based recommendation in online social networks. IEEE Transactions on Parallel and Distributed Systems 24(4): 642–651.

Zhang, C., Sun, J., Zhu, X. and Fang, Y. (2010). Privacy and security for online social networks: Challenges and opportunities. IEEE Network 24(4).

CHAPTER **6**

A Deep-dive on Machine Learning for Cyber Security Use Cases

Vinayakumar R.,[1,*] *Soman KP.,*[1] *Prabaharan Poornachandran*[2] *and Vijay Krishna Menon*[1]

Conventional methods, such as static and binary analysis of malware, are inefficient in addressing the escalation of malware because of the time taken to reverse engineer the binaries to create signatures. A signature for the malware is accessible if the malware might have made significant damages to the system. The ongoing malicious activities cannot be detected by the anti-malware system. Still, there is a chance of detecting the malicious activities by analyzing the events of DNS protocol in a timely manner. As DNS was not made by considering the security, it suffers from vulnerabilities. On the other hand, malicious uniform resource locator (URL) or malicious website is a primary mechanism for internet malicious activities, such as spamming, identity theft, phishing, financial fraud, drive-by exploits, malware, etc. In this paper, the event data of DNS and malicious URLs are analyzed to create cyber threat situational awareness. Previous studies have used blacklisting, regular expression and signature-matching approaches for analysis of DNS and malicious URLs. These approaches are completely ineffective at detecting variants of existing domain name/malicious URL or newly found domain name/URL. This issue can be mitigated by proposing the machine learning based solution. This type of solution requires extensive research on feature engineering and feature representation of security 'artifact type', e.g., domain name and URLs. Moreover, feature engineering and feature representation resources must be continuously reformed to handle the variants of existing domain name/URL

[1] Center for Computational Engineering and Networking (CEN), Amrita School of Engineering, Coimbatore, Amrita Vishwa Vidyapeetham, India.
[2] Center for Cyber Security Systems and Networks, Amrita School of Engineering, Amritapuri, Amrita Vishwa Vidyapeetham, India.
* Corresponding author: vinayakumarr77@gmail.com

or entirely new domain name/URL. In recent times, with the help of deep learning, artificial intelligent (AI) systems achieved human-level performance in several domains. They have the capability to extract optimal feature representation by itself taking the raw inputs. To leverage and to transform their performance improvement towards the cybersecurity domain, we propose a method named as Deep-DGA-Detect/Deep-URL in which raw domain names/URLs are encoded using character-level embedding. Character-level embedding is a state-of-the-art method amplified in natural language processing (NLP) research [1]. Deep learning layers extract features from character-level embedding followed by feed forward network with a non-linear activation function estimates the probability that the domain name/URL is malicious. For comparative study, various deep learning layers are used. The optimal parameters for network and network structure are selected by conducting experiments. All the experiments are run for 1,000 epochs with learning rate 0.001. The deep learning methods performed well in comparison to the traditional machine learning classifiers in all test cases. Moreover, convolutional neural network-long short-term memory (CNN-LSTM) has performed well in comparison to other methods of deep learning. This is due to the fact that the deep learning architectures implicitly obtain the hierarchical features and long-range dependencies in character sequence in the domain name/URL. The proposed framework is highly scalable and can handle 2 million events per second. This analyzes the large volume and variety of data to perform near real-time analysis which is crucial to providing early warning about the malicious activities.

1. Introduction

As the number of computer applications and the sizes of networks continue to grow, the potential harm which will be caused by attacks launched over the web keeps increasing dramatically. Attackers are constantly developing new security threats which can easily break the present detection and prevention mechanisms. Logs of different computing systems, networks, network packets and transactions contain useful information which can be used for Security Log Analysis. All types of log data can be used to detect the faults and malicious events and it can also be used to monitor the ongoing security threats. The log analysis application provides useful information to the network administrators to defend against existing and new types of security threats. The log analysis application allows the user to visualize the attacks, to establish surprising network activities [3] and to redefine the attack or malicious activity.

As modern computing paradigm is considerably more complex and connected, it generates extremely large volume of data, typically called 'big data'. This is primarily due to the advancement in ICT systems and rapid adoption of internet. The analysis of big data in a timely manner is considered as a significant task for malicious activity detection. Machine learning is a widely used approach to extract valuable knowledge from big data.

The idea of machine learning was introduced years back with the aim of developing machines/computers which could think like humans. This essentially means training them to think logically and take decisions as humans do. There are two popular approaches to this training—one is supervised learning and the other is unsupervised learning. In supervised learning, the machine is familiarized for an operation with a set of sample inputs and sample outputs such that after training, given a new data, the algorithm is able to generate the corresponding output accurately. In the case of unsupervised learning, the algorithm is not trained with sample outputs initially and the algorithm by itself learns some pattern within the data and manipulates the information. Deeply explored is the field of classification problems which addresses the issue of categorizing data based on earlier observations. These classes of algorithms try to learn the behavior of data and thus enable itself for future prediction, given similar data. Popular classification algorithms include decision trees, *k*-nearest neighbor, support vector machine (SVM), artificial neural networks (ANN), genetic algorithm, etc. Classical machine learning algorithms rely on feature engineering to extract the optimal data representations or feature representations of security artifacts. However, selection of optimal features requires extensive domain knowledge and considered a daunting task in cyber security. Though modern machine learning, typically called as deep learning, is a black box, it has the capability to extract features [5]. In a nutshell, even though the power of RNN and CNN in extracting time and spatial information is larger, adoption towards Cyber security tasks is in infancy stage [6–19]. A large study to measure the effectiveness of various CNN and RNN on both off-line and real-time for Cyber security tasks will be required.

The field of cyber security is an interdisciplinary field which now requires the knowledge of natural language processing (NLP), image processing (IP), speech recognition (SR) and many other areas. More importantly, the use cases of Cyber security domain are surrounded by large amounts of text data. These text data convey important information for security artifacts. Thus analysis of the text data to understand the syntactic and semantic information has been considered a significant task to enhance organizational security. In this work, we apply the techniques of NLP and text mining techniques with classical machine learning and deep learning approaches to cyber security use cases. The use cases considered in this study are (1) Domain Generation Algorithms (DGAs) (2) Malicious Uniform Resource Locator (URL).

The scope of the research has been on finding malicious activities by analyzing event data from DNS and URLs. The major contributions of the current research work are:

- **Situational awareness framework:** In this work, a sub module for collecting malicious URLs is added to the situational awareness framework [2]. Situational awareness framework is a robust distributed framework that can collect and process trillions of event data generated from DNS queries by the internet-connected hosts. The event data produced by DNS and URLs are correlated to detect the attack patterns. These identified patterns could be used to stop further damages by malware. The framework has been built for web-scale data analysis, processing billions of events per second in both real-time and on demand basis.

- **Early detection of malicious activities:** Whitelisting or blacklisting the domain names is a laborious task and needs to be updated manually on a regular basis. In a significant number of cases the additions about a malware will be made only after weeks or months of its propagation. By this time, most the systems might have been affected with bots that are controlled by the bot master. To find malicious activities at recursive DNS level, we perform multiple analysis on the DNS and URL logs. We employ highly scalable optimization and deep learning architectures to classify and correlate malicious events. To find an optimal method, the performance of various classical machine learning and deep learning architectures is shown over a large labeled dataset consisting of a mix of algorithmically generated domains and benign domains gathered from various sources for DGA and URLs. In addition to detecting the DGA generated domain name, we categorize them to the corresponding DGA-based malware family.

2. Domain Generation Algorithms (DGAs)

The attacks on the internet service keeps rising due to the wide availability of internet to the general public. One of the main internet services, DNS has also been attacked by using various new methods. Attackers are constantly developing new security threats which can easily break the DNS service. The security of DNS is concerning the entire internet. DNS log file provides the insights of the DNS security. The event data of DNS protocol are collected by using the 'port mirroring' method with setting a promiscuous mode for communication between the DNS server and the DNS clients. The event data consists of DNS queries. The sheer volume of DNS log information typically makes it tough to discover attacks and pinpoint their causes. This demands new frameworks and techniques to analyze, process and manipulate different DNS log files for preventing security issues [2]. In this work, we use a distributed, parallel and cluster computing framework, which permits organizations to analyze a vast number of DNS logs efficiently. By applying preprocessing to the DNS queries, Time, Date, internet protocol (IP) and Domain name information is extracted. The domain name information is analyzed for detecting malicious events. The proposed interactive visual analysis system for the DNS log files can detect the anomalies in DNS query logs.

3. Malicious Uniform Resource Locator (URL)

A uniform resource locator (URL) is a division of Uniform Resource Identifier (URI) that is used to identify the location of the resources and retrieve them from internet. This directs the user to a particular web page on a website. A URL is composed of two parts. The first part is the protocol, for example, Hypertext Transfer Protocol (HTTP), Hyper Text Transfer Protocol Secure (HTTPS) and the second part is the location of resources via domain name or internet protocol (IP) address. Both are separated by a colon and followed by two forward slashes. Most of the time a user by oneself is not known whether the URL belongs to either benign or malicious category. Thus unsuspecting users visits the websites through the URL presented in e-mail, web search results and others. Once the URL is compromised, an attacker imposes an attack. These compromised URLs are typically termed as malicious URLs. As a security mechanism, finding the nature of a particular URL will alleviate most of the attacks occurring frequently.

4. Deep Learning

Deep learning (DL) mechanism has evolved from Artificial Neural Network (ANN). Generally, DL can be considered as the classical Machine Learning (ML) of deep models. Convolutional Neural Network (CNN) and Recurrent Neural Network (RNN) are most commonly used deep learning methods. CNN are heavily used in the field of image processing (IP). CNN extracts a hierarchical set of features by composing features extracted in lower layers. The hierarchical feature representation allows CNN to learn the data in various levels of abstraction. A single or a set of convolution and pooling operations and a non-linear activation functions are primary building blocks of CNN. In recent days, the advantage of using ReLU as an activation function in deep architecture is widely discussed because ReLU as an activation function is easy to train in comparison to *logistic sigmoid* or *tanh* function [28]. RNN is mainly used for sequential data modeling in which the hidden sequential relationships in variable length input sequences is learnt by them. RNN mechanism has significantly performed well in the field of NLP [5]. In initial time, the applicability of ReLU activation function in RNN was not successful due to the fact that RNN results in large outputs. As the research evolved, authors showed that RNN outputs vanished and exploded gradient problem in learning long-range temporal dependencies of large-scale sequence data modeling. To overcome this issue, research on RNN progressed on three significant directions. One was towards on improving optimization methods in algorithms; Hessian-free optimization methods belong to this category [29]. Second one was towards introducing complex components in recurrent hidden layer of network structure [20–22] introduced long short-term memory (LSTM), a variant of LSTM network, gated recurrent unit (GRU) [25]. Third one was towards the appropriate weight initializations; recently [23, 24], authors have shown RNN with ReLU involving an appropriate initialization of identity matrix to a recurrent weight matrix in order to perform closer in as compared to LSTM. This was substantiated with evaluating the four experiments on two toy problems, language modeling and SR. They named the newly formed architecture of RNN as identity-recurrent neural network (IRNN). The basic idea behind IRNN is that, while in the case of deficiency in inputs, the RNN stays in same state indefinitely in which the RNN is composed of ReLU and initialized with identity matrix.

5. Big Data Analysis and Management for Cyber Security

Two types of Apache Spark cluster setup has been done to train deep learning architectures at scale (1) distributed deep learning on GPUs (NVidia GK110BGL Tesla k40), (2) distributed deep learning on CPUs. Due to the confidential nature of the research, the scalable framework details cannot be disclosed. Each system has specifications (32 GB RAM, 2 TB hard disk, Intel(R) Xeon(R) CPU E3-1220 v3 @ 3.10 GHz) running over 1 Gbps Ethernet network. On top of Apache Hadoop Yet Another Resource Negotiator (YARN), Apache Spark cluster setup is done. This framework allows end users to expeditiously distribute, execute and harvest jobs. Each machine, such as physical or virtual device in Apache Spark cluster, is defined as a node. The framework contains three types of nodes. They are (1) master node (2) slave nodes and (3) data storage node.

1. Master node (N_M) supervises all other nodes and provides an interface to the end-user to make communication to the other nodes in the cluster. Moreover, the workloads are distributed to the slave nodes and aggregate the output from

all other slave nodes. In offline evaluation, it receives data from the data storage node and performs preprocessing followed by segmentation.

2. Slave nodes (N_S) facilitate actual computation. In offline system, it retrieves the preprocessed data from the master node and performs the actual computation.

3. Data storage node (N_{DS}) acts as a data base to store the data in offline evaluations. A master node can issue command to the data storage node to act as a slave node. This usually keeps track of the data on daily basis and aggregates and stores in NoSQL data base. This data can retrieve data on daily, weekly and monthly basis.

Distributed deep learning framework has been constructed by considering the following advantages:

1. It helps to increase the speed of training deep learning models and as a result hyper parameter selection task becomes easier.

2. Moreover, this type of system can be deployed on a large scale in real time to analyze large amounts of data with the aim to detect and issue an alert in a timely manner.

The proposed scalable architecture employs distributed and parallel algorithms with various optimization techniques that make it capable of handling the very high volume of Cyber security data. The scalable architecture also leverages the processing capability of the General Purpose Graphical Processing Unit (GPGPU) cores for faster and parallel analysis of Cyber security data. The framework architecture contains two types of analytic engines—real-time and non real-time analytic engines. The purpose of analytic engine is to generate an alert in case of attacks and detailed analysis and monitoring of Cyber security data to estimate the state in running condition of Cyber security system. Merely using a commodity hardware server, the developed framework is capable of analyzing more than 2 million events per second. The developed framework can be scaled out to analyze even larger volumes of network event data by adding additional computing resources. The scalability and real-time detection of malicious activities from early warning signals makes the developed framework stand out from any system of a similar kind.

6. Background

This section discusses the concepts behind transforming the text to numerical representation, deep learning architectures along with its mathematical details.

6.1 Text Representation

Natural language processing (NLP) has roots in many Cyber security use cases, such as domain generation algorithms, malicious URL detection, intrusion detection, android malware detection, spam detection, log analysis, insider threat detection and many others. Representation of texts is considered as a predominant task in NLP. Text representation is a way of making computers understand the meaning, contexts and the syntactic and semantic relationship of words or characters. Generally, the performance of machine learning algorithms relies on the text representations. Thus discarding of the syntactic and semantic similarity between words leads to loss of explanatory factors. Traditional word representations used in text mining also fail to preserve the word order. Recent advancements in distributed word representation methods have the capability to learn the rich linguistic regularities between words. Word embedding follows distributed representations which can handle the high dimensionality and sparsity issues efficiently. In simple terms, word embedding is a method that converts texts into numbers. It has remained an active area of research to find out better word representations than the existing ones. There are two types of text representation most commonly used. They are:

- Representation of texts in character level
- Representation of texts in word level

Initially, preprocessing for texts is done by creating a dictionary in which a unique key is assigned to each word/character in the text corpus. Moreover, the unknown words or characters are assigned to a default key, 0. Preprocessing includes converting all texts into lower case to avoid regularization issue and removal of non-word objects. Non-sequential and sequential representations are two most commonly used methods to convert word/character to numerical representations.

- *Non-sequential input representation*: Non-sequential input representation of domain names/URLs fail to preserve the sequential information among the characters and disregard the spatial correlation between characters. Bag of words and tf-idf (term frequency-inverse document frequency) belong to non-sequential input representation.

- *Sequential input representation*: Initially, the dictionary D is formed by assigning a unique key for each character in the dataset. Based on the frequently occurring character statistics, the characters are placed in an ascending order in a dictionary D. Each character of a domain name/URL is assigned to an index in accordance with dictionary D. The length of domain name/URL character vectors are transformed to the same length by choosing the desired sequence length. The character vectors that are longer than the desired sequence length are cut down and zero padding is done to the short vectors.

- *Approach to extend the dictionary*: N-grams is a typically used approach in NLP. It provides a way to extend the dictionary. It belongs to non-sequential input representation. In the field of NLP, extracting gram is a classifier which does according to size of n, for example if n is 1, its unigram, 2 is bigram and 3 is trigram. If a model of a data is been high, four and five grams have been used. Total number of N-grams in a sentence can be found as follows:

$$Ngrams_S = S - (NW - 1) \tag{1}$$

where S denotes a sentence and NW is the number of words in a sentence S.

- *Character embedding*: [71] was the first person to present the dense representation of words in the year 2003. For language modelling, they have mapped the data to look up table and use the resulting dense vectors. In [72] showed an enhanced word embedding and demonstrated the flexibility of word embedding in various NLP errands. After the improvement of more productive word embeddings [73, 74], word embedding turned into an essential part of neural system-based NLP and text mining. Additionally, character level embeddings picked up prevalence among scientists. [75] utilized character embedding to manage the 'out of vocabulary' problem. [76] utilized character embeddings with a convolutional neural network for named entity recognition. In this work, Keras character embedding is utilized for DGA and URL detection. Initially, random weights are assigned and weight will learn embedding for all the characters in the corpus. It coordinatively works with the other layers and optimizes weights to minimize the loss during the back propagation.

6.2 Logistic Regression

It is one of the largely used machine learning algorithms for both classification and prediction. It is generally considered as a statistical tool used to analyze data when there are one or more independent variables determining the output. It is a special case of linear regression, wherein the logistic regression predicts the probability of outcome by fitting the data to a logistic function given as

$$\sigma(z) = \frac{1}{1 + e^{-z}} \qquad (2)$$

6.3 Random Forest

Random Forest (RF) combines a subset of observations and variables and builds a decision tree (DT). RF is like an ensemble tool which unites many decision trees. This logic of combining decision trees ensures a better prediction than individual DT. The concept of bagging is extensively used in RF to create several minimal correlated DT.

6.4 Artificial Neural Networks (ANNs)

Artificial neural networks (ANNs) represent a directed graph in which a set of artificial neurons generally called 'units' in mathematical model are connected together with edges. This is influenced by the characteristics of biological neural networks, where nodes represent biological neurons and edges represent synapses. Feed-forward networks (FFNs) and recurrent neural network (RNN) are two main types of ANNs.

A feed-forward network (FFN) consists of a set of units that are connected together with edges in a single direction without formation of a cycle. They are simple and most commonly used algorithms. Multi-layer perceptron (MLP) is a subset of FFN that consist of three or more layers with a number of artificial neurons, termed as units in mathematical terms. The three layers are input layer, hidden layer and output layer. There is a possibility to increase the number of hidden layers when the data is complex in nature. So, the number of hidden layers is parameterized and relies on the complexity of the data. These units together form an acyclic graph that passes information or signals in forward direction from layer to layer without the dependence of past input. MLP can be written as $HI : \mathbb{R}^p \times \mathbb{R}^q$ where p and q are the size of the input vector $x = x_1, x_2, \ldots, x_{p-1}, x_p$ and output vector $O(x)$ respectively. The computation of each hidden layer HI_i can be mathematically formulated as follows:

$$HI_i(x) = f(w_i^T \dot{x} + b_i) \qquad (3)$$

$HI_i : \mathbb{R}^{d_{i-1}} \to \mathbb{R}^{d_i}$
$f : \mathbb{R} \to \mathbb{R}$

$w_i \in \mathbb{R}^{d_i \times d_{i-1}}$, and $b_i \in \mathbb{R}^{d_i}$, f is the elementwise non-linearity function. This can be either *logistic sigmoid* or *hyperbolic tangent* function. *sigmoid* have values between [0, 1] whereas [1, −1] range of values for *hyperbolic tangent*. If we want to use MLP for multiclass classification problem, then the output usually has multiple neurons. For this, *softmax* function can be used. This provides the probabilities of each class and selects the highest one resulting in crisp value.

$$sigmoid = \sigma(x) = \frac{1}{1 + e^{-z}} \qquad (4)$$

$$hyperbolic\ tangent = \tanh(z) = \frac{e^{2z} - 1}{e^{2z} + 1} \qquad (5)$$

$$softmax = SF(Z)_i = \frac{e^{z_i}}{\sum_{j=1}^{n} e^{z_j}} \qquad (6)$$

Three-layer networks with a *softmax* function $SF(Z)$ in output layer of MLP is identical to multi-class logistic regression. An MLP network with l hidden layers together can be generally defined as:

$$HI(x) = HI_l\ (HI_{l-1}(HI_{l-2}(\dots (HI_1(x))))) \qquad (7)$$

6.5 *Loss Functions, Steepest Descent or Gradient Descent and Back Propagation*

The aforementioned description of FFNs shows that generally MLPs are parameterized functions. This requires an approach to find the optimal model parameters for good prediction and conclusion. In that, defining a loss function is the initial step. Loss function *loss(pr, tr)* takes the predicted value (*pr*) and a target value (*tr*) as input and gives the inconsistency between the predicted (*pr*) and target value (*tr*). One of most commonly used loss functions is the squared Euclidean distance:

$$d(tr, pr) = \|tr - pr\|_2^2 \qquad (8)$$

Negative log probability is used for multi-class classification with the *tr* as the target class and *p(pr)* as the probability distributions that result from the neural networks

$$d(tr, p(pr)) = -\log p(pr)_{tr} \qquad (9)$$

A few experiments conducted in this paper are multi-class problems. However, the network receives a list of corrected input-output set $inop = (p_1, q_1), (p_2, q_2), \dots, (p_n, q_n)$ in the training process. Then we aim to decrease the mean of losses as defined below:

$$loss(p_n, q_n) = \frac{1}{n} \sum_{i=1}^{n} d(q_i, f(p_i)) \qquad (10)$$

This type of loss function will be used in the following discussed deep learning algorithms.

Loss function has to be minimized to get better results in the neural network. A loss functions is defined as:

$$Train_{inop}(\theta) \equiv J_{inop}(\theta) = \frac{1}{n}\sum_{i=1}^{n}d(q_i,f_\theta(p_i)) \tag{11}$$

where $\theta = (w_1, b_1, ..., w_k, b_k)$

Our aim is to minimize the loss function $J_{inop}(\theta)$ by choosing the appropriate value of $\theta \in \mathbb{R}^d$. This involves the estimation of $f_\theta(p_i)$ and $\nabla f_\theta(p_i)$ at the cost $|inop|$

$$\min_\theta J(\theta) \tag{12}$$

To do this, we can follow various optimization techniques; typically gradient descent is one among them. This is one of the most commonly used and itself establishes as a good method for optimizing the loss functions. The repeated calculation of a gradient and a parameter update is done to do a single update for the whole training dataset by the following update rule:

$$\theta^{new} = \theta_{old} - \alpha\nabla_\theta J(\theta) \tag{13}$$

α is called learning rate. To find derivative of J backpropagation or backward propagation of errors algorithm is adopted.

6.6 Recurrent Neural Network (RNN)

Recurrent neural network (RNN) is very similar to multi-layer perceptron (MLP). The only difference is that in RNN, the units in hidden layer contain a self-recurrent connection. This facilitates to concatenate the output of the hidden layer from the previous input samples 'forward pass' with the next input sample, and is jointly used as the input to the hidden layer [4]. This has showed remarkable performance in long-standing artificial intelligence (AI) tasks related to the field of machine translation, language modeling and SR [5]. Generally, RNN takes an input sequence x_T of arbitrary length and maps into hidden state vector hi_T recurrently with the help of a transition function tf. The hidden state vectors at each time step t are calculated as a transition function tf of current input sequence x_t and previous hidden state vector hi_{t-1}, as shown below:

$$hi_t = \begin{cases} 0 & t = 0 \\ tf(hi_{t-1},x_t) & otherwise \end{cases} \tag{14}$$

where tf is made up of affine transformation of x_t and h_{t-1} with the element wise non-linearity. This type of transition function tf results in vanishing and exploding gradient issue while training for long sequences; a gradient vector can grow or decay exponentially over time-steps [20]. Long short-term memory (LSTM), is a variant of RNN, which has a special unit typically called a 'memory block'. A memory block contains a memory cell, set of input and output gate. The states of a memory cell are controlled by input and output gate. Forget gate [21] and peephole connections [22] are two important units in LSTM. Forget gate performs forget or reset operation of a memory cell and a peephole connection facilitates to learn the precision timings of output.

Generally, a memory block contains input gate (*it*), forget gate (*ft*), output gate (*ot*), memory cell (*m*) and hidden state vector (*hi*) at each time step *t*. The transition function for a memory block is defined as follows:

$$it_t = \sigma(w_{it}x_t + P_{it}\, hi_{t-1} + Q_{it}m_{t-1} + b_{it}) \tag{15}$$

$$ft_t = \sigma\,(w_{ft}x_t + P_{ft}\, hi_{t-1} + Q_{ft}\, m_{t-1} + b_{ft}) \tag{16}$$

$$ot_t = \sigma\,(w_{ot}x_t + P_{ot}\, hi_{t-1} + Q_{ot}\, m_{t-1} + b_{ot}) \tag{17}$$

$$m1_t = \tanh(w_{m}x_t + P_{m}hi_{t-1} + b_{m}) \tag{18}$$

$$m_t = ft_t^i \odot m_{t-1} + it_t \odot m1 \tag{19}$$

$$hi_t = ot_t \odot \tanh(m_t) \tag{20}$$

where x_t is the input at time step *t*, σ is *sigmoid* non-linear activation function, $\odot$ denotes element-wise product.

[23, 24] proposed identity-recurrent neural network (IRNN) and [25] gated recurrent unit (GRU). IRNN contains minor changes in comparison to RNN. This has significantly performed well in capturing long-range temporal dependencies. The minor changes are related to initialization tricks; initialize the appropriate RNNs weight matrix using an identity matrix or its scaled version and use ReLU as non-linear activation function. Moreover, this method performance is closer to LSTM in four important tasks; two toy problems, language modeling and SR. GRU is a minimal set of LSTM units.

6.7 Convolutional Neural Network (CNN)

Convolutional Neural Network (CNN) can be considered a supplement to the classical neural network. This has been largely used in the field of image processing (IP). Generally, one dimensional CNN is composed of one or more convolution 1D and pooling 1D layers [5], commonly used in texts and signals.

A domain name/URL contains a set of characters which can be defined as $du = \{c_1, c_2, ...,c_l\}$, where *c* denotes a character and *l*, the length of domain name/URL. Each character is taken from a dictionary *D*. Each character is represented by distributional vectors $c \in R^{1 \times d}$ looked up in a character embedding's matrix $C \in R^{d \times |D|}$. This matrix is constructed by concatenating embeddings of all characters in *D*. For facilitating look-up operations, the characters are mapped into indices in *D*.

For each domain name/URL *du*, we form a domain name/URL matrix $DU \in R^{d \times |du|}$, where each column *i* represents a character embedding matrix c_i at the corresponding position *i* in a domain name/URL. Convolutional neural network (CNN) applies a series of transformations to the input matrix *DU* using convolution1d, pooling1d operations to learn optimal feature representation of each character in a domain name/URL from lower-level word embeddings. This optimal feature representation captures the higher-level semantic concepts.

Convolution1d focuses on learning the discriminative character sequences found within the domain names/URLs. These are common throughout the training corpus.

A convolution1D layer applies a set of m filter operations to a sliding window of length h over a domain name/URL. Let $DU_{[i:i+h]}$ denote the concatenation of character vectors DU_i to DU_{i+h}. A new feature f_i for a given filter $F \in R^{d \times h}$ is estimated by

$$f_i := \sum_{k,j}(DU_{[i:i+h]})_{k,j}.F_{k,j} \tag{21}$$

A new feature vector $f_i \in R^{l-h+1}$ is formed by concatenating all possible vectors in a domain name/URL. Then the vectors f_i from all m filters are aggregated to form a feature map matrix $F \in R^{m \times (n-h+1)}$.

Convolutional1D layer output is passed via non-linear activation $ReLU$ function to pooling layer. Pooling 1D is a down-sampling operation. This takes a maximum element over a fixed set of non-overlapping intervals, aggregates and forms a feature map matrix $F_{pooled} \in {}^{m \times \frac{n-h+1}{s}}$, where s is the length of each interval.

Generally, penultimate layer of CNN network is a fully-connected layer. This has connection to all activations in a previous layer and contains $sigmoid$ non-linear activation function in binary classification and $softmax$ non-linear activation function in multi-class classification.

Max-pooling operation output can also be passed to recurrent structures, such as RNN, LSTM, GRU and IRNN network to capture the sequence information, as defined below:

$$FM = CNN(x_t) \tag{22}$$

where, CNN is formed of convolution1D and max-pooling1D layers. The feature map FM is passed to recurrent structures to learn the long-range temporal dependencies; x_t denotes the input feature vector with a class label.

7. Software Frameworks

Many powerful and ease-of-use software frameworks have been built to facilitate the implementation of deep learning architectures. The comparative study of these deep learning frameworks on various deep learning architectures executed on devices (CPU and GPU) were discussed in terms of extensibility, hardware utilization, and speed [30]. The deep learning architectures of this research are experimented using the most recent software framework Google's open source data flow engine, TensorFlow [31]. TensorFlow allows programmers to build numerical systems as unified data flow graphs. All experiments of RNNs approaches are trained using BPTT technique.

8. Deep-DGA-Detect: Applying Deep Learning to DGA-Based Malware Detection at Scale

8.1 Introduction

Nowadays, internet has turned out to be the largest and most important global communication medium and infrastructure. It interfaces billions of hubs, empowering them to communicate with each other. The Domain Name System (DNS) (or Service or Server) is of dominant importance within the operation of the internet. DNS is a distributed, scalable, reliable, dynamic and globally available database which maps domain name to IP address and vice-versa. These characteristics of DNS have provided an opportunity for cybercriminals to host malicious contents or facilitate the management of phishing websites to steal users' information [41]. These days, the behavior and operation of such threats are undergoing pivotal changes and advancements. One of the extensively utilized attacking strategies is to utilize botnets [42]. Botnets are considerably used for malicious purposes, such as distributed denial of service (DDoS) attacks, large-scale spam campaign, identity theft, sniffing traffic, key logging, spreading new malware, etc. A botnet is a set of compromised hosts that are controlled by a bot master by issuing the commands or instructions to a command-and-control server (C2C). Blacklisting and sinkholing are the most commonly used methods by security organizations to block the communication point between the C2C and a botmaster [43, 44]. Blacklisting is considered a daunting task mainly because the data base has to be updated manually on a regular basis. Moreover, botnets are consistently improved and new methods to evade detection are persistently being created. One of the widely used methods by attackers to evade blacklisting is DNS agility or fluxing methods, which is mostly an IP flux or domain flux service [45]. In this paper, only domain flux service is studied. A domain flux service uses domain generation algorithm (DGA). It generates domain names pseudo-randomly on a large scale and tries to contact their C2C server using hit-and-trial method. To successfully defeat an attack, all domain names must be blacklisted or sinkholed before they go to work. Sinkholing means reverse engineering the malware and its DGA family, to find the seed. A seed is a union of numeric, alphabet, date/time, etc. Using the seed, the respective botnets can be hijacked by registering the upcoming domain names and using them as an impersonator C2C server. Intuitively, an adversary must redeploy botnets with updated seeds to resume their functionality. This defense turns increasingly more complex because the degree of dynamically-generated domains increases. Overall, both blacklisting and sinkholing approaches are productive only when both the algorithm and the seed used by botnet campaign is known.

A detailed study of the effectiveness of blacklisting in extensive reportage of DGA-generated domain was done by [43]. The study was considered with both the public and private blacklists. DGA reportage of public blacklists was considerably less in comparison to the DGAs analyzed by way of the authors being contained in any of the blacklists. Private blacklists progressed well, but the performance of them varied from 0 per cent to 99.5 per cent over various malware families. This study claims that the blacklists are needed. They can act as an initial shelter and provide adequate level of protection; the other method has to be incorporated.

The creation of a DGA classifier to attack DGA-based malware is another most commonly used approach. This classifier can stay within the network, sniffing out DNS requests and seeking out DGAs. Whenever DGA classifier detects the DGAs, it informs the network administrators or other automated tools to further inspect the roots of a DGA. Previous work in DGA analysis and detection can be divided into retrospective detection and real-time detection. Retrospective detection acts as a reactionary system because it splits large sets of domain names into categories, using clustering with the aim to produce statistical properties of each category [46–48]. Pleiades [46] was the primary system that was capable of detecting DGA-based domains without reverse engineering the bot malware. This includes DGA discovery, DGA classification and C&C detection module. DGA discovery module detects and clusters botnet queries based on NXDomains. DGA classification module uses Alternating Decision Tree learning approach for discovering and differentiating the NXDomain clusters and C&C detection uses Hidden Markov Models, each DGA family using a unique model. Generally, during training, a model is created and DGAs are categorized based on the statistical tests, e.g., Kullback-Leibler divergence. This method is mainly based on analyzing the bunch of traffic, so DGA possibly establishes communication before being detected. Moreover, these types of detection systems make use of contextual information, such as NXDomains, HTTP headers and passive DNS to further enhance the DGAs detection rate. These kinds of DGAs detection systems cannot be adopted for real-time detection and prevention in real-time security applications [49]. Moreover, using the contextual information in detecting DGA's in many of the real-time applications, for example, endpoint detection and response may not be a suitable approach. Mainly due to that, the endpoint detection and response system has uncompromisable requirements on usage of memory, network and processing.

Real-time detection methods aim at detecting and classifying domain as either DGA generated or benign on a per domain basis by using only the domain name to block any C2C server communication. This is often considered to be complex in comparison to the retrospective method because achieving the optimal performance in real-time deployment system is difficult. In [49] the performance of the retrospective methods was very less. Many of the previous real-time methods are mostly based on machine learning. Classical machine learning-based DGA detection and classification rely on the feature engineering phase, e.g., string length, vowel to consonant ratio, entropy, etc. As a result, their performance implicitly depends on the domain-specific features. The domain-specific feature extraction is not always feasible, mainly due to the magnitude of data and features accessible. Deep learning is a feasible approach to obtain the optimal feature representation in accessing the big data. Moreover, feature engineering methods are easy to bypass. For the very first time, [46] proposed Hidden markov model (HMM) based real-time DGA detection system and the performance was really poor. Unlike the feature engineering mechanism in classical machine learning, deep learning architectures take the raw domain names as input and non-linearity in each hidden layer facilitates to learn the more complex and abstract feature representations. The objective of this subsystem is set as follows:

- The authors introduce Deep-DGA-Detect, a scalable DGA-based botnet identification framework that efficiently analyses the DNS traffic in a network, detecting and alerting the presence of DGA-generated domain.

- The performance of various classical machine learning and deep learning architectures is shown over a large labeled dataset in the context of detecting the DGA-generated domain name and classifying them to their DGA family. These labelled datasets consist of a mix of algorithmically-generated domains and benign domains gathered from the Alexa, OpenDNS and 40 well-known DGA-based malware families [33].

- A full-fledged implementation of Deep-DGA-Detect was performed and evaluated over a large labeled dataset consisting of a mix of algorithmically-generated domains and benign domains gathered from the network activity of private sources and public source, OSINT.

8.2 Experiments

To train deep learning architectures at scale, the TensorFlow[1] framework is used with Apache Spark.[2] TensorFlow with Apache Spark on top of Hadoop[3] stack enables more efficient and accurate analysis of big data.

8.2.1 Description of Data Set

To avoid the issues that are discussed in [32], we have used three different datasets as shown in Table 1. The detailed statistics of the Dataset 1 is displayed in Table 2. All are completely disjointed to each other. Dataset 1 is from [33]; each domain name is labeled as benign or DGA-generated in classifying whether the domain name as benign or DGA-generated. In the case of categorizing the DGA-generated domain name to their respective botnet family, domain names are labeled with their family. The benign domain names are from Alexa [34], OpenDNS [35] in all three datasets. Data-set 2 is collected from OSINT real time DGA feeds [36]. Dataset 3 is gathered from real-time DNS traffic in our LAB network. The three datasets are gathered from entirely different environments and each has different ratio of benign and DGA-generated domain names.

8.2.2 Hyper-parameter Selection

All deep learning architectures are trained using back propagation through time (BPTT) technique. The deep networks such as RNN, IRNN, LSTM, GRU, CNN and CNN-LSTM are parameterized functions and finding the best parameter is an important task. This is primarily due to the fact that the DGA detection rate implicitly depends on the optimal network parameters and network structures [37]. All experiments were run for 1000 epochs with batch size 128 and learning rate 0.001.

In RNN/LSTM/GRU/IRNN, two trial of experiments are run for the number of units/memory blocks 32, 64, 96, 128, 256 with moderately sized RNN/LSTM/GRU/IRNN network. A moderately sized RNN/LSTM/GRU/IRNN network contains a single RNN/LSTM/GRU/IRNN layer followed by dropout layer 0.1 and fully-connected layer with one neuron and *sigmoid* non-linear activation function. An RNN/LSTM/GRU/IRNN layer with 64 units/memory blocks performed well and moreover the performance of 32 units/memory blocks is comparable to the 64 units/memory blocks. To provide further level of freedom in feature extraction, the number of units/memory blocks is set to 128 for the rest of the experiments. In the second set of experiments,

Table 1: Statistics of data-set.

Source	Training	Validation	Testing
Data set 1	810000	414736	205000
Data set 2	–	–	94500
Data set 3	–	–	82000

---

[1] https://www.tensorflow.org/

[2] https://spark.apache.org/

[3] http://hadoop.apache.org/

Table 2: Detailed statistics of dataset, Data set 1 for multi-class classification.

Domain Family	Training	Validation	Testing
Alexa and OpenDNS	80000	50000	20000
bamital	20000	10000	5000
banjori	20000	10000	5000
bedep	10000	5188	2000
conficker	20000	10000	5000
corebot	20000	10000	5000
cryptolocker	20000	10000	5000
dnschanger	20000	10000	5000
dyre	20000	10000	5000
emotet	20000	10000	5000
gameover	20000	10000	5000
gameover p2p	20000	10000	5000
gozi	20000	10000	5000
locky	20000	10000	5000
matsnu	20000	10000	5000
murofet	20000	10000	5000
murofetweekly	20000	10000	5000
necurs	20000	10000	5000
nymaim	20000	10000	5000
oderoor	20000	10000	5000
padcrypt	20000	10000	5000
proslikefan	20000	10000	5000
pushdo	20000	10000	5000
pykspa	20000	10000	5000
pykspa2	20000	10000	5000
qadars	20000	10000	5000
qakbot	20000	10000	5000
ramnit	20000	10000	5000
ranbyus	20000	10000	5000
rovnix	20000	10000	5000
simda	10000	6000	2000
sisron	20000	10000	5000
suppobox	20000	10000	5000
sutra	20000	10000	5000
symmi	20000	10000	5000
szribi	10000	9536	4000
tinba	20000	10000	5000
torpig	20000	10000	5000
urlzone	20000	4012	2000
virut	20000	10000	5000
Total	**810000**	**414736**	**205000**

we pass bigrams as inputs to RNN/LSTM/GRU/IRNN. Their performances were considerably poor in comparison to the character-level inputs.

In CNN, three trial of experiments are run for filters 4, 16, 32, 64 and 128 and filter length 3 with a moderately sized CNN. The moderately sized CNN is composed of convolution1d, maxpooling1d with pooling length 2, fully-connected layer with 128 neurons, dropout layer with 0.2 and fully-connected layer with 1 neuron and *sigmoid* non-linear activation function. CNN network with 64 filters outperformed the other filters. Thus for the rest of the experiments, we have set the number of filters in convolution1d layer as 64.

8.2.3 Proposed Architecture

An intuitive overview of the proposed deep learning model is shown in Fig. 1. This is composed of three sections, such as domain name character embedding, feature representation and a classifier. This contains several parameters that are learned by the deep neural network during training. During training, embedding, deep learning layers and fully-connected layers, all work in concert to make an accurate and robust DGA detection. This model is evaluated on three various experimental designs:

- Binary classification, i.e., classifying the domain name as DGA or non-DGA.
- Multi-class classification, i.e., addition to detecting the DGA domain, categorizing into their corresponding DGA family.
- Deep learning models trained on Dataset 1 is evaluated on the Dataset 2 and Dataset 3 (both are collected in different environment) in the context of classifying the domain name as DGA or non-DGA.

The computational flow of the deep learning model is explained below:

- **Domain name character embedding:** By using the aforementioned approach (Section 6.1), a matrix is constructed for training (810000*81), validation (414736*81) and testing (205000*81) for Dataset 1, Dataset 2 (94500*81) and Dataset 3 (82000*81). These matrices are passed to an embedding layer with

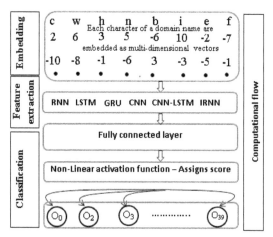

Fig. 1: An intuitive overview of proposed deep-learning approach. All connections and inner units are not shown and can be considered as representative of proposed deep-learning model.

batch size 128. An embedding layer maps each character on to a 128 length real valued vector. This can be considered as one of the hyper parameters. We chose 128 to provide further level of freedom to the deep learning architectures. This collaboratively works with the other layers in the deep network during back propagation. It facilitates to learn the similar characters that come close together. This kind of domain name character clustering facilitates to other layers to easily detect the semantics and contextual similarity structures of domain names. To visualize, the 128 high-dimensional vector representation is passed to the t-SNE, as shown in Fig. 2. It shows that the characters, numbers and special characters have entirely appeared in separate clusters. This reflects that the embedding representation has learnt the semantics and contextual similarity of domain names.

- **Feature representation:** We adopt various deep layers RNN, LSTM, GRU, IRNN, CNN and a hybrid network, such as CNN-LSTM for feature representation. Recurrent layers, such as RNN, LSTM, GRU and IRNN extract sequential information and CNN helps to extract spatial information among the characters. All the architectures are not too deep and that's the reason batch-normalization is not employed [38]. The pattern-matching technique followed by deep learning layers is more effective in comparison to the other techniques, like regular expressions. Regular expressions output binary value whereas deep learning layers outputs a continuous value that represents how much the pattern matched. Additionally, regular expressions are written by hand whereas deep learning models implicitly capture the optimal feature representation. Like this, deep learning layers automatically apprentice to accurate complex, abstruse patterns in a down-covered way.

- **Recurrent layers:** We use the following recurrent layers, RNN, LSTM, GRU and IRNN. Based on the knowledge obtained from hyper parameter selection, the number of units (memory cells in LSTM and GRU) is set to 128. Recurrent layer captures the sequential information from the embedding layer output, which is in the form of a tensor and passes into dropout layer. A unit in RNN uses *hyperbolic*

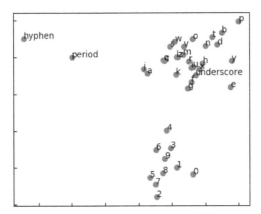

Fig. 2: Embedded character vectors learned by CNN-LSTM classifier are represented using two-dimensional linear projection (PCA) with t-SNE. Consider the CNN-LSTM group characters by similar effect on its states and the consecutive loss.

tangent as an input and output activation function that has values in range [–1, 1]. An LSTM memory cell uses *hyperbolic tangent* as an input and output activation function that has values in range [–1, 1] and gates uses *logistic sigmoid* that has values in range [1, 0].

- **Convolution layers:** To learn the spatial co-relationships or local features among the characters, CNN is used. It has two layers—in first layer with the help of multiple convolution1d operations local features are extracted and passed to the maxpooling1d. A convolution1d layer uses 64 filters with filter length 3 that slide over the character embedding vector sequence and output a continuous value at each step. This is actually a representation of how much the pattern has matched in the character-embedding vectors sub-sequence. This contains a pool length of 2 that divides the convolution feature map into two equal parts. Next, the maxpooling1d output is flattened to a vector and passed to the fully-connected network in CNN. For CNN-LSTM, maxpooling1d output, i.e., consolidated features of CNN network is again passed to the LSTM to learn the sequential information among the spatial features that are obtained from the CNN. This LSTM layer includes 70 memory blocks.

- **Regularization:** A dropout layer with 0.1 (0.2 in CNN) is used between deep layers and fully-connected layer that acts as a regularization parameter to prevent overfitting. A dropout is a method for removing the neurons randomly alongwith their connections during training a deep learning model. In our alternative architectures, the deep networks could easily overfit the training data without regularization, even when trained on a large number of DGA corpus.

- **Classification:** The newly formed features from deep layers are passed to fully-connected layers for classification. A fully-connected layer is an inexpensive way of learning the non-linear combination of the learnt features by deep layers. In fully-connected layers, each neuron in one layer is connected to every other neuron in another layer. The fully-connected neural network is composed of two layers, such as fully-connected layer with 128 units and followed by fully-connected layer with one unit. It contains *sigmoid* activation function in binary classification, *softmax* activation function in multi-class classification (fully-connected layer with 40 units). In classifying the domain name as either benign or DGA generated, the prediction loss of deep learning models is estimated using binary-cross entropy:

$$loss(p,e) = -\frac{1}{N}\sum_{i=1}^{N}[e_i \log p_i + (1-e_i)\log(1-p_i)] \tag{23}$$

where p is a vector of predicted probability for all domain names in testing dataset, e is a vector of expected class label, 0 for benign and one for DGA generated.

In classifying the domain name as either benign or DGA generated and categorizing them to their family, the prediction loss of deep learning models is estimated using categorical-cross entropy:

$$loss(p,e) = -\sum_{x} p(x)\log(e(x)) \tag{24}$$

where p is true probability distribution, q is predicted probability distribution. To minimize the loss of binary-cross entropy and categorical-cross entropy we have used *Adam* optimizer [39] via backpropagation.

8.3 Other Alternative Architectures Attempted

There are other architectures that we tried out while selecting the proposed deep learning model, i.e., CNN-LSTM:

- Stacked RNN/LSTM/GRU/IRNN layers
- Stacked embeddings followed by RNN/LSTM/GRU/IRNN/CNN/CNN-LSTM layer
- Stacked convolution1d layer to capture non-linear convolution activations
- Stacked convolution1d layer followed by RNN/LSTM/GRU/IRNN layer in hybrid architecture

8.4 Evaluation Results

The performance of proposed deep learning and machine learning models are evaluated on three different datasets, such as Dataset 1, Data set 2 and Dataset 3. Moreover, the performance of trained model on Dataset 1 is evaluated on Dataset 2 and Dataset 3 in classifying the domain name as benign or DGA generated. The results are shown in the form of an ROC curve by comparing two operating characteristics, such as true positive rate and false positive rate across varying thresholds in the range [0.0–1.0]. This metric is independent of the proportion of benign to malicious samples in the entire corpus. The ROC curve for Dataset 1, displayed in Fig. 3 and the other statistical metrics for all the three datasets are reported in Tables 3 and 4. Deep learning models outperform the N-gram or domain-expert derived features with a large size of domain name corpus. This suggests that the embeddings with deep learning layers can be a robust approach for automatic feature extractor. Moreover, the performances of N-grams are better than the domain-expert-derived features. This is due to the advantage with the large size of domain name corpus and bag of N-grams is an effective representation in the domain of NLP [40].

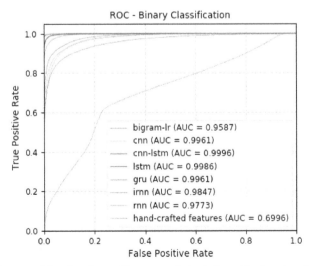

Fig. 3: ROC curve of deep-learning architectures and classical machine-learning classifier in the context of classifying the domain name as either non-DGA or DGA (TPR vs FPR) for Dataset 1

Table 3: Summary of test results in classifying the domain name into either benign or DGA generated.

Method	Accuracy	Precision	Recall	F-score
RNN	0.957	0.975	0.977	0.976
LSTM	0.989	0.990	0.998	0.994
I-RNN	0.967	0.978	0.985	0.982
GRU	0.984	0.988	0.994	0.991
CNN	0.960	0.969	0.987	0.978
CNN-LSTM	0.994	0.994	0.999	0.997
bigram-logistic regression	0.906	0.906	0.999	0.950
Expert level hand designed features				
Random Forest	0.902	0.902	1.000	0.949
Data set 2-CNN-LSTM	0.959	0.957	0.992	0.974
Data set 3-CNN-LSTM	0.936	0.931	0.989	0.959

Table 4: Summary of test results in classifying the domain name into either benign or DGA generated and categorizing DGA generated domain name to their family for Data set 1 (Classwise true positive rate and false positive rate is available in Fig. 1.5).

Method	Accuracy	Precision	Recall	F-score
RNN	0.920	0.922	0.920	0.918
LSTM	0.963	0.964	0.963	0.963
I-RNN	0.934	0.938	0.934	0.932
GRU	0.959	0.960	0.959	0.958
CNN	0.924	0.925	0.924	0.922
CNN-LSTM	0.977	0.978	0.977	0.976
bigram logistic regression	0.880	0.882	0.880	0.875
Expert level hand designed features				
Random Forest	0.855	0.859	0.855	0.849

The overall results for all the deep learning models for Dataset 2 and Dataset 3 is considerably less in comparison to Dataset 1. This is because the Dataset 2 and Dataset 3 are collected in an entirely different environment. Thus the domain names are unique and not seen in the trained model with Dataset 1. However, the results are considerable.

8.5 *Conclusion, Future Work and Limitations*

In this subsystem, the performance of deep learning architectures was assessed to detect and classify the DGA-generated domains as used by modern botnets. In addition, the other traditional supervised machine learning classifiers are evaluated. Deep learning methods are preferable over other traditional machine learning methods as they have the capability to obtain optimal feature representation themselves by taking the raw domain names as their input. Handcrafted feature engineering methods require

Domain Family	RNN		I-RNN		LSTM		GRU		CNN		CNN-LSTM		Bigram Logistic Regression		hand-crafted Features (RF)	
	TPR	FPR	TPR	FPR	TPR	FPR	TPR	FPR	TPR	FPR	TPR	FPR	TPR	FPR	TPR	FPR
Alexa and OpenDNS	0.9581	0.0028	0.9692	0.0017	0.9963	0.0011	0.9937	0.0009	0.9842	0.0074	0.9939	0.0004	0.8805	0.0067	0.8113	0.0058
bamital	1.0	0.0	1.0	0.0	1.0	0.0	1.0	0.0	1.0	0.0	1.0	0.0	1.0	0.0	1.0	0.0
banjori	1.0	0.0	1.0	0.0	1.0	0.0	0.9998	0.0	1.0	0.0	1.0	0.0	1.0	0.0	1.0	0.0
bedep	0.694	0.0009	0.71	0.0007	0.8925	0.0006	0.859	0.0006	0.716	0.001	0.9605	0.0003	0.6595	0.0013	0.655	0.0019
conficker	0.3808	0.0042	0.4262	0.0045	0.7116	0.0083	0.5402	0.004	0.5134	0.0099	0.654	0.002	0.346	0.0058	0.3532	0.0093
corebot	0.9994	0.0	0.999	0.0001	0.9964	0.0001	0.9988	0.0	0.9994	0.0	1.0	0.0	0.979	0.0001	0.9594	0.0001
cryptolocker	0.7926	0.0057	0.8158	0.0044	0.9244	0.002	0.9712	0.0057	0.7476	0.0041	0.9654	0.0007	0.7246	0.0062	0.698	0.0064
dnschanger	0.9836	0.004	0.9884	0.0036	0.9618	0.0015	0.9938	0.0022	0.9874	0.0042	0.9942	0.001	0.9794	0.0053	0.9854	0.0082
dyre	1.0	0.0	1.0	0.0	1.0	0.0	1.0	0.0	1.0	0.0	1.0	0.0	1.0	0.0	1.0	0.0
emotet	0.9998	0.0001	0.9998	0.0002	0.999	0.0	0.9994	0.0001	0.9998	0.0001	0.9996	0.0	0.9998	0.0004	0.9998	0.0006
gameover	0.9998	0.0	1.0	0.0	0.9986	0.0	1.0	0.0	0.9998	0.0	0.9998	0.0	1.0	0.0002	0.9994	0.0005
gameover_p2p	0.969	0.0005	0.9942	0.0004	0.9914	0.0	0.997	0.0	0.9884	0.0006	0.998	0.0	0.9088	0.004	0.9322	0.0052
gozi	0.9998	0.0001	0.9998	0.0001	1.0	0.0	0.9992	0.0	0.9994	0.0001	1.0	0.0	0.9954	0.0002	0.9898	0.0004
locky	0.8278	0.0017	0.8672	0.0013	0.9808	0.0006	0.9736	0.0006	0.8642	0.002	0.9884	0.0003	0.7724	0.0026	0.6794	0.0046
matsnu	0.9994	0.0012	1.0	0.0004	0.9994	0.0	0.9994	0.0	0.9566	0.0002	0.9994	0.0001	0.9988	0.0045	0.9916	0.0056
murofet	1.0	0.0	1.0	0.0	0.9996	0.0	1.0	0.0	1.0	0.0	1.0	0.0	1.0	0.0	0.9994	0.0
murofetwekly	0.9232	0.0032	0.9684	0.0026	0.9952	0.0005	0.9786	0.0002	0.969	0.0051	0.9922	0.0001	0.9006	0.0057	0.8672	0.0054
necurs	0.803	0.0008	0.835	0.0007	0.912	0.0004	0.8972	0.0002	0.7872	0.0002	0.9636	0.0007	0.7274	0.0005	0.6084	0.0013
nymaim	0.5988	0.0158	0.4968	0.0082	0.713	0.0056	0.7256	0.0084	0.4762	0.0099	0.9064	0.0072	0.3568	0.0078	0.3362	0.0091
oderoor	0.9314	0.0068	0.9648	0.001	0.993	0.0003	0.9908	0.0001	0.8944	0.0018	0.992	0.0002	0.4566	0.0051	0.3678	0.0027
padcrypt	1.0	0.0	0.9998	0.0	0.9994	0.0	1.0	0.0	1.0	0.0	1.0	0.0	1.0	0.0	0.9998	0.0
proslikefan	0.9306	0.0022	0.9862	0.0024	0.9924	0.0016	0.9954	0.0018	0.9776	0.0028	0.9978	0.0014	0.9558	0.0057	0.958	0.0066
pushdo	0.999	0.0003	0.9996	0.0005	0.9992	0.0001	0.9996	0.0001	0.9686	0.0	0.999	0.0	0.9662	0.0013	0.8852	0.0021
pykspa	0.9988	0.0002	0.9998	0.0003	0.9988	0.0	0.9984	0.0	0.9994	0.0002	0.9998	0.0	0.9938	0.0018	0.9828	0.0049
pykspa2	0.657	0.009	0.891	0.017	0.823	0.0046	0.8742	0.008	0.7256	0.01	0.9256	0.0035	0.7682	0.0227	0.6404	0.0222
qadars	0.9968	0.0001	0.9998	0.0	0.9988	0.0	0.999	0.0	0.9964	0.0	0.9998	0.0	0.9778	0.0013	0.9758	0.0014
qakbot	0.5648	0.0025	0.5866	0.0012	0.7278	0.0014	0.6884	0.0011	0.5642	0.0015	0.8006	0.0011	0.3638	0.004	0.3272	0.0039
ramnit	0.717	0.0101	0.8178	0.0105	0.8954	0.0064	0.8434	0.0053	0.792	0.0116	0.9424	0.0031	0.6754	0.0129	0.7062	0.0155
ranbyus	0.9996	0.0029	0.9744	0.0008	0.9982	0.0006	0.9948	0.0003	0.9584	0.0007	0.994	0.0002	0.9984	0.0037	0.8892	0.0029
rovnix	0.9952	0.0	0.9926	0.0	0.999	0.0001	0.9988	0.0	0.9964	0.0	0.9984	0.0	0.9918	0.0001	0.9868	0.0001
simda	0.9995	0.0001	1.0	0.0001	0.999	0.0	1.0	0.0001	0.9985	0.0001	0.9995	0.0	0.998	0.0004	0.974	0.0022
sisron	1.0	0.0	1.0	0.0	1.0	0.0	1.0	0.0	1.0	0.0	1.0	0.0	1.0	0.0	1.0	0.0
suppobox	0.9996	0.0003	0.9994	0.0001	0.9988	0.0	0.9982	0.0	0.9968	0.0001	0.9998	0.0002	0.9954	0.0016	0.988	0.0027
sutra	1.0	0.0	0.9998	0.0	1.0	0.0	1.0	0.0	0.9994	0.0	0.9998	0.0	1.0	0.0001	0.999	0.0003
symmi	1.0	0.0	0.9998	0.0	1.0	0.0	1.0	0.0	1.0	0.0	1.0	0.0	1.0	0.0	1.0	0.0001
szribi	1.0	0.0001	0.9995	0.0	0.9998	0.0	1.0	0.0	1.0	0.0001	0.9995	0.0	0.997	0.0001	1.0	0.001
tinba	0.994	0.0039	0.9886	0.0018	0.997	0.0006	0.994	0.0004	0.9888	0.0028	0.9992	0.0002	0.997	0.0075	0.9756	0.009
torpig	1.0	0.0001	0.998	0.0	1.0	0.0	1.0	0.0	1.0	0.0	1.0	0.0	1.0	0.0002	1.0	0.0006
urlzone	0.909	0.0001	0.9065	0.0	0.934	0.0001	0.938	0.0003	0.9115	0.0001	0.961	0.0	0.7835	0.0002	0.755	0.0004
virut	0.9814	0.0026	0.991	0.0027	0.9826	0.0009	0.9922	0.0011	0.9482	0.0018	0.9968	0.0007	0.9704	0.0035	0.992	0.0056
Accuracy	0.920		0.934		0.963		0.959		0.924		0.977		0.880		0.855	

Fig. 4: Classwise TPR and FPR.

domain-level knowledge of the subject and prove ineffective at handling the drifting of domain names in the scenarios of adversarial machine learning setting. Notably, deep learning architectures have performed well in comparison to the traditional machine learning classifiers in all kinds of datasets. Moreover, the LSTM and hybrid network which are composed of CNN followed by LSTM have significantly performed well in comparison to other deep learning approaches. In addition to detecting the DGA-generated domain, domain names are classified to corresponding DGA family, which can offer the context concerning the origin and intent of the domain-generating malware.

We have christened Deep-DGA-Detect that analyzes and explores the event data generated by core Internet protocol, i.e., Domain Name Systems (DNS) to combat the Cyber security threats. This can be used as cyber threat situational awareness platform at Tier-1 Internet Service Provider (ISP) level. This framework can be made more robust by adding malware binary analysis module, facilitating detailed information of the malware. Moreover, this completely fails at detecting the malicious communications that circumvent DNS by using direct IP addresses for their communication. This limitation can be alleviated by incorporating the NetFlow. With concern on computational cost of training on longer domain names, we are unable to train more complex architecture. The reported results can be further improved by using high computational cost architecture. The complex network architectures can be trained by using advanced hardware followed by distributed approach in training that we are incompetent to try.

9. Deep-URL: A Deep Dive to Catch Malicious URLs using Deep Learning with Character Embeddings

9.1 Introduction

In recent years, internet has progressed from simple static websites to billions of dynamic and interactive websites. This advancement has absolutely reworked the paradigm of communication, trading and collaboration as a good thing for humanity. However, these precious benefits of the web are adumbral by cyber-criminals who use the web as a medium to perform malignant exercises with the aid of illegitimate advantages. Generally a malicious author uses compromised uniform resource locator (URL) as a primary tool to point to the malicious websites [50]. Compromised URLs are termed as malicious URLs, distributed via communication channels like e-mail and social media resources. In [51] acclaimed that one-third of all websites are potentially malicious in nature, demonstrating aggressive use of malicious URLs to perpetrate cyber-crimes. Thus detecting compromised URLs in a timely manner avoids significant damages. Development of an appropriate solution to tackle malicious URLs has been considered a significant work in internet security intelligence.

Most commercial systems existing in the market are based on blacklisting, regular expression and signature-matching methods [52]. All of these methods are reactive that suffer from delay in detecting the variants of existing malicious URLs and entirely new ones at that. By that time, a malicious author can get benefits from the end users. Both systems require a domain expert in which they constantly monitor the system and create signatures and push out updates to the customer. As a solving measure, over a decade researchers have proposed several machine learning-based URL detection systems [52]. These systems require domain-level expert knowledge for feature engineering and feature representation of security artifact type, e.g., URLs and finding the accuracy of machine learning models using those representations. In case of real-time system deployment, the machine learning-based URL systems encounter issues like large labeled training URLs corpus and analyzing the systems when patterns of URL keep continuously changing. Deep learning is a subdivision of machine learning [37] that is a prominent way to reduce the cost of training and operates on raw inputs instead of relying on manual feature engineering. Towards this, we propose deep-URL that takes raw URLs as input, character-level embedding, deep layers and feed-forward network with non-linear activation function to detect whether the URL is malicious or benign. For comparative study, the performances of the other machine learning classifiers are evaluated.

9.2 Related Work

Researchers have used blacklisting and machine learning-based solutions as their primary techniques to attack the malicious URLs. Blacklisting is one of the non-machine learning mechanisms widely used to block malicious URLs. It uses honeypots, web crawlers and manual reporting through human feedback as their tools to update the repository of malicious URLs. While a user attempts to visit URL, blacklisting will be triggered automatically. This does the pattern matching to know whether the URL is in their repository. If so, the request is blocked. This has been employed in web browsers, such as PhishTank [53], DNS-BH [54] and jwSpamSpy [55]. Commercial malicious URL detection systems are Google Safe Browsing [56], McAfee SiteAdvisor [57], Web of Trust (WOT) [58], Websense ThreatSeeker Network [59], Cisco IronPort Web Reputation [60] and Trend Micro Web Reputation Query Online System [61]. Blacklisting mechanisms are simple, easy-to-implement and are completely ineffective at finding the new malicious URL and additionally, always require human input to update the malicious URL repository.

Machine learning methods rely on feature engineering to extract lexical, host-based features and a combination of both to distinguish between the benign and malicious URLs. Host-based features include information related to IP address, domain name, connection speed, WHOIS data and geographical properties. Lexical features include domain name and properties such as length, number of tokens and hyphens, IP address or port number, directory, file name and parameters. Initially, [62] used various machine learning classifiers with URL features to classify the URLs as either malicious or benign. In feature engineering, they used recursive entropy reduction-based mechanism to obtain tokens and in second step they extracted a set of features from the collected tokens. They claimed that the URL-based features performed well in comparison to the page content features. In [63] they studied hidden fraudulent characteristics of URL by using only the features of lexical, header and time information. Their reported results showed higher malicious detection rate and additionally they also made a statement such that the methods can perform well when large number of samples of malicious and benign URLs exist in training phase. [64] proposed scalable mechanism towards handling large-scale malicious URLs using an online learning algorithm. They reported that the performance of online learning algorithm were good in comparison to batch-learning algorithms by demonstrating it through a real-time URL classification system. Online learning mechanisms used feature set from a combination of blacklist host-based information and lexical information and showed malicious detection rate up to 99 per cent over a balanced data-set. In [65] they compared the effectiveness of Artificial Neural Network (ANN) approach over static classifiers, such as Support Vector Machine (SVM), Decision Tree (DT), Naive Bayes (NB) and K-Nearest Neighbors (KNN) to malicious web page detection by using the static feature sets from lexical in URL and page contents. ANN approach performed well by reporting highest accuracy at 95.08 in comparison to the other static classifiers. Additionally, the importance of each feature towards identifying attacks and thereby reducing the false positive rate was discussed in detail. In [66] they discussed the performance of SVM and ANN approaches to detect malicious URLs using lexical and host-based features. As a result, ANN mechanism performed well over SVM due to its capability to store the past information of visited URLs over a long time and update and delete when it sees a change in the hidden layer of the MLP network. Unfortunately, identifying the optimal features by following manual feature engineering is a daunting task. Moreover, the features have to be constantly updated to evade the evolving obfuscation techniques. This work automates the feature engineering process instead of relying on manual feature engineering.

9.3 Experiments

9.3.1 Description of Dataset

It was necessary to collect a variety of different URLs. Over time, as malicious URLs have grown, it has become increasingly difficult for people even after collaborating with one another to keep track of the growing samples of malicious URL. For Dataset 1, unique samples of legitimate and malicious URLs were acquired from Alexa. com and DMOZ directory and OpenPhish.org, Phishtank.com, MalwareDomainlist. com and MalwareDomains.com. Dataset 2 is from [68]. There are different ways to split the dataset into training and testing. For Dataset 2, two types of splits are used. One is random split using scikit-learn [67] and second one is using the time-split. In the case of malware detection, time split is good and helps to improve zero-day malware detection. The training dataset could be URL samples first visible in a time period earlier than the test dataset. The test set might be samples, that the model has and which were not visible earlier than playing the function as zero-day malware. The detailed statistics of Dataset 1 and Dataset 2 are displayed in Table 5.

Table 5: Statistics of URL corpus.

Dataset	Training	Testing
Data set 1	150793	73229
Data set 2 random split	87201	37373
Data set 2 time split	86315	38259

9.3.2 Hyper-parameter Selection

All experiments of deep learning models are trained using the backpropagation through time (BPTT) approach. The deep learning models, such as RNN, IRNN, LSTM, GRU, CNN and CNN-LSTM are parameterized functions. Thus the malicious detection rate of deep learning models implicitly depends on the optimal parameters [37]. Finding a best optimal parameter for the network and the network structure is considered an important task. All experiments are run for 1000 epochs with batch size 256 and learning rate 0.001.

We run three trial of experiments for the number of units/memory blocks—32, 64, 128, 256, 512 with a basic LSTM network. A basic network contains a single RNN/ LSTM/GRU/IRNN layer followed by dropout 0.2 and fully-connected layer with one neuron and *sigmoid* non-linear activation function. Experiments with 256 units/ memory blocks in RNN/LSTM/GRU/IRNN performed well in comparison to the others. Even though the experiments with 128 units/memory blocks are comparable to 256, we have set the number of units/memory blocks to 256 for the rest of the experiments to provide additional degrees of freedom in feature learning. We have also made an attempt by passing the bigrams as input to RNN/LSTM/GRU/IRNN model. The performance of bigram is lesser than the character-level inputs.

We run three trial of experiments for filters 4, 16, 32, 64, 128, 256 and 512 with filter length 4 in a basic CNN network. The basic CNN network is composed of convolution1d, maxpooling1d with pool length 2, fully-connected layer with 256 neurons, dropout layer with 0.3 and fully-connected layer with one neuron and *sigmoid* non-linear activation function. CNN network with 256 filters performed well in comparison to the other filters. Moreover, the performance of CNN network with 128 filters is comparable to 256. Thus, we decided to set the number of filters at 256 for the rest of the experiments.

9.3.3 Proposed Architecture

An intuitive overview of the proposed deep learning model is shown in Fig. 5. This contains three sections, such as URL character encoding, feature learning and a classifier. The deep learning architecture contains several parameters that are learned by the deep neural network during training. During training, embedding, deep learning layers and fully-connected layers all work in concert to make an accurate malicious URL detection model. The model is evaluated on three experimental designs given below:

- Binary classification, i.e., classifying the URL as legitimate or malicious on random split of Dataset 1.
- Binary classification, i.e., classifying the URL as legitimate or malicious on random split of Dataset 2.
- Binary classification, i.e., classifying the URL as legitimate or malicious on time split of Dataset 2.

The computational flow of the deep learning model is explained below:

1. **URL character encoding:** By following the aforementioned mechanism, a matrix is formed for training (150793*1666), testing (73229*1666) for Dataset 1, training (87201*1666), testing (37373*1666) for Dataset 2 and training (86315*1666), testing (38259*1666) for Dataset 3. These matrices are passed to an embedding layer with batch size 256. This maps each character on to a 256 length real valued vector. This serves as one of the hyper parameters, in order to provide an additional degree of freedom to the deep learning architectures where 256 is chosen as the embedding vector length. During backpropagation, embedding layer collaboratively works with other layers of the deep network. An embedding layer makes an attempt to group the similar characters in a separate cluster. This type of representation facilitates other layers to easily capture the semantics and contextual similarity structures of the characters of URLs. To visualize, the 256 high-dimensional vector representation is transformed to 2-dimensional vectors by passing into t-SNE, as shown in Fig. 6, where the characters, numbers and special characters appear in separate clusters. This demonstrates that the embedding layer has captured the semantics and contextual similarity of characters in URL.

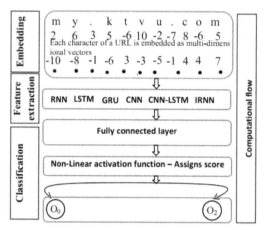

Fig. 5: An overview of deep-learning architecture; all layers and its connections are not shown.

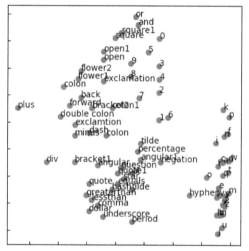

Fig. 6: Embedded character vectors learned by CNN-LSTM classifier is represented using 2-dimensional linear projection (PCA) with t-SNE.

2. **Feature learning via deep layers:** We use various deep layers, such as RNN, LSTM, GRU, IRNN, CNN and a hybrid network, such as CNN-LSTM for feature learning. Recurrent structures, such as RNN, LSTM, GRU and IRNN capture the sequential information and CNN extracts spatial information among the characters in URL. All the deep learning architectures contain only one deep layer. Thus the batch normalization technique is not used [38]. Unlike the pattern-matching technique followed by regular expressions, deep network layers match the patterns in a continuous value, i.e., degree of matching pattern rather than a binary value.

3. **Recurrent layers:** We employ the various recurrent layers, such as RNN, LSTM, GRU and IRNN. Based on the knowledge acquired during the hyper parameter

selection, the number of units/memory blocks is set to 256. Recurrent layers learn the sequential information from the character-level embedding output. This results best features in the form of tensor and passes them into dropout layer. A unit in RNN uses *hyperbolic tangent* as an input and output activation function that has values in range [–1, 1]. An LSTM memory cell uses *hyperbolic tangent* as an input and output activation function that has values in range [–1, 1] and gates use *logistic sigmoid* that has values in range [1, 0].

4. **Convolution layers:** We used CNN network to capture the spatial co-relationships among the characters. It is composed of two layers, such as convolution1d and maxpooling1d. A convolution1d layer includes more than one convolution operations that facilitate to extract the local features. Generally, a convolution1d layer contains 256 filters with filter length 4 that slide over the character embedding vector sequence and result in a continuous value at each step. This value is the representation of how much the pattern has matched in the character embedding vectors subsequence. This convolution1d output is passed to maxpooling1d. This contains a pool length of 4 that divides the convolution feature map into four equal parts. Again, the maxpooling1d output is flattened to a vector and passed to the fully-connected layer. To learn the sequential information from the max-pooling1d output, we pass into the LSTM layer, which contains 50 memory blocks.

5. **Regularization:** A dropout layer with 0.2 (0.3 in CNN) is placed between deep layers and fully-connected layer. This acts as a regularization parameter to prevent from overfitting. A dropout is an approach for randomly removing the neurons along with their connections during the training of a deep learning model. When we experiment with a deep learning model without dropout, the network overfits the data in most of the experiments (even with large number of URL corpus).

6. **Classification:** As a next step to compute a probability that the URL is malicious, the features that are captured by deep learning layers are passed to fully-connected layer. A fully-connected layer is a reasonable way of learning the non-linear combination of the captured features by deep layers. In fully-connected layer, each neuron in the previous layer has connection to every other neuron in the next layer. It has two layers, such as fully-connected layer with 128 units followed by fully-connected layer with one unit and non-linear activation function in classifying the URL as legitimate or malicious. The prediction loss of deep learning models is computed using binary-cross entropy as follows:

$$loss(pr,ep) = -\frac{1}{N}\sum_{i=1}^{N}[ep_i \log pr_i + (1-ep_i)\log(1-pr_i)] \quad (25)$$

where *pr* is true probability distribution, *ep* is predicted probability distribution. To minimize the loss of binary-cross entropy and categorical-cross entropy, we have used *Adam* optimizer [39] via backpropagation.

9.4 Evaluation Results

The efficacy of various deep learning and machine learning models are evaluated on two different datasets, such as Dataset 1 and Dataset 2. Dataset 1 is split into training, validation and testing using random split with scikit-learn [67]. Two designs of splits are used for Dataset 2. They are random split and time split. The results for Dataset 1, random split of Dataset 2 and time split of Dataset 2 are shown in the form of a ROC curve by comparing two operating characteristics, such as true positive rate and false positive rate across varying threshold in the range [0.0–1.0] in Figs. 7, 8 and 9 respectively. This metric is independent of the percentage of legitimate to malicious URLs in the entire corpus. The detailed statistical measures for Dataset 1 and Dataset 2 are reported in Table 6. Machine learning classifiers rely on features, N-grams, specifically three-gram is calculated. To translate the three-gram representation into numerical representation, hashing mechanism is used. This creates 1000 length long vector and hashes each three-gram using hashing algorithm. These vectors are passed as input to the deep neural network. A deep neural network contains five layers, the first two layers contain 256 units, third layer contains 128 units, fourth layer contains 64 units and fifth layer contains 32 units and final output layer contains one unit with *sigmoid* non-linear activation function. In between the deep layers in the deep neural network, we added ReLU as an activation function, followed by batch normalization layer and dropout layer with 0.1. The performance of both the deep learning models and the machine learning classifiers are less for Dataset 2 in comparison to the Dataset 1. The overall performance of the deep learning model is better than the N-gram or expert designed features in both Dataset 1 and Dataset 2. Thus, we claim that the character level URL embedding with deep learning layers can be a powerful method for automatic feature extractor. Additionally, N-grams outperform the expert designed features in both Dataset 1 and Dataset 2. The primary reason is that the advantage of large size of URL corpus and also the bag of N-grams is an effective representation in the field of NLP [40].

Understanding the internal dynamics of deep learning layers have remained as a black-box. These layers include a lot of information which could be seen by unrolling them. Generally, the embedded character-level vectors coordinately work with the

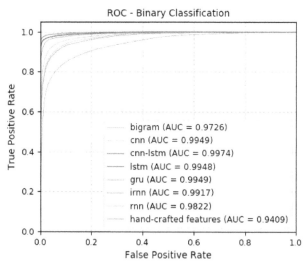

Fig. 7: ROC curve of deep-learning architectures and classical machine-learning classifier in the context of classifying the domain name as either benign or malicious (TPR vs FPR) for Dataset 1.

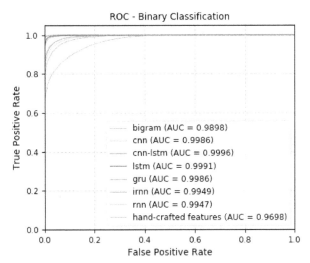

Fig. 8: ROC curve of deep-learning architectures and classical machine-learning classifier in the context of classifying the domain name as either benign or malicious (TPR vs FPR) for Dataset 2 (random-split).

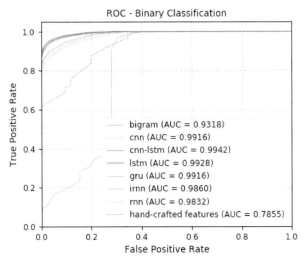

Fig. 9: ROC curve of deep-learning architectures and classical machine-learning classifier in the context of classifying the domain name as either non-DGA or DGA (TPR vs FPR) for Dataset 2 (time-split).

deep learning layers to obtain optimal feature representation. Implicitly, this kind of feature representation learns the similarity among characters in the URL. Finally, the feature representations of deep learning layers are passed to the fully-connected network. The non-linear activation function in the fully-connected layer facilitates distinguishing the feature vectors as legitimate or malicious. To visualize, the feature vectors of penultimate layer of CNN-LSTM are passed to the t-SNE, as shown in

Table 6: Summary of test results.

Method	Accuracy	Precision	Recall	F-score
Data set 1				
RNN	0.941	0.947	0.910	0.928
LSTM	0.975	0.982	0.957	0.970
I-RNN	0.965	0.971	0.945	0.958
GRU	0.974	0.978	0.960	0.969
CNN	0.921	0.893	0.926	0.909
CNN-LSTM	0.983	0.988	0.971	0.979
DNN (N-gram)	0.918	0.947	0.855	0.898
Expert level hand crafted features				
Random Forest	0.878	0.926	0.774	0.843
Data set 2 (random split)				
RNN	0.959	0.949	0.970	0.959
LSTM	0.986	0.990	0.982	0.986
I-RNN	0.961	0.975	0.947	0.961
GRU	0.992	0.997	0.987	0.992
CNN	0.943	0.951	0.936	0.943
CNN-LSTM	0.992	0.997	0.987	0.992
DNN (N-gram)	0.943	0.951	0.936	0.943
Expert level hand crafted features				
Random Forest	0.894	0.925	0.860	0.891
Data set 3 (time split)				
RNN	0.936	0.958	0.877	0.916
LSTM	0.959	0.963	0.933	0.948
I-RNN	0.939	0.964	0.880	0.920
GRU	0.954	0.951	0.933	0.942
CNN	0.929	0.963	0.855	0.905
CNN-LSTM	0.964	0.979	0.931	0.954
DNN (N-gram)	0.843	1.0	0.606	0.755
Expert level hand crafted features				
Random Forest	0.795	0.666	0.975	0.791

Fig. 10, which shows a clear separation between the malicious and legitimate URLs. This shows that the CNN-LSTM model has captured the behaviors of the legitimate and malicious URLs.

9.5 *Alternative Architectures Attempted*

During evaluating the deep learning approaches for malicious URL detection, we tried out the following architectures:

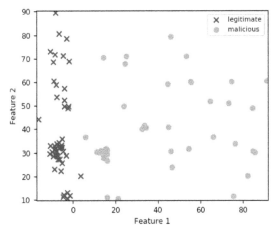

Fig. 10: 50 URLs are randomly chosen from legitimate and malicious and their activation values in penultimate layer are displayed using t-SNE.

- Stacked convolution1d layer followed by RNN/LSTM/GRU/IRNN layer in hybrid architecture.
- Stacked embeddings followed by RNN/LSTM/GRU/IRNN/CNN/CNN-LSTM layer.
- Stacked convolution1d layer to capture non-linear convolution activations.
- Stacked RNN/LSTM/GRU/IRNN layers.

9.6 Conclusion, Future Work Directions and Discussions

In this subsystem, we investigated the application of featureless, deep learning-based detection method for the purpose of malicious URLs detection. Additionally, the traditional machine learning algorithms with feature engineering are used. Deep learning architectures have performed well in comparison to the traditional machine learning classifiers. Moreover, this can easily handle the drifting of malicious URLs and serves as a robust mechanism in adversarial environment. Based on the results, we claim that the deep learning-based malicious URL detection obviate the need for blacklisting, using regular expression, signature matching method and machine learning-based solutions. We propose christened Deep-URL and it can be made more robust by adding auxiliary information like website registration services, website content, network reputation, file paths and registry keys.

10. Proposed Architecture for Organizational Security

The architectural details of the developed framework are provided in Fig. 11. This is composed of three modules (1) Data collection (2) Deep learning module (3) Continuous monitoring.

Passive sensor collects data of DNS and URLs from various sources from entirely five different places, discussed in [2]. Each sensor can handle data from multiple sources. Collected data contains the DNS and URL logs. These logs contain the details

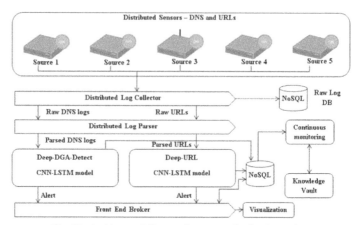

Fig. 11: Architectural diagram for an organizational security.

of user queries. Parsing unit parses the logs and extracts domain names and URLs exclusively which have more than one IP address. The domain names and URLs are passed to the deep learning module. The detected domain names/URLs are then fed into the continuous monitoring system. Here, the system continuously monitors the targeted domains/URLs with time duration of five seconds. This system will actively query to check for any change in IP of the detected domain/URLs.

11. Conclusion, Future Work and Limitations

In this work, we investigate and explore the event data of DNS and URLs for the purpose of Cyber security threats situational awareness. To handle large volumes of DNS and URL logs, a highly scalable and distributed big data framework is developed. The proposed framework leverages the concepts of NLP and uses deep learning architectures for automatically detecting Cyber security threats. The developed framework is capable of analyzing more than 2 million events per second and it could detect the malicious activities within them in real-time. The developed framework can be scaled out to analyze even larger volumes of network event data by adding additional computing resources. The scalability and real-time detection of malicious activities from early warning signals makes the developed framework stand out from any system of similar kind.

In the current research, the functionality for the malware binary analysis is not provided. This can provide an additional information of the characteristics of the malware. Additionally, adding web page content analysis to the URL detection module can enhance the malicious detection rate. These two tasks will remain as a room for future enhancement of malicious activities detection rate.

Acknowledgment

This research was supported in part by Paramount Computer Systems and Lakhshya Cyber Security Labs Pvt. Ltd. We are also grateful to NVIDIA India for the GPU hardware support and to Computational Engineering and Networking (CEN) department for encouraging the research.

References

[1] Collobert, R., Weston, J., Bottou, L., Karlen, M., Kavukcuoglu, K. and Kuksa, P. (2011). Natural language processing (almost) from scratch. Journal of Machine Learning Research 12: 2493–2537.

[2] Vinayakumar, R., Poornachandran, P. and Soman, K.P. (2018). Scalable framework for cyber threat situational awareness based on domain name systems data analysis: Big data in engineering applications. Studies in Big Data. Springer (Inpress).

[3] Snyder, M.E., Sundaram, R. and Thakur, M. (2009, June). Preprocessing DNS log data for effective data mining. pp. 1–5. *In*: Communications, 2009. ICC'09. IEEE International Conference on, IEEE.

[4] Elman, J.L. (1990). Finding structure in time. Cognitive Science 14(2): 179–211.

[5] LeCun, Y., Bengio, Y. and Hinton, G. (2015). Deep learning. Nature 521(7553): 436.

[6] Vinayakumar, R., Soman, K.P. and Poornachandran, P. (2018). Evaluating deep learning approaches to characterize and classify malicious URLs. Journal of Intelligent & Fuzzy Systems 34(3): 1333–1343.

[7] Vinayakumar, R., Soman, K.P. and Poornachandran, P. (2018). Detecting malicious domain names using deep learning approaches at scale. Journal of Intelligent & Fuzzy Systems 34(3): 1355–1367.

[8] Vinayakumar, R., Soman, K.P., Poornachandran, P. and Sachin Kumar, S. (2018). Detecting android malware using long short-term memory (LSTM). Journal of Intelligent & Fuzzy Systems 34(3): 1277–1288.

[9] Vinayakumar, R., Soman, K.P., Poornachandran, P. and Sachin Kumar, S. (2018). Evaluating deep learning approaches to characterize and classify the DGAs at scale. Journal of Intelligent & Fuzzy Systems 34(3): 1265–1276.

[10] Vinayakumar, R., Soman, K.P. and Poornachandran, P. (2017, September). Deep encrypted text categorization. pp. 364–370. *In*: Advances in Computing, Communications and Informatics (ICACCI), 2017 International Conference on IEEE.

[11] Vinayakumar, R., Soman, K.P. and Poornachandran, P. (2017, September). Secure shell (ssh) traffic analysis with flow based features using shallow and deep networks. pp. 2026–2032. *In*: Advances in Computing, Communications and Informatics (ICACCI), 2017 International Conference on IEEE.

[12] Vinayakumar, R., Soman, K.P. and Poornachandran, P. (2017, September). Evaluating shallow and deep networks for secure shell (ssh) traffic analysis. pp. 266–274. *In*: Advances in Computing, Communications and Informatics (ICACCI), 2017 International Conference on IEEE.

[13] Vinayakumar, R., Soman, K.P., Velan, K.S. and Ganorkar, S. (2017, September). Evaluating shallow and deep networks for ransomware detection and classification. pp. 259–265. *In*: Advances in Computing, Communications and Informatics (ICACCI), 2017 International Conference on IEEE.

[14] Vinayakumar, R., Soman, K.P. and Poornachandran, P. (2017, September). Applying deep learning approaches for network traffic prediction. pp. 2353–2358. *In*: Advances in Computing, Communications and Informatics (ICACCI), 2017 International Conference on IEEE.

[15] Vinayakumar, R., Soman, K.P. and Poornachandran, P. (2017, September). Long short-term memory based operation log anomaly detection. pp. 236–242. *In*: Advances in Computing, Communications and Informatics (ICACCI), 2017 International Conference on IEEE.

[16] Vinayakumar, R., Soman, K.P. and Poornachandran, P. (2017, September). Deep android malware detection and classification. pp. 1677–1683. *In*: Advances in Computing, Communications and Informatics (ICACCI), 2017 International Conference on IEEE.

[17] Vinayakumar, R., Soman, K.P. and Poornachandran, P. (2017, September). Applying convolutional neural network for network intrusion detection. pp. 1222–1228. *In*: Advances in Computing, Communications and Informatics (ICACCI), 2017 International Conference on IEEE.

[18] Vinayakumar, R., Soman, K.P. and Poornachandran, P. (2017, September). Evaluating effectiveness of shallow and deep networks to intrusion detection system. pp. 1282–1289. *In*: Advances in Computing, Communications and Informatics (ICACCI), 2017 International Conference on IEEE.

[19] Vinayakumar, R., Soman, K.P. and Poornachandran, P. (2017). Evaluation of recurrent neural network and its variants for intrusion detection system (IDS). International Journal of Information System Modeling and Design (IJISMD) 8(3): 43–63.

[20] Hochreiter, S. and Schmidhuber, J. (1997). Long short-term memory. Neural Computation 9(8): 1735–1780.

[21] Gers, F.A., Schmidhuber, J. and Cummins, F. (1999). Learning to forget: Continual prediction with LSTM.

[22] Gers, F.A., Schraudolph, N.N. and Schmidhuber, J. (2002). Learning precise timing with LSTM recurrent networks. Journal of Machine Learning Research 3: 115–143.

[23] Le, Q.V., Jaitly, N. and Hinton, G.E. (2015). A simple way to initialize recurrent networks of rectified linear units. arXiv preprint arXiv: 1504.00941.

[24] Talathi, S.S. and Vartak, A. (2015). Improving performance of recurrent neural network with relu nonlinearity. arXiv preprint arXiv: 1511.03771.

[25] Cho, K., Van Merrinboer, B., Gulcehre, C., Bahdanau, D., Bougares, F., Schwenk, H. and Bengio, Y. (2014). Learning phrase representations using RNN encoder-decoder for statistical machine translation. arXiv preprint arXiv: 1406.1078.

[26] Nair, V. and Hinton, G.E. (2010). Rectified linear units improve restricted boltzmann machines. pp. 807–814. *In*: Proceedings of the 27th International Conference on Machine Learning (ICML-10).

[27] Martens, J. (2010, June). Deep learning via Hessian-free optimization. In ICML 27: 735–742.

[28] Bahrampour, S., Ramakrishnan, N., Schott, L. and Shah, M. (2015). Comparative study of deep learning software frameworks. arXiv preprint arXiv: 1511.06435.

[29] Abadi, M., Barham, P., Chen, J., Chen, Z., Davis, A., Dean, J. and Kudlur, M. (2016, November). TensorFlow: A system for large-scale machine learning. In OSDI 16: 265–283.

[30] Hillary Sanders and Joshua Saxe. Garbage in, garbage out: How purport-edly great ML models can be screwed up by bad data. Available at https://www.blackhat.com/docs/us-17/wednesday/us-17-Sanders-Garbage-In-Garbage-Out-How-Purportedly-Great-ML-Models-Can-Be-Screwed-Up-By-Bad-Data-wp.pdf.

[31] Plohmann, D., Yakdan, K., Klatt, M., Bader, J. and Gerhards-Padilla, E. (2016, August). A comprehensive measurement study of domain generating malware. pp. 263–278. *In*: USENIX Security Symposium.

[32] Alexa, Available at https://support.alexa.com/hc/en-us/articles/.

[33] OpenDNS, Available at https://umbrella.cisco.com/blog.

[34] Osint, Available at http://osint.bambenekconsulting.com/feeds/.

[35] Goodfellow, I., Bengio, Y., Courville, A. and Bengio, Y. (2016). Deep Learning (Vol. 1). Cambridge: MIT Press.

[36] Ioffe, S. and Szegedy, C. (2015, June). Batch normalization: Accelerating deep network training by reducing internal covariate shift. pp. 448–456. *In*: International Conference on Machine Learning.

[37] Kingma, D.P. and Ba, J. (2014). Adam: A method for stochastic optimization. arXiv preprint arXiv: 1412.6980.

[38] Shazeer, N., Pelemans, J. and Chelba, C. (2015, September). Sparse non-negative matrix language modeling for skip-grams. pp. 1428–1432. *In*: Proceedings Interspeech.

[39] He, Y., Zhong, Z., Krasser, S. and Tang, Y. (2010, October). Mining dns for malicious domain registrations. pp. 1–6. *In*: Collaborative Computing: Networking, Applications and Work-sharing (CollaborateCom), 2010 6th International Conference on IEEE.

[40] Feily, M., Shahrestani, A. and Ramadass, S. (2009, June). A survey of botnet and botnet detection. pp. 268–273. *In*: Emerging Security Information, Systems and Technologies, 2009. SECUR-WARE'09. Third International Conference on IEEE.

[41] Khrer, M., Rossow, C. and Holz, T. (2014, September). Paint it black: Evaluating the effectiveness of malware blacklists. pp. 1–21. *In*: International Workshop on Recent Advances in Intrusion Detection. Springer, Cham.

[42] Stone-Gross, B., Cova, M., Gilbert, B., Kemmerer, R., Kruegel, C. and Vigna, G. (2011). Analysis of a botnet takeover. IEEE Security & Privacy 9(1): 64–72.

[43] Antonakakis, M., Perdisci, R., Nadji, Y., Vasiloglou, N., Abu-Nimeh, S., Lee, W. and Dagon, D. (2012, August). From throw-away traffic to bots: detecting the rise of DGA-based malware. *In*: USENIX Security Symposium (Vol. 12).

[44] Yadav, S., Reddy, A.K.K., Reddy, A.L. and Ranjan, S. (2010, November). Detecting algorithmically generated malicious domain names. pp. 48–61. *In*: Proceedings of the 10th ACM SIGCOMM Conference on Internet Measurement. ACM.

[45] Yadav, S., Reddy, A.K.K., Reddy, A.N. and Ranjan, S. (2012). Detecting algorithmically generated domain-flux attacks with DNS traffic analysis. IEEE/Acm Transactions on Networking 20(5): 1663–1677.

[46] Krishnan, S., Taylor, T., Monrose, F. and McHugh, J. (2013, June). Crossing the threshold: Detecting network malfeasance via sequential hypothesis testing. pp. 1–12. *In*: Dependable Systems and Networks (DSN), 2013 43rd Annual IEEE/IFIP International Conference on IEEE.

[47] Hong, J. (2012). The state of phishing attacks. Communications of the ACM 55(1): 74–81.

[48] Liang, B., Huang, J., Liu, F., Wang, D., Dong, D. and Liang, Z. (2009, May). Malicious web pages detection based on abnormal visibility recognition. pp. 1–5. *In*: E-Business and Information System Security, 2009. EBISS'09. International Conference on IEEE.

[49] Sahoo, D., Liu, C. and Hoi, S.C. (2017). Malicious URL detection using machine learning: A survey. arXiv preprint arXiv: 1701.07179.

[50] Malware URL, Available at http://www.phishtank.com/.

[51] Malware URL, Available at http://www.malwaredomains.com.

[52] Malware URL, Available at http://www.jwspamspy.net.

[53] Malware URL, Available at https://developers.google.com/safe-browsing/.

[54] Malware URL, Available at http://www.siteadvisor.com.

[55] Malware URL, Available at http://www.mywot.com/.

[56] Malware URL, Available at http://www.websense.com/content/threatseeker.asp.

[57] Malware URL, Available at http://www.ironport.com.

[58] Malware URL, Available at http://reclassify.wrs.trendmicro.com/.

[59] Kan, M.Y. and Thi, H.O.N. (2005, October). Fast webpage classification using URL features. pp. 325–326. *In*: Proceedings of the 14th ACM International Conference on Information and Knowledge Management. ACM.

[60] Liang, B., Huang, J., Liu, F., Wang, D., Dong, D. and Liang, Z. (2009, May). Malicious web pages detection based on abnormal visibility recognition. pp. 1–5. *In*: E-Business and Information System Security, 2009. EBISS'09. International Conference on IEEE.

[61] Ma, J., Saul, L.K., Savage, S. and Voelker, G.M. (2009, June). Identifying suspicious URLs: an application of large-scale online learning. pp. 681–688. *In*: Proceedings of the 26th Annual International Conference on Machine Learning. ACM.

[62] Sirageldin, A., Baharudin, B.B. and Jung, L.T. (2014). Malicious web page detection: A machine learning approach. pp. 217–224. *In*: Advances in Computer Science and its Applications. Springer, Berlin, Heidelberg.

[63] Gandhi, J. An ANN approach to identify malicious URLs.

[64] Pedregosa, F., Varoquaux, G., Gramfort, A., Michel, V., Thirion, B., Grisel, O. and Vanderplas, J. (2011). Scikit-learn: Machine learning in Python. Journal of Machine Learning Research 12: 2825–2830.

[65] Schiappa, Madeline. Machine Learning: How to Build a Better Threat Detection Model. Available at https://www.sophos.com/en-us/medialibrary/PDFs/technical-papers/machine-learning-how-to-build-a-better-threat-detection-model.pdf?la=en.

[66] Bengio, Y., Ducharme, R., Vincent, P. and Jauvin, C. (2003). A neural probabilistic language model. Journal of Machine Learning Research 3: 1137–1155.

[67] Collobert, R., Weston, J., Bottou, L., Karlen, M., Kavukcuoglu, K. and Kuksa, P. (2011). Natural language processing (almost) from scratch. Journal of Machine Learning Research 12: 2493–2537.

[68] Mikolov, T., Sutskever, I., Chen, K., Corrado, G.S. and Dean, J. (2013). Distributed representations of words and phrases and their compositionality. pp. 3111–3119. *In*: Advances in Neural Information Processing Systems.

[69] Pennington, J., Socher, R. and Manning, C. (2014). Glove: Global vectors for word representation. pp. 1532–1543. *In*: Proceedings of the 2014 Conference On Empirical Methods in Natural Language Processing (EMNLP).

[70] Kim, Y., Jernite, Y., Sontag, D. and Rush, A.M. (2016, February). Character-aware neural language models. In AAAI pp. 2741–2749.

[71] Chiu, J.P. and Nichols, E. (2015). Named entity recognition with bidirectional LSTM-CNNs. arXiv preprint arXiv: 1511.08308.

CHAPTER 7

A Prototype Method to Discover Malwares in Android-based Smartphones through System Calls

B.B. Gupta,[1,*] *Shashank Gupta,*[2] *Shubham Goel,*[1] *Nihit Bhardwaj*[1] and *Jaiveer Singh*[1]

1. Introduction

Ever since the smartphones came into existence, the mobile industry has been revolutionized by a number of applications made available by different operating systems. These applications are available in the application store, which are called differently in different operating systems, like in windows, it is OVI store, or in android, it is Google playstore. The android OS provides users with utmost comfort, while providing thousands of apps in the Google playstore, many of which have served more than 10 million downloads. These android applications provide users with different functionality depending upon the requirements of the user. All these applications are readily available to the user and can be easily downloaded and installed on smartphones. Like downloading and installing, hosting an application by a developer on the app store is also very simple and straightforward [33–39]. These uploading and downloading have become very efficient with the availability of good mobile data connectivity, powerful smartphones, an application store and an integrated billing system. The developers, who host their applications, collect their payments easily through this billing system and the users, who download some of the paid apps, get charged through the same [40–47].

[1] Department of Computer Engineering, National Institute of Technology, Kurukshetra, Haryana, India.
[2] Department of Computer Science and Information Systems, Birla Institute of Technology and Science, Pilani, Vidhya Vihar, Pilani-333031, Rajasthan, India.
* Corresponding author: gupta.brij@gmail.com

Users download different applications based on their specific needs. These applications before downloading ask for certain permissions from the user. The user can either allow all these permissions or can simply reject the application from being downloaded. There is no provision to allow certain permissions while rejecting some permission. These permissions allow the application to access many of the core services of the smartphone, e.g., camera, contacts, calendar, phone location, text messaging, etc. As is evident, these core services provided by a smartphone can become a major privacy concern if utilized in a wrong way by some malicious application.

The number of applications in the Google playstore has been escalating at a tremendous rate. With this, the number of malicious applications and all the genuine developers and Google, which owns Android, are aware of this fact. Therefore, only those applications that contain digital signatures of a trusted key are allowed in the playstore and made available for download. No application without the digital signature is allowed to be hosted on the Google playstore, but these apps can be downloaded from other non-trustworthy sources, which are never advised. Although these digital signatures can help in getting rid of malicious applications in the initial stage itself, but malicious developers have many other ways to get round this, like repackaging or update-attack to name a few. These attacks utilize an existing application on the playstore, modify those applications, make them malicious and again host them on playstore. When legitimate users download and install such applications, their devices get corrupt and data as well as personal information get compromised.

In this paper, we have proposed a prototype method, Droid-Sec, which aims at detecting malicious applications if the user tries to download and install them. Droid-Sec gives the user a fair warning that the application is malicious and can corrupt the device as well as leak personal information to unknown sources. Droid-Sec utilizes behavior-based malware detection system to discover the malicious applications. The main idea is to compare the system calls used by any android application with standard system calls to be used by that application that is stored on the remote server. If system calls used by applications match with standard permissible system calls, then that android application is legitimate; otherwise it is malicious and once an application is deemed malicious, then Droid-Sec has a provision to disable the internet connection of the corresponding application.

Our approach is similar in nature with other anti-viruses available out there in the market for smartphones. However, conventional antiviruses employ the technique of signature-based comparison; Droid-Sec utilizes the technique of system calls-based comparison. This modification is put in use because digital signature-based filtering can be easily by-passed by attackers. In addition to this, Droid-Sec aids the user of smartphones in selectively disabling the internet connection of those malicious applications.

Contributions: Droid-Sec is a dynamic analysis-based approach, which examines the behavior of an android application at run-time and on the basis of this behavior and system calls used by the applications, our approach can determine the malicious nature of the application. We introduced Droid-Sec to effectively detect and block the malicious applications. The main contributions of this paper are as follows:

- A lightweight detection method, Droid-Sec is presented, which traces the system calls of the applications and transmits the files to the remote server for comparison purposes, thus making it very lightweight. It requires less CPU usage, battery and other resources than traditional antivirus.

- Droid-Sec was tested on indigenous applications such that applications were modified to show some malicious nature. Droid-Sec detected malicious nature of the applications with an accuracy of about 99 per cent.

- Droid-Sec not only detects the malicious nature of the applications, but also blocks the internet access of the application.

- We have provided experimental evidence and demonstrated that this approach can be feasible and usable in real-time environment.

2. Related Work

Basically, there are two ways in which the android malware can be detected—static analysis and dynamic analysis. In static analysis, we generally check the source code of android applications to find out its maliciousness; if the source code comes out to be different from the original source code, then the application is considered to be malicious. In static analysis, applications need not be executed, while in the case of dynamic analysis, malicious nature is found at run-time. Some parts of code might be missed in this approach but it provides better accuracy than static analysis.

In static analysis, permissions requested by an application are generally checked to find out the nature of the application. Simple Naïve Bayes were applied to get requested permissions by Peng et al. [1] to develop a hierarchical model of classification for android applications. AdDroid [2] also proposed a permission-based mechanism, which separates permissions from advertising framework and then the sensate information is kept safe from advertising library. Sarma et al. [3] determined the malicious nature of the application on the basis of permissions given to the actual application and malicious application. Permission-based approaches generally showed low accuracy rate.

Many approaches have also been developed on the basis of dynamic analysis which checks the malicious nature of the android application at dynamic run-time. CrowDroid [4] is one of the approaches that use dynamic analysis. It is machine learning-based approach, which can detect malware in the system on the basis of the behavior of malicious application. It tracks the system calls used by the applications during run-time and apply machine-learning algorithm to make clusters of benign and malicious applications. Andromaly [5] is also a dynamic analysis-based machine learning approach that monitors the behavior of the smartphone and the user on the basis of several parameters, ranging from sensor activities to CPU usage. Another approach developed like this is MADAM [6] that detects the malicious nature of applications on the basis of some utilized features. MADAM showed good results on the malware found in the wild. In [7], the authors discuss an approach that uses machine learning algorithm to extract features and find out the anomaly. In [8], a behavior-based malware detection system is discussed that relates to the input given by the user with the system call to determine the malicious nature of applications related to SMS/MMS sending.

2.1 Existing Solutions

Smartphones are important but dangerous devices. They contain so much sensitive and private information such that all the above-mentioned vulnerabilities and more of these devices must be tackled and handled accordingly. Keeping that in mind, many counter-measures are being developed and added to smartphones to keep them secure and safe from attackers or intruders. These counter-measures are in various forms, ranging from development of security layers in the operating system itself to downloadable applications in the application market, depending on the threats to the system.

Android Sandboxing: The operating system of android contains a feature which makes the core of security architecture in android. It is android sandboxing, which essentially creates virtual boxes for all the applications installed on the smartphone device and these applications cannot access or manipulate any resource outside of their sandbox. This key feature helps in securing the android at the most basic level.

Antivirus: Above the operating system security is a layer of security software, which contains different components for the purpose of securing a smartphone device. Antivirus can be deployed on a smartphone as that component and secures the device from malicious apps and different malwares. It is a software which detects any malicious app lurking in the smartphone by checking the signatures of malicious executable files and comparing them with the signatures stored in its own repository. If these signatures match, then it is malicious in nature, otherwise not. Thus, malwares and other malicious applications can be avoided, using a good antivirus like Avast antivirus [22].

Anti-theft Software: Theft is one of the oldest and most common crime we see happening in the world. And therefore, there have been many ways to keep the valuables safe from the thieves. From the invention of locks to the announcement and then introduction of biometric identification, the anti-theft system has grown significantly. Anti-theft software is a piece of code which helps in the prevention of unauthorized access to a smartphone device. They help in tracking of the device if the device is stolen or lost and can also accomplish remote wiping out of data [23].

Updates and Password Protection: It is always advisable to keep all the software updated because when a software is updated, old bugs are removed from the application and any loopholes in the coding of the application are managed and handled properly, making it all the more difficult for any hacker to hack into such apps. Keeping the entire installed applications' passwords protected is a good way to keep the data stored in these applications safe and secure from attackers or even when the device gets stolen or lost.

Secure Wi-Fi: Connecting to insecure or public Wi-Fi can prove to be one of the biggest threats to users and their data. As anyone can join these open networks and thus, it opens an easy door for the attacker. Hence, it is advisable to join only those networks where browsers indicate that they have SSL encryption active. When SSL encryption is active, the websites first verify themselves to the user's browser before opening. If any website fails to verify itself, then an 'untrusted' security certificate pops up.

Turn-off the Bluetooth: Sometimes users leave their blue-tooth feature open, either accidentally or on purpose, and fall prey to intruders who forcibly pair their devices with the user's device and corrupt the device. Once a device has been force-paire, sensitive and private information becomes accessible to the attacker without the user knowing. Therefore, it is always advisable to turn off the blue-tooth in the settings menu after its use and not left open.

2.2 Current Issues and Challenges

We presented some solutions to various mobile vulnerabilities in the earlier section. These solutions were very basic in nature. Unfortunately, attackers in today's world are far more skilled than they used to be and are well aware of the aforementioned solutions. The part of the skill set of these attackers includes new varieties of ways to exploit a device and corrupt it. Turning the bluetooth off or downloading and installing an antivirus are not enough to keep these attackers in check. These solutions contain many loopholes and the attackers are well aware of them. Thus, these methods are not sufficient to cope with the significant rise in the number of malwares lurking in the market. According to a recent report [24], malwares targeting the android platform is at an all-time high, with a 600 per cent increase in just last 12 months.

As stated in the previous section, software installed on the smartphone is needed to be updated on a routine basis so that it can deal with the latest threats in the world. But most of the times, the security patches are not always released at regular time intervals; thus, creating room for the attackers to find loopholes and corrupt the system.

Password protection enables users to authenticate themselves before accessing any application on the mobile phone. But, only password protection does not guarantee to a user that the applications will not contain any malware. Also, many users do not use password protection as it conflicts with their ease of use of the device. They don't want to make a random pattern or enter a pin again and again and find it a nuisance. This lack of passwords eases the use of attackers too in case of mobile thefts.

Mobile users install mobile security software to restrict malware, but it is not necessary that this software can detect all kinds of malwares [26]. Numerous mobile security software are not competent enough to secure mobile from latest malware. Mobile anti-viruses detect malicious apps, using digital signatures. They have a repository of various signatures of malwares and the ones matching those signatures are deemed malicious. These anti-viruses are needed to be updated on a timely basis in order to get the database for the latest malware. Moreover, by-passing the signature-based detection is not that difficult for today's attackers [25]. Sometimes, the software makes the device slow or corrupt in some way, so users don't always install such security software.

3. Background, Motivation and Statistics

Android Operating System, a Linux kernel based OS, was designed initially for touch screen devices, such as smartphones, etc., but these days, many wearables and television sets have come out in the market, sporting android OS. Android OS supports an open platform which has been made possible because of the robust security architecture and rigorous security program. It consists of a multi-layered architecture, which provides android with the flexibility it needs, combined with security. The main benefits which the secure architecture of android aims at in providing are: shielding the important and sensitive user information, protecting the various system resources and isolating each application from one another.

These objectives are achieved by android Operating System through the following security features which the Android team introduced in this platform—user-granted permissions before any application is installed, digital signing of each application and android sandboxing. Thus, android is a secure platform for application development in which applications are mainly written in Java language and run in the Dalvik Virtual Machine (DVM). It has a multi-layered architecture which not only provides flexibility but also the security, which is lacking in other Operating Systems.

3.1 Android Software Stack

Figure 1 shows the android software stack, which provides the android architecture with different drivers of the devices coupled with the kernel in its bottom-most layer. As we go up the stack, we encounter the middleware, android run-time and applications [9, 10].

Linux Kernel: The bottom layer of the android software stack is Linux kernel, which provides functionalities, like memory management, process management, device management, etc. All other things at which Linux is good, such as networking and device drivers that are used to make interface with the hardware, are also managed by kernel.

Libraries: Libraries reside above the kernel layer and contain library libc, open source web browser engine, web-kit and SQLite database used for storage of application data and SSL libraries for security.

Android Run-time: Android run-time is another part of middleware. It consists of Dalvik Virtual Machine (DVM), which is a Java Virtual Machine of android. DVM

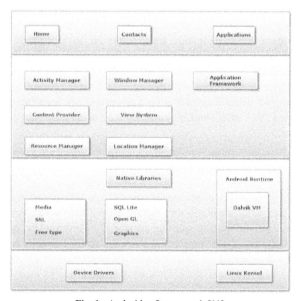

Fig. 1: Android software stack [11].

take advantage of Linux features, like memory management and multithreading. Android run-time also contains some libraries that help the android developer while developing the applications [12].

Application Framework: This layer is one of the higher layers in android software stack that provides services to applications which can be used by android developers to make applications in a most efficient manner.

Applications: The topmost layer is occupied by android applications, which are developed in this layer, for example, android games, browsers, etc.

3.2 Background and History

In this section, we present how the android operating system came into existence and became what it is today. As we chose android as our operating system upon which we are building our framework, it only seems prudent to discuss the history of android OS. Android, which started as a mobile operating system can now be seen in different tablets and has even extended to support smart televisions and wearables too. Android is also expanding its reach to various automobiles as they released the SDK for android auto last year.

The day is not far when android will start to support anything and everything tangible. It was a long journey which started long back in 2003, but has achieved tremendous milestones since then. Table 1 highlights all the different updates in the android API levels corresponding to the month and year of their releases.

- Android Inc., in 2003, was founded in Palo Alto and was later acquired by Google in 2005. Prior to the beta version, which was released in November 2007, the alpha version of android was released inside Google and OHA (Open Handset Alliance). The various internal releases were code-named 'Astro Boy' and 'R2-D2' [16].

- The first commercial version of the android was Android 1.0 (API level 1) which was released on September 23, 2008 and is also known as Android Alpha. HTC dream was the first commercially available android device.

- Android 1.1 (API level 2), on February 9, 2009, update was made available for general public, initially only for HTC dream. Internally, Android 1.1 was also known as 'Petit Four', but this name was never used officially. Now, this version is known as the Android Beta.

- Two months later, on April 27, 2009, another update was made and released. Android 1.5 (API level 3) was made available which was based on Linux kernel 2.6.27 and was the first version to have a confectionary-based code name which was later adopted for all releases henceforth. It was known as Android Cupcake.

- Later that same year, on September 15, 2009, Android 1.6 (API level 4) was released which is also known as Android Donut. It was based on Linux kernel 2.6.29 and brought numerous new features.

- On October 26, 2009, Android 2.0 (API level 5), popularly known as Android Eclair was released. Later two more minor updates were introduced—API level 6 and API level 7 under Android Eclair.

- Android 2.2–2.2.3 (API level 8) was released on May 20, 2010 based on Linux kernel 2.6.32. It is popularly known as Android Froyo, short for frozen yogurt.

Table 1: Android updates releases with month and year [17, 18].

Month, Year	Android Version	API Level
September, 2008	V1.0 (Alpha)	API 1
February, 2009	V1.1 (Beta)	API 2
April, 2009	V1.5 (Cupcake)	API 3
September, 2009	V1.6 (Donut)	API 4
October, 2009	V2.0 (Eclair)	API 5, 6, 7
May, 2010	V2.2 (Froyo)	API 8
December, 2010	V2.3 (Gingerbread)	API 9, 10
February, 2011	V3.0 (Honeycomb)	API 11, 12, 13
October, 2011	V4.0 (Ice-cream Sandwich)	API 14, 15
November, 2012	V4.1–4.3 (Jellybean)	API 16, 17, 18
June, 2014	V4.4 (Kitkat)	API 19, 20
March, 2015	V5.0 (Lollipop)	API 21, 22

- At the end of the same year, on December 6, 2010, Android 2.3–2.3.2 (API level 9) was released and is known as Android Gingerbread. It was called a major android release as it brought many key changes in the user interface and support for Near Field Communication (NFC) and also improved power management. Android 2.3.3–2.3.7 Gingerbread (API level 10) updates were also released the following year and which mainly introduced support for voice or video chat and open accessory library.

- On February 22, 2011, Android 3.0 (API level 11) was released as the first tablet—only android update-based on Linux kernel 2.6.36. It is known as Android Honeycomb. Motorola Xoom tablet was the first device to feature this version of Android. API level 12 and 13 were also later released for Android Honeycomb.

- Android 4.0–4.0.2 (API level 14), also known as Android Ice cream Sandwich, based on Linux kernel 3.0.1 was publicly released on October 19, 2011. Android 4.0.3–4.0.4 (API level 15) was also released under the same name.

- Android 4.1 (API level 16), JellyBean, based on Linux kernel 3.0.31 was released on June 27, 2012 and the first device to run this version was released in July 13, 2012. Android 4.2 (API level 17) was released on November 13, 2012 under the same Android Jelly Bean. The second generation Nexus 7 tablet was the first device to feature Android 4.3 (API level 18).

- On October 31, 2013, Android 4.4 (API level 19), KitKat was made available and it had its debut on Google's Nexus 5. Android 4.4 KitKat with wearable extensions (API level 20) was released on June 25, 2014. Besides having support for wearables, like smart watches, it had UI updates for Google Maps navigation and alarms.

- Android 5.0–5.0.2, Lollipop (API level 21) was unveiled under the code name Android L on June 25, 2014. Android Lollipop featured a new and redesigned user interface built around a responsive design language. Notifications, in Android L, could be accessed from the lock screen and displayed within other applications at the top of the screen. Internal changes were made in it by Google. Later, Android 5.1, Lollipop (API level 22) was released on March 9, 2015 [17, 18].

3.3 Statistics

In today's world, people have become indispensably dependent upon smartphones. The dependence is not limited to urban population only, but in rural households too. In short, smartphones have become a way of life these days. So, because of this rapid growth of smartphone usage, the number of applications available in the market is also increasing at an exponential rate. Downloading and installation of apps from the Google playstore takes only a few minutes. But most of the apps available on the play-store are open source, which attract many attackers. The attackers try to modify these open-sourced apps in various egregious ways and steal or leak personal or critical information from the uninformed user's phone.

The mindset of an intruder is hard to decipher as there can be any range of reasons behind the intruder's attacks. The first and foremost reason is money. Although, most hackers are not motivated by any financial profits, there are still many criminals who utilize their hacking techniques to make money. They mainly do this by putting up a fake e-commerce application in the market or a website on the internet to collect sensitive credit and debit card details. They then leak this information to their remote servers, which then use this information to steal money from the legitimate users. Many attackers or intruders attack a device just to spy on the user and his daily activities. In this case, the attacker can be a friend, a family member or even a business rival. Cyber espionage is not unconventional these days and many organizations indulge in this ill-practice in order to maintain their status and be ahead of other rivals. These organizations employ third-party specialists to do their dirty work for them. More often than not, the reason behind an attacker hacking into a device is to steal the intellectual property of the user. Some attackers are said to indulge in hacking practices just for some excitement and thrill or for some publicity stunt in order to earn limelight. Some attackers just want to disrupt the normal workings of a society. Their main aim is to cause as much trouble for the normal posterity as possible. Whereas, some attackers attack a device just to use it as an instrument for their large-scale hacking, like as a part of a DDoS attack or as a storage media for their illicit materials, like pirated software or pornography. Identity theft is also common by these attackers when they corrupt a device and compromise all the sensitive data on that device [19]. Therefore, the categorization of attacks is a little difficult to answer but attacking a device in order to steal information and harm the user in any way is a criminal offence and no reason can justify such acts. Some of the most recent attacks on smartphones are mentioned in Table 2.

The fact is, practically anyone having a smartphone device and connected to internet is susceptible to being hacked by these intruders, attacked by malwares or infiltrated by a malicious application. Thus, there is a requirement of some proactive method, when it comes to saving a device from being hacked. There is a need to fight against these attackers in any possible way and make the device safe and secure for use.

In this paper, we propose a prototype method, Droid-Sec, which aims at identifying any malicious application if the user, accidentally or on purpose, tries to download and install them from the app market. Our method gives the user a full warning when the

Table 2: Statistics of recent attacks on mobile devices.

S. No.	Name	Description
1.	SMSZombie	Hidden inside the application was a Trojan virus which used to send expensive messages to an online payment system and steal sensitive information.
2.	Angry Birds: Space	A fake app of the same name as the original Angry Birds: Space was found in an unofficial app store, which installed a malware on the user's devices.
3.	Super Battery Charger	This application was supposed to extend the battery life of a mobile smartphone when installed, but actually used to send expensive text messages away.
4.	FakeInst	This malware was found to belong to the FakeInst family and it also used to send text messages away to expensive numbers located mainly in Russia.
5.	LuckyCat	Chinese hackers created this spyware that hit both Windows and Macintosh OS which used to open backdoors when downloaded and installed.
6.	Unknown	User mobile car's network was hacked into and all the sensitive data belonging to the customers as well as the drivers was leaked to a remote database by a third party.

application is checked, that the application installed is malicious and can corrupt the device and leak personal and sensitive information to unknown sources.

Droid-Sec utilizes behavior-based malware detection system to detect any malicious applications lurking in the device. The method of Droid-Sec is to compare the system calls used by any android application with the standard system calls to be used by that same application stored on a remote server. After this comparison, we check whether the system calls used by applications match with standard permissible system calls stored on the server. If the calls match, then that android application is non-malicious; otherwise, if the calls do not match, that means a third-party person has modified those system calls and can be malicious. And finally, the users are provided with this feature through which they can selectively disable the internet connection of those applications.

4. Taxonomy of Mobile Vulnearabilities

Mobile smartphones have become an indispensable utility in today's world with everyone, ranging from a child to an adult. Millions of users, including various major businesses and organizations, use mobile phones as a tool to communicate and as a means of planning their schedules and organizing their lifestyle according to their needs. All the data regarding that is saved on their phones and any intrusion on their devices can compromise all that information. Therefore, mobile security has become increasingly important. All the personal data and sensitive information compiled and stored in mobile phones needs to be secure and safe from intruders [20].

In today's world, like computers, mobile phones have become a preferred target for attackers. As smartphones are capable of doing so much more than just calling a person, so these other features can be exploited by any skilled attacker, who can exploit the means of communication, like Short Message Service (SMS), Bluetooth,

WI-FI networks, GPS, etc. [21]. In addition to this, they can also exploit software vulnerabilities of the different web browsers and the underlying operating system. Moreover, attackers can attack through malwares or malicious software. These malwares or software, when downloaded, play havoc with the device and often leak sensitive and private data of the users to attackers. All these attacks can corrupt the smartphone and compromise any sensitive information stored on the phone.

As mentioned above, there are many ways through which a smartphone can be exploited by an attacker. Table 3 presents a brief description of various types of mobile vulnerabilities.

These vulnerabilities in smartphones open numerous doors for the attackers to illegally access a smartphone device and obtain sensitive information from the user, without their consent. As most of these come as in-built features in a smartphone in order to ease a user's access and help him function smoothly, they cannot be got rid of too.

Table 3: Description of mobile vulnerabilities.

S. No.	Mobile Vulnerability	Brief Description
1.	Exploiting SMS or MMS	Attackers can exploit this feature and send text messages with viruses as attachments to the unaware users and corrupt their devices.
2.	Exploiting Wi-Fi networks	Attackers can try to eavesdrop on Wi-Fi communications to obtain sensitive information, like username and password.
3.	Bluetooth based attacks	Attackers can exploit the blue-tooth feature by connecting himself/herself with the user's phone and take full control of the device.
4.	Data sync from corrupt PC	Malwares from the computer can get transferred to the user's phones while data sync takes place without the user's knowledge.
5.	Malicious webpage	Smartphones can fall victim to various hacking techniques like phishing, etc., used by attackers in order to illegally obtain information.
6.	Malicious applications	Many malicious apps lurk in application store which when downloaded and installed corrupt a smart phone accordingly.
7.	Mobile theft	Any data which is outside the scope of encrypted storage area is susceptible to get compromised if the device is stolen or lost.

5. Proposed Approach: DROID-SEC (Android Security)

As android applications are open source and widely available, these applications are susceptible to various kinds of attacks, especially repackaged attacks, update attacks as described in [27]. Droid-Sec focuses on detecting the malicious applications and after detecting the malicious nature of the application, it can be used to block the internet connectivity of that application. If the application is not able to connect to the internet, then it cannot send the private data to the outside world. Droid-Sec only blocks the internet access of the application while the application remains on the android device. So a smartphone user can enjoy all the desired features of malicious application without letting it to compromise with the user's privacy. We can take example of Barcode Scanner, a brilliant android application that is used for scanning and it may need permissions like contact data, internet access and browser history and it might possibly have been forged by some malicious attacker. This application can run properly without the internet access. If the user finds out that this application is showing some malicious nature, its internet connection can be blocked and still the user can access most of the services provided by this application.

Figure 2 illustrates the approach, which relies on the comparison of system call events used by normal as well as malicious applications. Our approach demands that the publisher of any android application provides the set of system calls that is used by android application and these system calls are stored on a remote server. These system calls set are standard system call set used by that particular android application. Any normal application will follow this set while the malicious application will tend to use the system calls different from this standard set, as the requested permissions from this malicious applications will be different. Droid-Sec tends to track the system call used by any application and comparing this set of system calls set that is actually used by the application with the standard system calls set. On the basis of this comparison, this approach finds which application is malicious and which is not. When this approach has detected whether the application is malicious or not, then it blocks the internet access of that application, thus making the malicious application harmless and useable for the user.

In this process, the most important step is to trace the system call, which is done by utilizing the capabilities of Linux utility strace, which is also called 'System Call Tracer'.

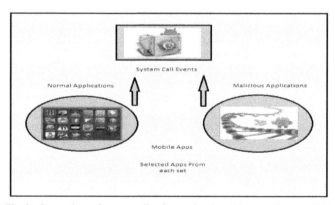

Fig. 2: Comparison of system calls of normal and malicious applications [28].

5.1 *Strace*

Strace is a Linux utility that is used to trace the system calls used by any program. In this way, we can say that it tracks how the application interacts with the Linux kernel. This interaction means that which Linux signals are received by the program and which system calls are used by the program [29]. In other words, we can say that Strace runs the specified program during its execution. It traces down and then records all the system calls used by the program and which signals are received during its execution. The name, return value and arguments of each system calls are either printed on the file whose name is specified by the –o option or standard error [30].

Strace command basic syntax is:

strace [command or program][arguments]

Here, in this approach, pid of any particular process is found out and then strace is applied on that process before the generated output is stored in straceOut.txt

strace -p pid –o straceOut.txt

Here the process, whose pid is used, will be traced and all system calls generated will be stored and sent to the remote server for comparison. The main concept behind blocking the internet access moves around IP tables. As in windows, a background firewall keeps running all the time and also keeps scanning the system for malicious applications in the system. This firewall decides which programs can have internet connectivity and which cannot. For similar tasks, IP tables are there in Linux which define which applications or process is allowed internet access or which is blocked. As we already know, android is a Linux-based Operating System, so the IP tables can be used to block malicious internet traffic requests.

5.2 IP Tables

IP tables are Linux-based command-line firewall utilities that generally use policy chains to block or allow internet traffic. If any connection has to be established in the system, IP tables check for a rule in the list for matching purpose. If no rule is found, it goes for the default rule [31].

Types of Chains: IP tables have three types of chains—input, forward and output.

Input: These types of chains are generally used to govern the behavior of all incoming connections. For, e.g., any data is coming to yours computer/server from the other user, the IP tables will try find out the rule for that particular IP address and port number and on the basis of that, accept or reject the incoming internet traffic [31].

Forward: If there are incoming connections and these connections are not supposed to be delivered locally, then forward policy chains are required. We can take the example of a router which forwards all incoming connections because all traffic sent to the router is meant to be forwarded to its target as the router is not any destination; it is only intermediary. If you are using nating, routing or some else mechanism of routing, then only use these types of policy chains, otherwise do not even use them.

Output: This chain is used for outgoing connections, for, e.g., if we try to ping facebook.com, IP tables will check the output chain to see what is the rule about the ping and facebook.com before making the decision to allow or deny the connection attempt.

5.2.1 IP Tables in Android

IP tables are default modules in AOSP; *c* code can be written to use net filter and handle it. We can use net filter to write *c* code to handle that. For example, we can create an android project and write a JNI file, use NDK-build to compile that and then ADB push the executable to the android file system to execute. And at the mobile end, you can ADB shell to it and directly use IP tables command as a root user, just like in Linux [31].

Most Common IP Tables Commands
Some common IP tables are described in [32].
Block from an IP
iptables -A INPUT -s 13.21.23.54 -j DROP
for an specific port
iptables -A INPUT -s 13.21.23.54 –p tcp –dport 22 -j DROP
Using a Network and not only one IP
iptables -A INPUT -s 13.21.23.0/24 -j DROP

5.3 Allowing or Blocking Specific Connections

By using IP tables policy chains, internet traffic can be allowed or blocked on specific ranges, ports and specific addresses. For example, we are going to set connections to DROP, but they can be switched to ACCEPT or REJECT; all depends on the requirement of the user and how the user wants to configure the policy chains.

We have used IP tables to block or unblock the internet access of android applications. IP tables are better than any type of firewall, like Windows firewall because it does not need to keep running in the background. Whenever the user wishes to block access of any application, then he can call any of the commands and change the access permission given to the particular android applications. Since IP tables do not run all the time, the resources, like CPU resource, memory battery can be saved a lot and then these resources can be used by other applications. Figure 9 illustrates the sequence of steps for detecting the malicious nature of applications.

5.4 Operation of Droid-Sec

Droid-Sec operates in two phases: firstlym it detects the malicious nature of the application and secondly, it blocks the internet access of the malicious application. Two modules have been incorporated in this approach for detecting malicious nature and then blocking the malicious application. In the first module, a basic web application has been developed, where any authorized developer can upload all legitimate system calls' file of his application. This will be used as standard set of system calls to compare with the traced out system calls of real applications. This web application uses PHP as backend, HTML, CSS, JavaScript, Jquery as frontend, Apache server and MySql database. While for the second module, we developed a real time android application. This application jams the internet access of malicious applications that were detected by the first module.

Figure 3 illustrates some of the system calls of malicious and normal applications, which are traced with the help of Droid-Sec client and the generated output files that contain the system calls are sent to the remote server. The main concept in detection of malicious nature of android application is to compare the standard system call set with the actually used system calls by the android application. Standard system call set is provided by the publisher and stored on the remote server. This remote server runs the comparison algorithms for stored standard system calls and recorded system calls.

So, basically in the first part, whenever the user wants to verify whether the application is malicious or not, the user starts application DroidSec to start tracing all running applications using strace. Whenever, strace is attached with any process, it will start recording all the system calls used by that particular process. All system

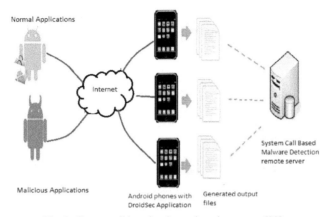

Normal Applications

Internet

Malicious Applications

Android phones with
DroidSec Application

Generated output
files

System Call Based
Malware Detection
remote server

Fig. 3: System call based malware detection system [28].

calls generated by this process are recorded in the output log files. Now the client/user side work is done. These generated output log files are now sent on remote web server which already contains the standard system call events for use by the applications. Now at remote server, there are two sets of system call events—one is standard system call events and second is recorded system call events. Now on the remote server, these sets of system calls are compared with each other. If the result of this comparison comes out to be same, then it means that set of system call events used by that application is according to the system call events allowed by the publisher. It also implies that permissions needed by applications are same as the genuine application. In that case, that instance of application is declared safe by the remote server and no further action is required. But if the comparison shows that the recorded system calls are different than permissible standard system call, then it means that the application is showing deviations from the normal functioning of the genuine application. It implies that this particular instance of the application has been modified by some attacker for his use. It is demanding permission that is different from the standard application. In that case, the server replies with a warning message to the user declaring that particular instance of the application as malicious application. The entire working of this module is explained in Fig. 4. Now, the user knows about the malicious nature of the application and can block the internet access of that application or completely uninstall from his device. To block internet access of malicious application, a separate module is designed which is explained in the next section.

In the first part, Droid-Sec has successfully detected which application is malicious in the system or which is not; the user has to block or unblock the internet access of the malicious applications according to his requirement. Now here we describe how Droid-sec is able to block internet access of malicious applications.

Whenever Droid-Sec is launched to block internet access, it looks in the cache directory and checks out if the binary files are present in it. Then it will check for the name of the applications installed in the system and look in the cache first. If Droid-Sec finds out that the applications list is not in the cache, then the application list will be loaded first before being displayed. Now a screen is shown to the user in which the list of applications is displayed to the user with two types of mode—black list mode and white list mode. The black list mode is utilized to provide the facility to block the

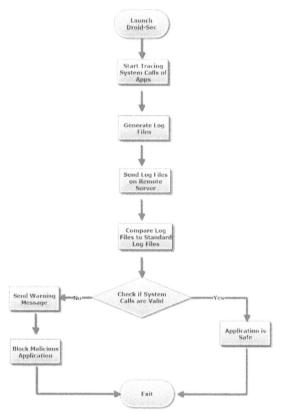

Fig. 4: Flow diagram of process of detection of malicious nature of application.

internet access to applications and white list mode is used to grant internet access to applications. From here, the user has to check his preferences, i.e., which applications he wants to block or to which applications he want to provide the internet access. Figure 5 illustrates the sequence of steps of blocking internet access to malicious application.

If the user checks his preferences for the black list mode, then the rule will be formed to block internet access of the application and the user goes for the white list for his preferences. The rule formed will be for providing internet access to blocked applications. As the user clicks on save rule, Droid-Sec will first check it has been granted the root access and if root access is not there for Droid-Sec, then root access is given to it. Root access is required because Droid-Sec needs to change permissions of other applications and it is known that in Linux only that the super user with root access has only the authority to do these kinds of permission task. For gaining root access, Droid-Sec uses the third party software, that is called KingoRoot. As the user provides root permissions to Droid-Sec, a script is formed that contains rules for applications. This script mainly contains all the necessary commands that have to be executed on Linux kernel, such as commands for blocking internet access of applications in black list mode. It also contain PID of applications on which the rule is to be applied.

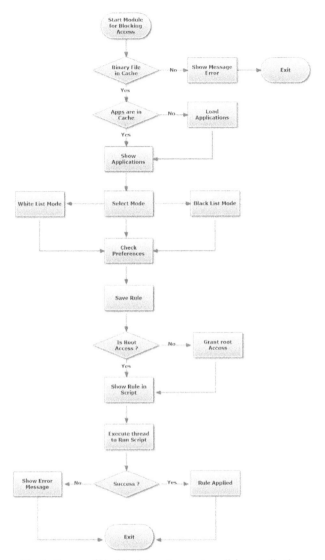

Fig. 5: Process of blocking internet access to malicious application.

Now internal threads are created and scripts are run on Linux kernel with these internal threads. These threads now hold the script that contain all the necessary information, such as PID of the applications selected and the Linux command to block internet access or to provide internet access and these internal threads are passed root access. These threads are executed on Linux kernel. The scripts contain the rules of IP tables which decide the internet connectivity of the process. The current rule of IP tables for the selected application is modified and as the IP tables rule is changed for the selected applications, it is either blocked or granted access, depending on the choice of the user. If any error condition arises while the execution process going on and rules are being modified, an exception is thrown and warning message is displayed on the screen and the thread is then terminated. If the thread is executed without any error, then IP tables rule is changed and the applied message is displayed on the screen. Now internet access of the selected malicious application is blocked.

6. Experimental Setup and Results

Numerous applications have been tested on a range of android smartphones. It was installed on various mobile smartphones which have different android versions, namely Jellybean, Kitkat, GingerBread. Firstly, Droid-Sec was tested on android applications, like the calculator. We made some changes in the source code of the calculator such that the modified calculator could take the contact details from the android device and send these contact details to a server without notifying the user. So the modified calculator made different system calls than the actual calculator. Droid-Sec compared the system calls generated by this modified calculator and actual calculator and declared the modified calculator as malicious. We blocked the internet access of the modified calculator and kept it in our device for use without any security risk. Table 4 lists some of the popular android smartphone devices with a variety of android versions on which our modified calculator application was installed and tested by DroidSec to provide successful results.

Similarly, like the calculator, we made some changes in a simple Sudoku puzzle android game such that it started taking the GPS location of the android device and sending it to the remote server. In this manner, this modified Sudoku compromises with the privacy of the user. It was also declared malicious. Then, Droid-Sec was tested for normal calculator and normal Sudoku and all traced system calls came out to be in normal system call events sets. So these were declared as normal applications. In the same way, Droid-Sec was tested with a few more modified malicious applications and normal applications on various android versions and in all the cases, Droid-Sec was able to differentiate between the normal applications and modified applications.

Table 4: Mobile and androids on which mockdroid was tested.

S. No.	Mobile	Android Version	Remarks
1.	Samsung Galaxy Grand 2	Android v4.3 (Jelly Bean)	Successful
2.	Moto G (2nd Gen)	Android v4.4.4 (KitKat)	Successful
3.	XOLO Q700	Android OS, v4.2 (Jelly Bean)	Successful
4.	Samsung Galaxy Y Plus S5303	Android v4.0 (Ice Cream Sandwich)	Successful
5.	Samsung Galaxy Note N-7000	Android v2.3 (Gingerbread)	Successful

6.1 Observed Outcomes

6.1.1 Detection of Malicious Nature of Applications

In the first step, a developer has to upload all legitimate system calls file of his application as shown in the Fig. 6. Now the developer has provided the standard set of system calls. This file containing the system calls will be stored at the server database for comparing with system calls to be used by real applications.

Now, Droid-Sec will trace out the system calls generated by every application, using strace and transmit these generated output log files on the server having the standard system call files. These output generated files will contain system calls used by any particular application as shown in Fig. 7.

Now the server has a standard set of system calls and traces out the set of system calls. The server will compare the two files and if the system calls from generated output file lie in the standard set provided by the developer, then it means the application has used valid system calls and is safe. But if the system calls present in generated output file are different from the standard set of system calls event, then it means that application has used invalid system calls and will be declared malicious by the server. The server replies with a warning message, declaring the application as malicious.

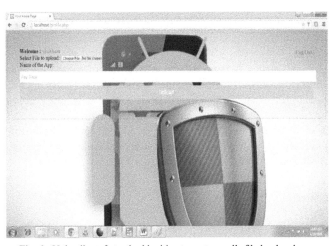

Fig. 6: Uploading of standard legitimate system calls file by developer.

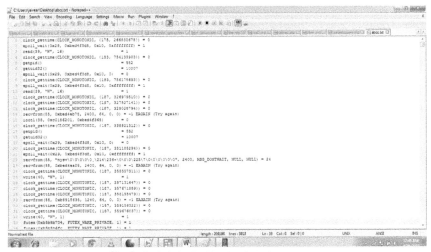

Fig. 7: List of standard legitimate system calls file.

6.2 Jamming the Internet Access of Malicious Applications

The functioning of Droid-Sec for blocking internet access is quite easy and efficient. When Droid-Sec has been installed in the system, the user has to provide root access to the Droid-Sec by using third-party software, KingoRoot. Root access is required because Droid-Sec has to modify other applications' permissions for internet. Firewall is in disabled mode by default. A user can enable the firewall by tapping the settings key on his device so that a menu is brought up and the user can touch the firewall-disabled button. When the firewall has been enabled, DroidSec asks for root access permission as shown in the Fig. 8.

Now that the firewall is enabled, the user can see the list of all applications in the system and knows which application is malicious as detected by the Droid-Sec. Smartphone user has to check the boxes that are next to the apps in order to block or allow internet access as shown in the Fig. 9.

The user has now selected the malicious applications and has to tap on save rules, so that rules for internet access of those malicious applications can be modified, as shown in Fig. 10.

Droid-Sec does not need to run all the time to restrict internet access of applications because Droid-Sec changes the IP tables rule for that application that has been blocked. The rule remains as it is until further modification. If the user wants to provide internet access again to blocked applications, then he can again use Droid-Sec to change the permissions of IP tables which again remain saved until further modification. In this manner, Droid-Sec can restrict the internet access of malicious applications such that they cannot send the private data outside the user's device.

7. Discussion

In this approach we have presented an elegant and efficient approach—Droid-Sec, which not only detects the malicious application but also restricts the internet access

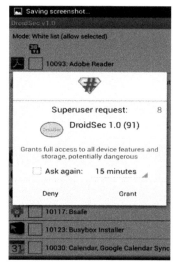

Fig. 8: Granting root access to DroidSec.

Fig. 9: Jamming of internet access of checked applications.

Fig. 10: Applying IP tables rules.

of the malicious application. Recently, numerous defensive approaches have been put forward for detecting the malicious applications. Many of these approaches utilized the concept of signature-based detection system. These techniques show low accuracy rate as the signature can be easily bypassed by the attackers. However, Droid-Sec shows a far better accuracy rate than these types of approaches.

In addition to this, there are wide varieties of other approaches that utilize system calls to detect malicious nature of the application. In addition, these approaches have used machine learning algorithms to make the clusters of non-malicious and malicious applications on the basis of system call events. However, these techniques have to use large database to store applications log files and are also time consuming, extensive machine learning algorithms to make clusters. In our approach, only the standard system calls files are needed to be stored, which makes it efficient in terms of storage and also the Droid-Sec does not require any machine learning algorithm to make it more efficient in terms of time complexity.

There are numerous other approaches which deal only with detection of malicious applications; however, they do not provide a solution to deal with malicious applications. Usually users of smartphones have only the desired option to delete the malicious application. But in our approach, we have presented an efficient mechanism to block the internet access of malicious applications and thus make the malicious application ineffective because it cannot leak user's data to the outside world.

8. Conclusion and Future Scope

In this paper, we presented a novel approach—the Droid-Sec, which aims at detecting any kind of malicious application in the device. Droid-Sec compares the system calls of a particular application against the system calls used by the standard, genuine application uploaded by the bonafide developer in the app market. If the system calls happen to be same, then the application is deemed innocuous; otherwise, if the system calls happen to be different, then the application is flagged as malicious and a warning is given to the user regarding the application. DroidSec also provides a provision to block the internet access of the flagged application. It might seem like any ordinary antivirus at first sight but it is totally different from one because where antiviruses use signature-based detection techniques to find malicious applications, our application uses system calls-based detection technique. Bypassing of the signature-based technique has become rather easy for today's attackers but simulating system calls is very difficult still. Moreover, it also provides a provision to block the internet connection of malicious applications.

Currently, Droid-Sec is able to block the internet connection access to malicious applications. But breach in security in smartphone devices can occur through other means without even using the internet, for example, some applications can take access over calendar database or phonebook or use the GPS of smartphone to track the user. Such a kind of suspicious application can send the location to third party, if not through internet, then maybe through the SMS feature. So, we are currently identifying many such problems and extending Droid-Sec to tackle all of these problems. Most importantly, we want our application to work pro-actively without any help or guidance needed from the user, so that in the near future we can extend DroidSec to achieve these characteristics.

References

[1] Peng, H., Gates, C., Sarma, B., Li, N., Qi, Y., Potharaju, R., Nita-Rotaru, C. and Molloy, I, (2012). Using probabilistic generative models for ranking risks of Android apps. pp. 241–252. *In*: Proceedings of the 2012 ACM Conference on Computer and Communications Security, ACM, Raleigh, North Carolina, USA.

[2] Pearce, P., Felt, A.P., Nunez, G. and Wagner, D. (2012). AdDroid: Privilege separation for applications and advertisers in Android. pp. 71–72. *In*: Proceedings of the 7th ACM Symposium on Information, Computer and Communications Security, ACM. Seoul, Korea.

[3] Sarma, B.P., Li, N., Gates, C., Potharaju, R., Nita-Rotaru, C. and Molloy, I. (2012). Android permissions: A perspective combining risks and benefits. pp. 13–22. *In*: Proceedings of the 17th ACM Symposium on Access Control Models and Technologies, ACM, Newark, New Jersey, USA.

[4] Iker Burguera, Urko Zurutuza and Simin Nadjm-Tehrani. (2011). Crowdroid: Behavior-based malware detection system for android. *In*: ACM, SPSM 11 Proceedings of the 1st ACM Workshop on Security and Privacy in Smartphones and Mobile Devices.

[5] Shabtai, A., Kanonov, U., Elovici, Y., Glezer, C. and Weiss, Y. (February 12). Andromaly: A behavioral malware detection framework for android devices. *In*: Journal of Intelligent Information Systems, Archive 38(1).

[6] Dini, G., Martinelli, F., Saracino, A. and Sgandurra, D. (2012). MADAM: A multi-level anomaly detector for android malware. *In*: MMM-ACNS'12 Proceedings of the 6th International Conference on Mathematical Methods, Models and Architectures for Computer Network Security.

[7] Schmidt, A.D., Peters, F., Lamour, F., Scheel, C., Camtepe, S.A. and Albayrak, S. (2008). Monitoring smartphones for anomaly detection. *In*: Mobileware '08 Proceedings of the 1st International Conference on Mobile Wireless Middleware, Operating Systems and Applications.

[8] Xie, L., Zhang, X., Seifert, J.P. and Zhu, S. (2010). PBMDS: A behavior-based malware detection system for cellphone devices. pp. 37–48. *In*: Proceedings of the Third ACM Conference on Wireless Network Security, WISEC 2010, Hoboken, New Jersey, USA, March 22–24.

[9] Software Stack. Available at: http://www.pcmag.com/encyclopedia/term/51702/software-stack [last accessed on 25 April 2015].

[10] Android Architecture. Available at: http://www.android-app-market.com/android-architecture. html [last accessed on 25 April 2015].

[11] Android Software Stack. Available at: http://www.javatpoint.com/android-software-stack [last accessed on 25 April 2015].

[12] Android Software Stack. Available at: http://java.dzone.com/articles/android-software-stack. [last accessed on 25 April 2015].

[13] Android Sandboxing. Available at: http://androiddls.com/wiki/index.php?title=Sandboxing [last accessed on 25 April 2015].

[14] Android Sandboxing. Available at: http://www.techopedia.com/definition/25266/sandboxing [last accessed on 25 April 2015].

[15] Android Malware Statistics. Available at: http://www.kaspersky.co.in/about/news/virus/2011/ Number_of_the_Week_at_Least_34_of_Android_Malware_Is_Stealing_Your_Data [last accessed on 25 April 2015].

[16] History of Android Operating System. Available at: http://en.wikipedia.org/wiki/Android_ (operating_system) [last accessed on 25 April 2015].

[17] Android Version History. Available at: http://en.wikipedia.org/wiki/Android_version_history. [last accessed on 25 April 2015].

[18] Android History. Available at: http://www.android.com/history/.[last accessed on 25 April 2015].

[19] Malware Attacks. Available at: http://www.fiercemobileit.com/story/malware-attacks-android-devices-see-600-increase-says-sophos/2014-03-04 [last accessed on 25 April 2015].

[20] Mobile Vulnerabilities. Available at: http://blog.trendmicro.com/trendlabs-security-intelligence/exploiting-vulnerabilities-the-other-side-of-mobile-threats/. [last accessed on 25 April 2015].

[21] Mobile Vulnerabilities. Available at: http://www.pcworld.com/article/2010278/10-common-mobile-security-problems-to-attack.html. [last accessed on 25 April 2015].

[22] Mobile Antivirus Software. Available at: http://www.avast.com/en-in/free-mobile-security [last accessed on 25 April 2015].

[23] Android Anti-theft Software. Available at: http://www.androidauthority.com/best-android-apps-to-help-find-a-lost-smartphone-565016. [last accessed on 25 April 2015].

[24] Android Malware Report. Available at: http://www.fiercemobileit.com/story/malware-attacks-android-devices-see-600-increase-says-sophos/2014-03-04. [last accessed on 25 April 2015].

[25] Mobile Antivirus Software. Available at: https://www.nccgroup.trust/en/learning-and-research-centre/white-papers/the-demise-of-signature-based-antivirus/ [last accessed on 25 April 2015].

[26] Mobile Antivirus Software. Available at http://www.darkreading.com/vulnerabilities-and-threats/does-mobile-antivirus-software-really-protect-smartphones/d/d-id/1106727? [last accessed on 25 November 2014].

[27] Zhou, Y. and Jiang, X. (May 2012). Dissecting android malware: Characterization and evolution. *In*: Proceedings of IEEE, 2012 Symposium on Security and Privacy (SP), IEEE.

[28] Ham, Y.J. and Lee, H.W. (2014). Detection of malicious android mobile applications based on aggregated system call events. International Journal of Computer and Communication Engineering 3(2): 149–154.

[29] Strace. Available at: http://mylinuxbook.com/linux-strace-command-a-magnificent-troubleshooter/ [last accessed on 25 April 2015].

[30] Strace. Available at: http://linux.die.net/man/1/strace. [last accessed on 25 April 2015].

[31] IP Tables. Available at: http://www.howtogeek.com/177621/the-beginners-guide-to-iptables-the-linux-firewall [last accessed on 25 April 2015].

[32] IP Tables. Available at: https://help.ubuntu.com/community/IptablesHowTo [last accessed on 25 April 2015].

[33] Gupta, Shashank, B.B. Gupta and Pooja Chaudhary. (2018). Hunting for DOM-based XSS vulnerabilities in mobile cloud-based online social network. Future Generation Computer Systems 79: 319–336.

[34] Gupta, Shashank and Brij Bhooshan Gupta. (2017). Cross-Site Scripting (XSS) attacks and defense mechanisms: classification and state-of-the-art. International Journal of System Assurance Engineering and Management 8(1): 512–530.

[35] Gupta, Shashank and Gupta, B.B. (2016). XSS-SAFE: A server-side approach to detect and mitigate cross-site scripting (XSS) attacks in JavaScript code. Arabian Journal for Science and Engineering 41(3): 897–920.

[36] Gupta, Shashank and Brij Bhooshan Gupta. (2016). JS-SAN: Defense mechanism for HTML5-based web applications against JavaScript code injection vulnerabilities. Security and Communication Networks 9(11): 1477–1495.

[37] Gupta, B.B., Shashank Gupta and Pooja Chaudhary. (2017). Enhancing the browser-side context-aware sanitization of suspicious HTML5 Code for halting the DOM-based XSS vulnerabilities in cloud. Application Development and Design: Concepts, Methodologies, Tools, and Applications, 216.

[38] Gupta, Shashank and Gupta, B.B. (2016). XSS-secure as a service for the platforms of online social network-based multimedia web applications in cloud. Multimedia Tools and Applications, 1–33.

[39] Gupta, Shashank and Brij B. Gupta. (2016). An infrastructure-based framework for the alleviation of JavaScript worms from OSN in mobile cloud platforms. pp. 98–109. *In*: International Conference on Network and System Security, Springer, Cham.

[40] Gupta, Shashank and Gupta, B.B. (2017). Detection, avoidance and attack pattern mechanisms in modern web application vulnerabilities: Present and future challenges. International Journal of Cloud Applications and Computing (IJCAC) 7(3): 1–43.

[41] Gupta, Shashank and Brij Bhooshan Gupta. (2016). XSS-immune: A Google chrome extension-based XSS defensive framework for contemporary platforms of web applications. Security and Communication Networks 9(17): 3966–3986.

[42] Nagpal, Bharti, Naresh Chauhan and Nanhay Singh. (2017). SECSIX: Security engine for CSRF, SQL injection and XSS attacks. International Journal of System Assurance Engineering and Management 8(2): 631–644.

[43] Gupta, Shashank, B.B. Gupta and Pooja Chaudhary. (2018). A client-server JavaScript code rewriting-based framework to detect the XSS worms from online social network. Concurrency and Computation: Practice and Experience, e4646.

[44] Gupta, Shashank and Brij Bhooshan Gupta. (2018). Robust injection point-based framework for modern applications against XSS vulnerabilities in online social networks. International Journal of Information and Computer Security 10(2-3): 170–200.

[45] Gupta, Shashank and Gupta, B.B. (2018). RAJIVE: Restricting the abuse of JavaScript injection vulnerabilities on cloud data centre by sensing the violation in expected workflow of web applications. International Journal of Innovative Computing and Applications 9(1): 13–36.

[46] Gupta, Shashank and Gupta, B.B. (2018). POND: Polishing the execution of nested context-familiar runtime dynamic parsing and sanitisation of XSS worms on online edge servers of fog computing. International Journal of Innovative Computing and Applications 9(2): 116–129.

[47] Gupta, Shashank and Gupta, B.B. (2018). A robust server-side javascript feature injection-based design for JSP web applications against XSS vulnerabilities. pp. 459–465. *In*: Cyber Security: Proceedings of CSI 2015, Springer, Singapore.

Additional References

Arshad, Saba, Munam A. Shah, Abdul Wahid, Amjad Mehmood and Houbing Song. (2018). SAMADroid: A Novel 3-Level Hybrid Malware Detection Model for Android Operating System. IEEE Access.

Bhattacharya, Abhishek and Radha Tamal Goswami. (2018). A hybrid community-based rough set feature selection technique in android malware detection. pp. 249–258. *In*: Smart Trends in Systems, Security and Sustainability, Springer, Singapore.

Fan, Ming, Jun Liu, Xiapu Luo, Kai Chen, Zhenzhou Tian, Qinghua Zheng and Ting Liu. (2018). Android malware familial classification and representative sample selection via frequent subgraph analysis. IEEE Transactions on Information Forensics and Security.

Hussain, Muzammil, Zaidan, A.A., Zidan, B.B., Iqbal, S., Ahmed, M.M., Albahri, O.S. and Albahri, A.S. (2018). Conceptual framework for the security of mobile health applications on android platform. Telematics and Informatics.

Kim, Duhoe, Dongil Shin, Dongkyoo Shin and Yong-Hyun Kim. (2018). Attack detection application with attack tree for mobile system using log analysis. Mobile Networks and Applications, 1–9.

Malik, Sapna and Kiran Khatter. (2018). Malicious application detection and classification system for android mobiles. International Journal of Ambient Computing and Intelligence (IJACI) 9(1): 95–114.

Wang, Chao, Qingzhen Xu, Xiuli Lin and Shouqiang Liu. (2018). Research on data mining of permissions mode for Android malware detection. Cluster Computing, 1–14.

Zhu, Hui-Juan, Zhu-Hong You, Ze-Xuan Zhu, Wei-Lei Shi, Xing Chen and Li Cheng. (2018). DroidDet: Effective and robust detection of android malware using static analysis along with rotation forest model. Neurocomputing 272: 638–646.

Zulkifli, Aqil, Isredza Rahmi A. Hamid, Wahidah Md Shah and Zubaile Abdullah. (2018). Android malware detection based on network traffic using decision tree algorithm. pp. 485–494. *In*: International Conference on Soft Computing and Data Mining, Springer, Cham.

CHAPTER 8

Metaheuristic Algorithms-based Feature Selection Approach for Intrusion Detection

Ammar Almomani,[1] Mohammed Alweshah,[2,] Saleh Al Khalayleh,[2] Mohammed Al-Refai[3] and Riyadh Qashi[4]*

1. Introduction

Most modern applications, especially with an increase in the amount of digital data, have become dependent on the feature-selection process as an important part of their work. So, feature selection is applied in several well-known fields, such as text mining, pattern recognition, image processing and computer vision, industrial applications, web page, bookmark categorization and so on [1–3].

Feature selection is a complex problem, requiring an artificial intelligence method to solve it. However, researchers for many years have been trying to find the best ways to increase the accuracy and efficiency in this scientific field. The point of research that scientists are now trying to work on for improvement of the feature-selection process is to use another algorithm with the learning model. One of these algorithms, which can be used with the learning model, is metaheuristic techniques as an optimizer [4].

Research in recent years show the ability of these algorithms to give very high accurate results and improve the process of selecting attributes. There are many metaheuristic algorithms used in feature-selection problem, such as Simulated

[1] Department of Information Technology, Al-Huson University College, Al-Balqa Applied University, P.O. Box 50, Irbid, Jordan; ammarnav6@bau.edu.jo
[2] Prince Abdullah Bin Ghazi Faculty of Information Technology, Al-Balqa Applied University, Salt, Jordan; Saleh.fayez@hotmail.com
[3] Department of Computer Science, Zarqa University, Zarqa, Jordan; refai@zu.edu.jo
[4] Leipzig University of Applied Sciences, Gustav-Freytag-Str. 42A, 04277 Leipzig, Germany; QashiRiyadh.qashi@htwk-leipzig.de
* Corresponding author: weshah@bau.edu.jo

Annealing, Ant colony Algorithm, Artificial Bee Colony, Differential Evolution, Genetic Algorithm, Harmony Search and many other algorithms which will be mentioned later in the book. Although many of these algorithms are used in this task, researches are ongoing to find new algorithms to provide solutions to this problem. The motive behind this is twofold: to find fewer selected features or to upgrade the accuracy [5].

Also, one area where feature selection and other data-mining techniques are applied is cyber security [6]. National and cyber security are now the focus of many researchers because of the development of internet and networks applications and the increase of users who access them in the world. Because of data mining, users can now make all kinds of correlations and this also raises privacy concerns. So the researchers are interested to find approaches which can deal with malware detection. These proposed approaches are effectively combined with the information gain method as a feature selector with the evolving clustering method as an evolving learning classifier [6].

In this paper, several topics will be addressed. In the beginning and after introduction, the problem statement is clarified in Section 2. Then, the concept of feature-selection process is illustrated in Section 3 and in Section 4, the concept of the metaheuristic algorithms is discussed as also the most important modern researches that use these algorithms in the feature selection are reviewed. Next, in Section 5, data-mining methods with cyber security are discussed. Finally, discussion and conclusion are presented in Sections 6 and 7 respectively.

2. Problem Statement

Features selection is the problem as to how to choose the least number of features from the original dataset which sometimes contains huge features. The need for feature selection has now become important with the advancement of time to improve and find solutions and give a clear luster to many problems [7]. For example, to find specific conclusions and relationships in a large set of data, some features are related to the problem and others are not. If all the features were selected, this would certainly affect the search result. In order to find the best solutions, it is necessary to choose the features that are associated with the research problem only and avoid choosing any one of them that may affect the outcome of the work, either by giving inaccurate results or taking a lot of time in the analysis process. Figure 1 shows the concept of reduction attributes.

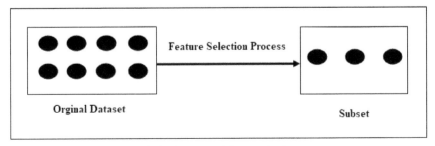

Fig. 1: A concept of feature selection

The objective of feature-selection process is to reduce the number of features as much as possible, but not at the expense of accuracy. So, the success of the selection process depends on two important factors: reducing the number of attributes and increasing the accuracy rate [8].

3. Feature-selection Process

Feature selection is one of the most important processes in the field of data mining and artificial intelligence as it aims to build a set of training data to apply in predictive modeling. The process is done by selecting a subset of the full dataset and this subset represents the minimum necessary features to reach the maximum possible variation in the data [9, 10].

Feature-selection process will eliminate redundant, irrelevant and misguiding features in order to get the best subset representing the best solution, where any feature is relevant if the decision depends on it; otherwise it is irrelevant. And it will be redundant if it is highly correlated with other features. All this will often improve and increase classification accuracy [11, 12]. Also, reducing features improves the performance model-training process efficiently, especially if the classifier is very expensive to train [13, 14].

The feature-selection process passes through several stages to reach the optimal solution [15, 16]. Figure 2 illustrates the stages of this process.

Figure 2 shows the four stages of features selection process and these include a generation procedure, an evaluation function, a stopping criterion and a validation procedure. In generation phase, a set of attributes is selected from the complete dataset for an evaluation process to determine whether it fits the solution or not [15]. This process can be done in two ways:

- *Forward selection*: Here the features are chosen one by one, when at the beginning, the generation subset is empty.
- *Backward elimination*: Here all features are chosen and then they are eliminated one by one.

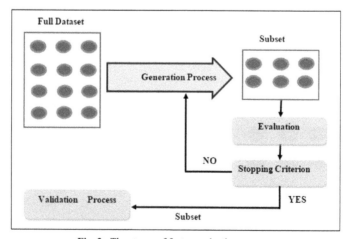

Fig. 2: The stages of feature-selection process.

In the evaluation phase, a certain standard is used to select the required features that are relevant to the problem under study. This criterion represents a threshold, where features are selected if they are equal or larger to the value of the threshold and the remaining features are ignored. There is a set of metrics that can be used as a threshold, including distance, information, correlation and consistency [17]. A stopping criterion is the control over the evaluation; it determines whether the subset of attributes achieves the desired solution to solve the problem or not. If it achieves the required solution to solve the problem in a suitable manner, it is presented to the classification algorithm (validation phase), else a new subset is generated to evaluate it [18].

After the stages of features selection process are clarified, the general algorithm for this process can be written as shown in Fig. 3.

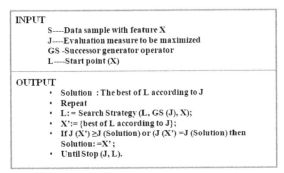

Fig. 3: The general algorithm of feature selection.

3.1 Feature-selection Approaches

There are several ways in which the process of feature selection is carried out. There is no better way than the other. Each method suits some kind of problems, so, the nature of the research problem determines which one of these methods is appropriate to give a better result. The common methods of feature-selection process are filter, wrapper, embedded and hybrid [19].

3.1.1 Filter Method

Filter method is a technique which selects features by examining and evaluating the intrinsic characteristics of the data independently of the classification algorithm [20]. A certain standard is used to select the required features that are appropriate to the problem under study. This criterion represents a threshold, where features are selected if they are equal or larger to the value of the threshold and the remaining features are ignored [18]. There is a set of metrics that can be used as a threshold, including distance, information, correlation and consistency [1].

The filter method is independent of the classification algorithm and therefore, it is fast and scalable. However, because it is not associated with the classifier to select attributes, it loses some accuracy. There are a lot of methods which use filter technique, including Information Gain (IG), Gain Ratio (GR), Symmetric Uncertainty (SU), Correlation based Feature Selection (CFS), Markov Blanket Filter (MBF), Fast

Correlation-based Feature Selection (FCBF) and Minimum Redundancy Maximum Relevance (MRMR) [21].

3.1.2 Wrapper Method

Wrapper method differs from the filter method as it uses the learning algorithm (classifier) as a part of evaluation phase (criterion measuring process). This approach creates an interaction between the search subset and the classification algorithm. Therefore, the results that will be obtained will be more accurate than in the previous filter method [22]. However, since the classifier will be used twice, this will lead to a longer time to reach the results than the previous method. Another drawback is overfitting which occurs here if the classification algorithm, which is used, learns the data perfectly and is unable to make generalization [23].

3.1.3 Embedded Method

Embedded method is similar to the wrapper method, where the same classifier is used in the attribute selection in the evaluation phase, but the use of the classifier here is done at a relatively less computational cost as compared to the wrapper method [24].

3.1.4 Hybrid Method

Hybrid method uses filter and wrapper methods sequentially. It is based on two phases to select features. At the first, filter approach is applied to generate a subset of features and then the wrapper approach is applied to select features from the subset which is achieved from the first phase [25].

4. Metaheuristic Algorithms

Optimization techniques are procedures of calculation used to solve certain problems and the results obtained by these methods are either optimal or suboptimal solution. Optimization models can be divided into two main methods—exact methods and approximate methods.

Exact methods ensure the optimal solution but the approximate methods obtain a near optimal solution. Heuristic algorithms and approximation algorithms are two types of approximate methods, where the approximation algorithms ensure a high-quality solution in bounded run-time, whereas heuristic algorithms obtain good solution within acceptable time. There are two types of heuristic algorithms: (i) specific heuristic, which are designed to solve specific cases and (ii) metaheuristics, which are top-level methodologies.

Metaheuristic algorithms are based on the principle of finding a specific mathematical technique that optimizes a particular problem. The improvement is done through numerous repeated attempts of implementation in order to connect to the best solution to the problem [26, 27].

Metaheuristic algorithms are used in a variety of fields, such as management administration, statistics and all engineering and informatics fields. They are capable of creating dynamic ways appropriate to the nature of the issue to be addressed and find the most appropriate solution among the range of possible solutions to this issue before improving the value of this solution to the maximum extent of possibility [28].

Metaheuristics take into count the information collected during the search to steer the search process. They are embedded with operators to scape a local optimum. They generate new solutions by combining one or more good solutions. As mentioned above, metaheuristics are usually incomplete methods; they do not guarantee finding a globally best solution. They often find approximation results [29].

When designing metaheuristic algorithms, both exploration and exploitation should be taken into account [30, 31]:

- *Exploration phase* is the movement to multiple sites and new location in order to discover and know all things about these places. It represents a new learning process. Population-based metaheuristic algorithms are exploration-oriented.
- *Exploitation phase* is the utilization of these sites and makes full use of existing resources therein. Single-based metaheuristic algorithms are exploitation-oriented.

The use of metaheuristic algorithms in various application fields has become widespread, as these smart algorithms have the ability to reach the best solutions. One of the areas in which these algorithms are applied is feature selection. Many researches have tried to find solutions to feature selection problems, using these algorithms and the results showed high efficiency of these smart algorithms in this domain.

In this review, the metaheuristic algorithms which were used in the review are 23 and they are Simulated Annealing, Tabu Search, Ant Colony Optimization, Ant Loin Optimization, Artificial Bee Colony, Bat Algorithm, Bees Algorithm, Binary Biogeography Based Optimization, Cat Swarm Optimization, Coco Search, Crew Search Algorithm, Differential Evolution, Dragonfly Algorithm, Firefly Algorithm, Fish Swarm Optimization, Grasshopper Optimization, Gravitational Search Algorithm, Grey Wolf Optimizer, Genetic Algorithm, Harmony Search, Moth-Flame Optimization, Particle Swarm Optimization, Whale Optimization Algorithm. This literature review is arranged by the algorithm name.

4.1 Single-based Metaheuristic Algorithms

These algorithms are based on the creation of an initial solution for the problem and then they move away from it to neighboring regions [28]. They are called 'trajectory algorithms' because they follow a specific path in the search process, starting with the first solution. In this method, every step is executed only if the outcome solution is better than the current one and the process continues until a local minimum is obtained [29].

4.1.1 Simulated Annealing

Simulated Annealing (SA) has been widely used in optimization problems. The idea of this algorithm comes from the steel process [32], which simulates the cooling of material in a heat bath. This algorithm is concerned with studying the properties of the material structure after cooling, as these properties are affected by the rate of cooling speed. If the cooling process is slow, large crystal beads will form, but if the opposite happens, the crystals will have flaws. Kirkpatrick et al. (1983) [33] used SA in the optimization problem, where the idea was to use this algorithm to find possible solutions and convergence for optimal solution.

Barbu et al. (2017) proposed a learning scheme for feature selection in high-dimensional data applications, using annealing for computer vision and big data learning. Many advantages were achieved, such as ruining much more efficiently and it did not introduce any undesired bias in estimation. Also, it solves the constrained optimization problem and has a performance guarantee in both estimation and selection. However, it can't handle object detection problems [34].

4.1.2 Tabu Search

Glover (1989) created Tabu Search (TS) as one of the local search algorithms that can be used for solving optimization problems. Tabu means the forbidden thing which cannot be touched or accessed; it is taken from the language which is used by the inhabitants of Polynesia Island [35].

Wang et al. (2016) proposed a hybrid method to select attributes using gene expression data by applying hybridization between Binary Imperialist Competition Algorithm (BICA) and Tabu Search. The results show that the proposed method reduced features with high classification accuracy [36].

Shi et al. (2018) proposed hybrid method by combining genetic algorithm and Tabu Search to select objects from high-resolution remote-sensing images. They aim to improve the premature convergence of genetic algorithm by Tabu Search. The proposed method increases the classification accuracy thought it cannot improve the number of reduced features [37].

4.2 Population-based Metaheuristic Algorithms

These algorithms deal with a set of solutions, rather than a single solution. At first, a set of solutions are initialized and then another set of solutions (population) are generated. Search steps are stopped after reaching a certain criteria [38]. In these algorithms, it should be noted that the reach to optimal solution depends on how the population is manipulated. Most of the methods associated with this type of algorithms are divided into Evolutionary Computation (EC) and Swarm Intelligence (SI) [29].

4.2.1 Ant Colony Algorithm

Ant colony optimization (ACO) was proposed by Dorigo and Di Caro (1999). The idea of the algorithm came from simulating the search for food in ants. It is one of the search algorithms that relies on experience and error rate to give an acceptable solution. So it is used to solve problems that take a long time in the computer or issues that need to test all the possibilities in order to reach the desired solution [39].

Sabeena and Sarojini (2015) used ACO and SVM classifier to reduce features and improve accuracy for two datasets—breast cancer and liver cancer, and predict the target class accurately for each dataset. This technique shows that the selected subset acquires higher accuracy and the implementation of feature-selection process can generally improve the accuracy of the classifier [40].

Sharma (2015) used Ant Colony Optimization for feature selection of keystrokes dynamics using SVM and KNN classifiers. In this approach, the accuracy of the result was not only maintained but it also improved, and the required time for authentication using subset of features takes very less time [41].

Wan et al. (2016) employed for feature selection is a Modified Binary Coded Ant Colony Optimization (MBACO) combined with genetic algorithm. SVM is utilized as a classifier, where there are several advantages of using this method, such as raising the average of fitness value and it is more convenient to reduce the data dimension based on SVM than GA, EGA, IGA, BPSO BACO, ABACO, BDE and GA-ACO. Interm of CPU time, it is fast and suitable to use with real-time applications and the heavy computation efficiency could be conquered at the maximum degree when it is combined with MBACO. As a result, it is able to maintain a good balance on the efficiency and classification accuracy [42].

Aghdam and Kabiri (2016) used ACO with Nearest Neighbor Classifier to select features and trained classifier to identify any kind of new attacks. This method reduces the number of features by approximately 88 per cent, and the detection error is reduced by around 24 per cent, using KDD Cup 99 test dataset. Also, it reduces both the memory size and the CPU time required and is very reliable for intrusion detection besides being a more robust representation of data. However, this method wasn't used by complicated classifier to improve its performance. Also, feature selection for the payload-based intrusion detection is not mature yet [43].

Mehmod and Rais (2016) presented anomaly detection feature-selection method, by using previous information as pheromones and then using the Support Vector Machine (SVM) to build anomaly detection model. Result shows improvements in the true positive rate (TPR) which uses this approach efficiency in networks as real-time intrusion-detection model [44].

Neagoe and Neghina (2016) presented two approaches for feature selection. The first one is the ACO-Band Selection (ACO-BS), which applies ant colony to decrease dimensionality of space images by preserving the most significant spectral bands for multi-spectral pixel classification. The second approach is the ACO-Training Label Purification (ACO-TLP), which removes contaminants from the training set of classifier by optimizing the quality of supervised classifier through minimizing the classification errors. Results show an increase in classification accuracy [45].

Behrouz Zamani et al. (2016) used ACO to search the solution space and calculate the correlation between selected-features subset that was chosen by ACO and the unselected subsets. ACO will build a solution as an affinity matrix in order to select the probability function from the ranked features. In this method, many classifiers were used to evaluate the performance; also the proposed work was compared with many supervised and unsupervised algorithms [46].

4.2.2 Ant Lion Optimizer Algorithm

Zawbaa et al. (2015) aimed to increase the performance of the ALO for feature selection by using chaos. This approach (CALO) is applied to data from biology and medicine to show the effectiveness of this method on the complex data. The advantages of this method are represented as—the accuracy of CALO is better than the other technique with the same test data. However, there are some limitations in using this technique, such as, the subset of features selected might differ and the running time may increase when made convert to another classifier. So in real time, application may not work efficiently [47].

Emary et al. (2016) proposed a wrapper subset feature-selection mechanism using binary variants of the ALO by transforming ALO to binary ALO, using two mechanisms—the V-shaped and the S-shaped functions, or creating an equal binary operator that is inspired from ant-lion optimization operators. The presented mechanism was applied on feature selection problem and compared with other algorithms, like (PSO), (GA), continuous ALO and binary bat algorithm (BBA). Results indicate the efficiency of the proposed algorithm (BALO). Also, the V-shaped showed better results than S-shaped [48].

Mafarja et al. (2017) used ALO as a wrapper feature selection method, presenting six transfer functions categorized in half in two categories—S-shaped and V-shaped. The presented approach is tested on 18 famous datasets and the results were compared with other approaches, like PSO, GSA and two previous Ant Lion optimization-based algorithms. Results indicate the efficiency of the proposed algorithm in term of accuracy [49].

4.2.3 Artificial Bee Colony

Karaboga (2005) proposed Artificial Bee Colony wherein bees in the colony are divided into three different groups—forager bees, observer bees and explore bees, and there is one artificial bee for each food source. This algorithm is often used in the field of optimization and achieves satisfactory results [50]. [50]

Ragothaman and Sarojini (2016) combined ABC with Pareto Optimization to select non-dominated features. There are many advantages of using this method and these are increase in classification accuracy for non-dominated features, and determining the individuality and independent nature of the features in the feature subset. However, clustering and some statistics methods can be used to select non-dominated and validated feature subsets [51].

Samsani and Suma (2016) presented a binary method of ABC to select features of thyroid disease dataset through the Decision Tree classifier, Naïve Bayesian and K-Nearest Neighbor. In this method, the classifiers' accuracy level for all the three classifiers were significantly improved and ABC gave small features subset. However, to achieve this result, the authors implemented three classifiers together rather than one [52].

Ghanem and Jantan (2016) employed multi-objective ABC to select features of Pareto fronts of non-dominated solutions, using a fitness function to reduce features, the error rate of classification and false alarm rate in IDSs. The good result in this method showed improvement of performance in selected features, false alarm rate and classification performance. However, the ABC algorithm was originally proposed as a single-objective technique [53].

Hancer et al. (2015) presented ABC with multi-objective methods to select features in classification by developing a multi-objective ABC framework (MOABC). Also, proposed a filter criterion to measure the dependence between a subset of features and the class labels. This is done by developing a fuzzy mutual information measure by integrating the dependence between two features and the class label in the original fuzzy mutual-information measure. The advantages are represented by removing irrelevant and redundant features and getting high performance in classification. On the other hand, there are some disadvantages of using this method; like the performance of classification sometimes doesn't reach a high level and the author didn't make a comparison with another approach. Also a novel multi-objective ABC to get better search was not developed, that is, the Pareto front of non-dominated solutions in large-scale feature selection problems [54].

4.2.4 Bat Algorithm

Yang (2010) proposed the Bat Algorithm (BA), which is a naturally inspired algorithm based on the idea of the prey being located by the bat through echoes even in darkness [55].[55]

Rozlini et al. (2015) proposed a framework for enhancement of the BA with Dumpster-Shafer Theory to find the optimum features to improve the classification performance. There are many advantages in this approach; such as high classifier accuracy indicates a high classification performance and the feature and classification performance are related. On the other hand, the irrelevant feature of classification process is not known and the authors did not make a comparison with other technique, like KNN and BaiyesNaives to ensure high accuracy [56].

Bansal and Sahoo (2015) presented a method to solve the problem of generating the set of not-dominated solutions. This problem is solved by investigating the performance of two FMS through BA-based multi-objective algorithms. The advantages of this method represent maximization of the classification and minimization of the feature vectors dimension. However, this approach must be tested with large spaced datasets [57].

Kumar (2017) proposed an approach for face recognition using the extracted and reduced features through Discrete Cosine Transform combined with Principle Component Analysis DCT-PCA and BA. The result shows a higher recognition rate of the proposed approach with less features selected as compared by other feature selection algorithms, like (PSO), (GA), and cuckoo search (CS) [58].

Jeyasingh and Veluchamy (2017) presented an enhancement of the BA to develop the Modified Bat Algorithm (MBA) in order to apply it in feature selection on the Wisconsin Diagnosis Breast Cancer (WDBC) dataset. This was done to achieve precise diagnosis of breast cancer. Authors chose random instances in the local random walk in order to detect the best solutions. The best solutions thus obtained are ranked and applied in the training process of Random Forest Classifier (RF). The result shows a higher accuracy in the classification process than other techniques, like Gain Ratio and Correlation-Based Feature Selection [59].

4.2.5 Bees Algorithm

Pham et al. (2006) presented the Bees Algorithm which is inspired by bee behavior and can be used in many applications, like models of smart systems and social behavior [60]. In order to apply this algorithm, the distance between the solutions must be determined as this is a condition for application of the algorithm [61].[61]

Eesa et al. (2015) proposed a combination model based on ID3 and Bees Algorithm to select features for network intrusion-detection system. The advantages of this method are that it shows a high performance when compared with the results obtained in all the features. However, it can be made more efficient by controlling the number of bees that are responsible for searching the possible location for food recovery [62].

4.2.6 Binary Biogeography-based Optimization

Boukra (2017) proposed a wrapper feature-selection approach for intrusion detection using a modified method, called Guided Adaptive Binary Biogeography-based Optimization (GAB-BBO). In this approach, the modifications are done to BBO by using new migration and mutation operators. Also, the Evolutionary State Estimation (ESE) is modified for adapting the algorithm behavior through Hamming distance and fuzzy logic consecutively. Furthermore, a balance between global and local search is achieved by adjusting the GAB-BBO parameters in exploration phase and using a Weighted Local Search (WLS) to enhance the exploitation ability. The experiments are applied in two phases. At first, the GAB-BBO performance is evaluated, using benchmark functions. The second experiment is applied, using the Kdd'99 intrusion detection dataset. After the comparisons with other methods, the results demonstrate the ability of the proposed approach to provide good competitive solutions in feature-selection problems [63].

4.2.7 Cat Swarm Optimization

Chu et al. (2006) proposed Cat Swarm Optimization (CSO) which relies on observing the behavior of cats, and it has two phases—tracing mode and seeking mode, where the integration of these two parts gives this algorithm great strength [64].[64

Lin et al. (2016) presented improvements in the original cat swarm optimization for feature selection problem, called ICSO. Authors presented two ways to enhance the current CSO by using the term Frequency-Inverse Document Frequency (TF-IDF). The presented approach proved to be more accurate than the native CSO [65].

4.2.8 Cuckoo Search Algorithm

Yang and Deb (2009) proposed the Cuckoo Search Algorithm (CSA) inspired from the parasitic behavior of some cuckoo birds. This algorithm showed great success in the field of optimization [66]. [66]

Sujana et al. (2017) [67] used CS to develop the parallel cuckoo search optimization (PCSO) algorithm and applied it in feature selection domain. The results show better accuracy as well as the selection size is reduced with greater efficiency than other methods which were compared [63].

4.2.9 Crow Search Algorithm

Askarzadeh (2016) proposed the Crow Search Algorithm inspired by the clever behavior of the crow, as it is known that the crows store the excess food in a different storage capacity so that they can return to these in case they are hungry [68]. [68]

Sayed et al. (2017) proposed a new algorithm, called Chaotic Crow Search Algorithm (CCSA) in order to enhance the native CSA in the field of feature selection. They used 10 chaotic maps to replace the random movement parameters of the native CSO. The result indicates that (CCSA) achieved a higher value in the best and mean fitness value test and it outperforms the native CSA [69].

4.2.10 Differential Evolution

Storn and Price (1997) proposed Differential Evolution (DE) algorithm. Through its use in many applications, this algorithm has proved its ability to perform tasks efficiently and is a reliable optimization algorithm. The strength of this algorithm lies in its use of only three parameters during the search process (166). However, it has some limitations, like stagnation of population as well as slow convergence speed [70]. [70]

Noghabi et al. (2015) proposed a new method for DE which can generate new mutations and extraction in gene selection for cancer classification capabilities. Subsequently, the trade off empty places are filled. Here, the lack of exploitation in DE is solved; consequently, the accuracy of classification becomes better. However, there is a problem due to the effect of population size and dimensionality on this method [71].

Wang et al. (2016) used DE for data preprocessing to solve the problem of reducing data or increase it with the same accuracy of classification. The advantages of this method that were achieved represented reducing features of the dataset when instant selection is used. However, the accuracy of the classification in the search space sometimes becomes worse when using instant selection. Also, it needs a novel representation of solutions as well as a computationally cheap fitness measure [72].

Shahbeig et al. (2016) applied the binary differential evolution optimization algorithm to reduce attributes and determine the purity of sponge iron. The advantages of this method were reduction in plant random conditions and improvement in the robustness of the algorithm [73].

4.2.11 Dragonfly Algorithm

Mirjalili (2016) proposed the DA algorithm which is inspired from dragonfly behavior and has been applied in several areas of optimization problems. In the feature-selection field, Mafarja et al. (2017) [74] proposed a wrapper feature-selection approach using Binary Dragonfly Algorithm (BDA) with KNN classifier. The results show the ability of BDA to avoid falling in local minima when it searches the features' space. It shows a superior performance compared to other approaches [75].[75]

4.2.12 Firefly Algorithm

Firefly Algorithm (FA) was proposed by Yang (2009) and is inspired by the behavior of fireflies in how they attract the other fireflies by flashing signals, and it is currently used for a wide range of optimization problems. Zhang et al. (2017) [76] improved the capability of FA in feature-selection domain by proposing a Return-Cost-based Binary Firefly Algorithm (RC-BBFA). After applying the proposed method on 10 datasets, the results demonstrate that the RC-BBFA reduced the number of attributes with high classification accuracy [77, 78].

4.2.13 Fish Swarm Optimization Algorithm

Li (2002) proposed Fish Swarm Optimization (FSO) to express the intelligent behavior of fish during their lifetime in terms of food search, migration and other collective behavior [79]. [79]

Manikandan and Kalpana (2017) enhanced the technique of native FSO in artificial fish swarm optimization (AFSO) and applied it with the Classification and Regression Tree (CART) to increase the accuracy of the feature-selection process. Results indicate that AFSO performs better in terms of measure, accuracy, precision and recall [80].

4.2.14 Grasshopper Optimization Algorithm

Saremi et al. (2017) proposed Grasshopper Optimization Algorithm (GOA), which introduces solutions for optimization problems by providing mathematical models that simulate the behavior of grasshoppers and insects in Nature [81].

Mafarja et al. (2017) improved the efficacy of the basic GOA for feature-selection problems by using Evolutionary Population Dynamics (EPD) with GOA. After applying the proposed method on 22 benchmark datasets, the results and analysis showed that the proposed method improved the feature-selection process in classification accuracy, number of selected features, consumed CPU time and convergence behaviors of all hybrid methods when compared [82].

4.2.15 Gravitational Search Algorithm

Rashedi et al. (2009) introduced Gravitational Search Algorithm (GSA) as one of the optimization techniques based on the law of gravity according to Newton's physical laws, where light masses are attracted to the heavy one [83].

Nagpal et al. (2017) used GSA to implement a wrapper-based method to select features in medical datasets using k-Nearest Neighbor's Classifier. The results show the ability of GSA to reduce attributes efficiently with an average of 66 per cent. Also, a better performance of classification accuracy was achieved by GSA in comparison to Particle Swarm Optimization and Genetic Algorithm [84].

Bostani and Sheikhan (2017) proposed a hybrid filter and wrapper approach based on mutual information (MI) and binary gravitational search algorithm (BGSA) to reduce attributes in intrusion detection systems. MI-BGSA method was compared to other wrapper and filter methods, such as BGSA and BPSO, ReliefF and Chi-square. The results showed that the MI-BGSA improved the classification accuracy and reduced the attributes with most relevance to the target class [85].

4.2.16 Genetic Algorithm

Genetic Algorithm (GA) was proposed by John Holland and his students at the University of Michigan in 1970's [86]. Genetic algorithm is an important technique in the search for the best option of a set of solutions available for a specific design. Genetic manipulation passes the optimal advantages through successive breeding processes and strengthens these traits. These traits have the greatest ability to enter breeding and produce an optimal generation by repeating the genetic cycle; the quality gradually improves [87].

Cerrada et al. (2015) used GA for designing fault diagnosis models in a multi-stage form to select in each stage the best features from a subset of candidate features. The study showed that GA is still a good tool for optimizing search processes and can be implemented under new approaches to improve its performance. However, this method leads to a study of the physical meaning of relevant condition parameters for fault diagnosis in spur gears [88].

Hong et al. (2015) selected a set of optimized features from the corpus using GA to minimize computational complexity and to increase performance of the classification. In this approach, the classification performance was better when all the features were used than when FSGA is used with the exception of the result of recall. Also, classification time can be reduced as well and it is capable with real-time data processing. However, the authors recommended using parallel GA to improve the performance of its repetition and the proposed algorithm regarding duplicate feature selection and fitness function. Also, this approach cannot deal with big data environment [89].

Soufan et al. (2015) presented DWFS—a web tool, using parallel genetic algorithm on the wrapper approach and mixed it with a randomized search algorithm to enhance the feature-selection process. The advantages of using this technique represented achievement of the capability of reducing the number of features while preserving the classification performance. However, it sometimes is not feasible with complex biological and biomedical data complementing or large problems [90].

Chen et al. (2016) used a hybrid CGPGA-based model by combining the Coarse-Grained Parallel Genetic Algorithm and SVM to solve the problem of feature selection in a classifier. There are many advantages of using this technique, such as minimizing the size of selected feature subset, the number of support vectors and training time. Also, it optimized SVM model parameters and correctly obtained the discriminating feature subset in an efficient manner. Furthermore, it makes the low level maintain the support vectors proportion. However, it didn't minimize the number of SVM models constructed through optimization instruction. So it does not work efficiently if the problem gets very large [91].

Aalaei et al. (2016) selected effective features for breast cancer diagnosis by genetic algorithm and PS-classifier, using wrapper approach. This technique improves accuracy, specificity and sensitivity of classifiers [92].

4.2.17 Grey Wolf Optimizer

Mirjalili and Lewis (2014) proposed Grey Wolf Optimizer (GWO) in which the fishing mechanism of the grey wolves is simulated through the leadership hierarchy using four types of wolves—their names are alpha, beta, delta and omega. The optimization process is done through three steps—searching for prey, encircling prey and attacking prey [93]. [93]

Emary et al. (2016) proposed a binary version of GWO to be used in the wrapper mode feature selection. The original GWO was enhanced by applying two mechanisms—the first aims to find the new position of the grey wolf by performing crossover on the simulated steps toward the best first three solutions; the second mechanism is used to find updated binary grey wolf position. The present work was compared with PSO and GA. Results showed the efficiency of BGWA based on 18 datasets tests [94].

4.2.18 Harmony Search

Geem et al. (2001) proposed the Harmony Search (HS) which is not considered to be an algorithm inspired by Nature, but it is simulates the musical process to reach an ideal state of harmony according to specific criteria. HS has achieved great success in the fields of control, signal processing, mechanical engineering and others [95].[95]

Bagyamathi and Inbarani (2015) used the hybridization of the Rough Set Theory and the Improved Harmony Search Relative Redact Algorithm (IHSRRA) to classify the protein sequences and reduce the dataset by removal of irrelevant features. This study showed that the proposed method can help to solve the problem of high-dimension, and the classification of protein sequence which was very accurate. Also, it could be very useful for predicting the function of protein. As a result, this paper proved the efficiency and effectiveness of HS and RST-based approaches [96].

4.2.19 Moth Flame Optimization

Mirjalili (2015) proposed Moth Flame Optimization (MFO) which is inspired by the movement of moths in Nature, where the moths fly at night in a fixed angle for the moon and in a straight line for very long distances. This simulation method was used effectively for optimization problems [97]. [97]

Zawbaa et al. (2016) presented a new mechanism to solve the feature-selection problem and reach the optimal subset by applying the Moth-Flame Optimization (MFO) using wrapper approach. The proposed algorithm was compared with other feature-selection algorithms, like the GA and PSO based on famous datasets. The proposed algorithm proved its performance to be higher than other feature selection [98].

4.2.20 Particle Swarm Optimization

Eberhart and Kennedy (1995) proposed Particle Swarm Optimization (PSO) which is a simulation of the flocking behavior of birds. The application of this algorithm in the field of optimization has yielded very accurate results [99]. [99]

Wu (2015) implemented two-round feature selection for text categorization, combining both the filter and wrapper approaches. The filter approach is Information Gain (IG) method, and the wrapper approach is BPSO based on 5NN and Rocchio. The advantages of this technique were achieved by reducing the number of features and the classification performance was improved. Also, it has much short computation time. On the other hand, it can't be used efficiently in distributed systems [100].

Butler-Yeoman, Xue et al. (2015) proposed a method to find a target size which roughly indicates the smallest size of feature subsets by using PSO based on wrapper feature-selection approach by developing a sequential random target-size optimization method. The advantage of this method is that it reduces the number of features and achieves a similar or even better classification than using all features or standard PSO. However, the consistency analysis for stochastic feature-selection algorithm has a problem but is relatively simple and lacks theoretical proof [101].

Aghdam and Heidari (2015) applied PSO to text features of a bag of words model in which a document is considered as a set of words or phrases and each position in the feature vector corresponds to a given term in a document. One of the advantages of using this approach is achieving a strong search capability on the problem space and this can efficiently find minimal feature subset. Also, features can be selected without any previous knowledge of the original dataset [102].

Javidi and Emami (2016) introduced a combinational search method that can balance between discovering the new region and the local search using PSO. CBPSOL has a strong search capability in the problem space and can efficiently find minimal feature subsets in real and high-dimensional synthetic datasets. But if the filter-based method is used before CBPSOL, it cannot show a high performance [103].

4.2.21 Whale Optimization Algorithm

Mirjalili and Lewis (2016) proposed this algorithm, which simulates humpback whales fishing process in the sea [104].[104]

Sharawi, Zawbaa et al. (2017) presented a feature-selected approach in order to get the subset of features that are optimal by using a wrapper-based Whale Optimization Algorithm (WOA). The approach was compared to other feature-selection algorithms, like PSO and GA using 16 datasets and the results indicate that the WOA achieved a higher accuracy than other algorithms [105].

Mafarja and Mirjalili (2017) proposed two hybrid techniques as a feature selection models that are based on WOA. The first technique improved the top solutions found at the end of each iteration by combining Simulated Annealing (SA) with the WOA. The second approach improved the exploitation process by searching the best zones that were found by the WOA. Test results over 18 well-known datasets are compared with other feature-selection algorithms—PSO, ALO and GA. Results show the enhancements on the accuracy of classification [106].

Mafarja and Mirjalili proposed a wrapper-based feature-selection approach that uses WOA. A new binary version of WOA was presented by the authors. The idea of the proposed approaches is to change the way of selecting operators for the search process by guiding randomization of operators. The first two methods used Tournament and Roulette Wheel selection mechanisms. The third optimization is done using crossover and mutation operators to enhance the exploitation of the WOA algorithm. Approaches tested and compared with other feature-selection wrapper-based algorithms, like the PSO, GA, and ALO. Results show the efficiency of the proposed approaches in terms of accuracy of classification and average selected features [107].

After the most important works were reviewed, the most important strengths of metaheuristic algorithms are derived as below:

- The ability to deal with the problem of constrained engineering optimization.
- The ability to deal with the problem of truss structures with discrete variables.
- The ability to deal with the problems of real world optimization.
- The ability to provide a fast convergence rate.
- The ability to achieve high accuracy results.
- The computational cost is very low, so the time complexity is reduced.
- The ability to be a global algorithm for optimization problems.
- This algorithm is adjustable and to be optimized even more.
- The ability to hybridize with other optimization algorithms.

However, these algorithms, like other optimization algorithms had the ability to deal with very complex problems, but the results depend on the nature and extent of the problem.

5. Data Mining Methods with Cyber Security

This part of the research discusses the importance of data-mining techniques for cyber and national security, and then a brief overview is provided. Currently, this field of research takes researchers' attention widely, to find the best methods for data protection from intrusions. The most important data-mining techniques are classification, cluster and feature selection. In this review, the genetic algorithm as metaheuristic algorithm and fuzzy data mining-based Intrusion Detection System are discussed.

Thuraisingham et al. (2008) presented a study of cyber security applications based on data-mining techniques, such as link analysis and association rule mining. These techniques can detect threats of abnormal patterns that may affect the system. The static and dynamic analyses provide a continuous dynamic process to detect new threats and monitor the adverse. These methods enable network intrusion protection by mining network traffic and data stream mining, which contribute to the detection of malignant patterns with high efficiency [108].

Kshirsagar et al. (2012) presented a study focused on several models for Intrusion Detection System based on genetic algorithm and data-mining techniques [109]. The study was concerned with the signature method to protect data from misuse, where this process requires special rules to be set up and stored in Intrusion Detection System in order to identify attacks through the matching process between the stored rules and any new anomalies.

Harshna and Navneet Kaur (2014) used the fuzzy logic with genetic algorithm for Intrusion Detection System, generating more flexible patterns in detecting anomalies by selecting a set of features available from audit data. The experiment was executed on the numeric dataset. The results showed the efficiency of the integration of fuzzy logic and genetic algorithm in the detection of real threats. However, the drawback of this method is that it requires considerable time to reach high efficiency results, particularly, in the large number of iterations, population size and correlation ratio [110].

As noted above, the problem of detecting threats across the network presents a complex challenge, especially with the increasing number of users and the volume of data. This research is still the focus of attention of many researchers in recent years. Data-mining techniques offer highly efficient solutions to reduce the risk of these threats. The genetic algorithm also provided solutions if used with data-mining applications.

Al-Qurishi et al. (2017) developed a method to expose the malicious user and counterfeit software across social networks by using a centralized central management protocol to prevent such damage and provide a secure means of communication across social networks. This protocol creates a roadblock for all users of these networks and ensures that the communications is human only. It thus identifies the accounts of the robot in an effective manner [111].

Moreover, in the field of encoding images transmitted over the network, the Quaternion Fresnel Transforms (QFST) method is used in [112] to encrypt images, and also to improve security and link degradation. The encrypted 3-D image is encrypted on a 2D basis. This proposed approach proves the validity and effectiveness of the encryption used to keep the content of the images confidential.

In terms of the Internet of Things (IOT) concept, the security field of risk detection has become an important space in the field of research in recent years. The users in the IOT store a lot of sensitive data in the cloud and as a result, the security concerns may increase, where this data must be protected from damage or that are accessed illegally. The sensitive data encryption process which encrypts sensitive data stored in the electronic clouds by internet users is the first solution to deal with these security concerns and all researches expect that within the next few years, all the data companies that provide Internet service to clients will be protected by encryption services [113].

Companies provide internet services that the internet of things adopts increasingly data security in cloud computing. Sensitive data needs to be protected at its inception, before it moves out of the cloud or enters. It requires the right strategy for data protection—data-centric approach to the cloud [114], which is applicable and adaptable to several different systems. Figure 4 shows data encryption before being sent either from the device or from the cloud.

This research aspect has attracted the attention of many researchers. Recently, Tewari and Gupta (2017) suggested a new mechanism for mutual authentication between the device and the server in the IOT environment. The mechanism depends on the method of encoding the Oval Curve (ECC) [115] and through the results, the method used is an effective solution to resist any potential attack on the network. In another paper [116], NS simulation technique was used to design a wireless video surveillance system by ensuring optimal frequency distribution and minimizing video distortions during transmission over the network.

Memos et al. (2017) described the security challenges facing IOT by analyzing the most important researches related to the security and protection of media exchanged over the network. Also, a study of most algorithms used to reach to the smart city concept in the future, such as the Media-based Monitoring System (EAMSuS), and High Efficiency Video Encoding (HEVC) [117].

Another area in which security techniques are applied and confidentiality is maintained is mobile cloud computing system. Zkik et al. (2017) proposed a scheme to encrypt the homogeneous information to preserve the confidentiality of the information as well as to retrieve it remotely. The results proved that after implementing the proposed approach, it was found to be robust, highly efficient and capable of security analysis in this field [118].

Ibtihal and Hassan (2017) used a new method to get a secure structure through the use of two clouds—the first to encrypt and decrypt and the second to store the data.

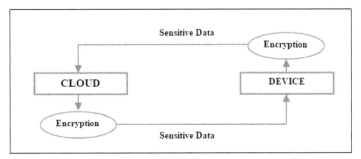

Fig. 4: Data encryption in the IOT.

202 *Machine Learning for Computer and Cyber Security*

To achieve this, Openstack was used and the PAILLIER as the encryption system. Also, watermarking, DWT algorithm was applied [119].

6. Discussion

As noted in this research, the problem of feature selection is a complex problem which requires intelligent methods to deal with the datasets for selecting the best attributes associated with the study problem. After submitting literature, there are many algorithms used to solve this problem. Many of these algorithms have been used more than once, either alone or in a hybrid way with other algorithms or by making some improvements to the same algorithm for providing better solutions in this field. Furthermore, it is clear that no single method can be applied successfully to solve every kind of problem. So no method is best for all the cases because each technique has advantages in specific areas of interest and the results depend on the nature and extent of the problem. Also, the problem of detecting threats across the network presents a complex challenge, especially with the increasing number of users and the volume of data. This research is still the focus of attention of many researchers in recent years. Data-mining techniques offer highly efficient solutions to reduce the risk of these threats. The genetic algorithm also provided solutions if used with data-mining applications as well as cryptographic algorithms that have been mentioned in the literature above.

7. Conclusion

The main objective of the research is to propose a new method that has the ability to find good solutions and optimize the process of feature selection. Metaheuristic algorithms are used to generalize subsets in feature-selection process randomly. It collects the information during the search to steer the search process, then generates new solutions by combining one or more good solutions. Metaheuristic algorithms have been widely used in feature selection problem as they are able to provide solutions to this problem with high efficiency. The positive results that have been achieved through these algorithms have made researchers continue to apply these algorithms in this field, even if new algorithms have not yet been applied or to hybridize different types of these algorithms to get better results. Furthermore, it suggests some future studies which could improve the feature-selection problems in the field of cyber security.

References

[1] Jović, A., Brkić, K. and Bogunović, N. (2015). A review of feature selection methods with applications. pp. 1200–1205. *In*: Information and Communication Technology, Electronics and Microelectronics (MIPRO), 38th International Convention on.
[2] Jensen, R. and Shen, Q. (2004). Fuzzy-rough attribute reduction with application to web categorization. Fuzzy Sets and Systems 141: 469–485.
[3] Forman, G. (2003). An extensive empirical study of feature selection metrics for text classification. Journal of Machine Learning Research 3: 1289–1305.
[4] Yusta, S.C. (2009). Different metaheuristic strategies to solve the feature selection problem. Pattern Recognition Letters 30: 525–534.

[5] Talbi, E.-G., Jourdan, L., Garcia-Nieto, J. and Alba, E. (2008). Comparison of population based metaheuristics for feature selection: Application to microarray data classification. pp. 45–52. *In*: Computer Systems and Applications, AICCSA 2008. IEEE/ACS International Conference on.

[6] Altaher, A., ALmomani, A., Anbar, M. and Ramadass, S. (2012). Malware detection based on evolving clustering method for classification. Scientific Research and Essays 7: 2031–2036.

[7] Pereira, R.B., Plastino, A., Zadrozny, B. and Merschmann, L.H. (2018). Categorizing feature selection methods for multi-label classification. Artificial Intelligence Review 49: 57–78.

[8] Yu, L. and Liu, H. (2004). Efficient feature selection via analysis of relevance and redundancy. Journal of Machine Learning Research 5: 1205–1224.

[9] Martin-Bautista, M.J. and Vila, M.-A. (1999). A survey of genetic feature selection in mining issues. pp. 1314–1321. *In*: Evolutionary Computation, CEC 99, Proceedings of the 1999 Congress on, 1999.

[10] Abdelaziz I. Hammouri, Mohammed Alweshah, Issa A. Alkadasi and Asmaran, M. (2017). Biogeography-based optimization with guided bed selection mechanism for patient admission scheduling problems. International Journal of Soft Computing 12: 103–111.

[11] Molina, L.C., Belanche, L. and Nebot, À. (2002). Feature selection algorithms: A survey and experimental evaluation. pp. 306–313. *In*: Data Mining, ICDM 2003, Proceedings. 2002 IEEE International Conference on.

[12] Alshareef, A.M., Bakar, A.A., Hamdan, A.R. Abdullah, S.M.S. and Alweshah, M. (2015). A case-based reasoning approach for pattern detection in Malaysia rainfall data. International Journal of Big Data Intelligence 2: 285–302.

[13] Sun, Y., Todorovic, S. and Goodison, S. (2010). Local-learning-based feature selection for high-dimensional data analysis. IEEE Transactions on Pattern Analysis and Machine Intelligence 32: 1610–1626.

[14] Alweshah, M., Hammouri, A.I. and Tedmori, S. (2017). Biogeography-based optimisation for data classification problems. International Journal of Data Mining, Modelling and Management 9: 142–162.

[15] Dash, M. and Liu, H. (1997). Feature selection for classification. Intelligent Data Analysis 1: 131–156.

[16] Alshareef, A., Ahmida, S., Bakar, A.A., Hamdan, A.R. and Alweshah, M. (2015). Mining survey data on university students to determine trends in the selection of majors. pp. 586–590. *In*: Science and Information Conference (SAI).

[17] Langley, P. (1994). Selection of relevant features in machine learning. pp. 245–271. *In*: Proceedings of the AAAI Fall Symposium on Relevance.

[18] Liu, H. and Setiono, R. (1997). Feature selection and classification—A probabilistic wrapper approach. pp. 419–424. *In*: Proceedings of 9th International Conference on Industrial and Engineering Applications of AI and ES.

[19] Hoque, N., Bhattacharyya, D. and Kalita, J.K. (2014). MIFS-ND: A mutual information-based feature selection method. Expert Systems with Applications 41: 6371–6385.

[20] Cateni, S., Colla, V. and Vannucci, M. (2017), A fuzzy system for combining filter features selection methods. International Journal of Fuzzy Systems 19: 1168–1180.

[21] Guyon, I. and Elisseeff, A. (2003). An introduction to variable and feature selection. Journal of Machine Learning Research 3: 1157–1182.

[22] Bermejo, P., Gámcz, J.A. and Puerta, J.M. (2011). A GRASP algorithm for fast hybrid (filter-wrapper) feature subset selection in high-dimensional datasets. Pattern Recognition Letters 32: 701–711.

[23] Visalakshi, S. and Radha, V. (2017). A hybrid filter and wrapper feature-selection approach for detecting contamination in drinking water management system. Journal of Engineering Science and Technology 12: 1819–1832.

[24] Chandrashekar, G. and Sahin, F. (2014). A survey on feature selection methods. Computers & Electrical Engineering 40: 16–28.

[25] Naqvi, S. (2014). A Hybrid Filter-wrapper Approach for Feature Selection.

[26] Crainic, T.G. and Toulouse, M. (2003). Parallel strategies for meta-heuristics. pp. 475–513. *In*: Handbook of Metaheuristics, Springer.

[27] Alweshah, M. (2018). Construction biogeography-based optimization algorithm for solving classification problems. Neural Computing and Applications 29: 1–10.

[28] Yang, X.-S. (2010). Nature-inspired Metaheuristic Algorithms, Luniver Press.

[29] Boussaïd, I., Lepagnot, J. and Siarry, P. (2013). A survey on optimization metaheuristics. Information Sciences 237: 82–117.

[30] Talbi, E.-G. (2009). Metaheuristics: From Design to Implementation, Vol. 74, John Wiley & Sons.

[31] Alweshah, M. and Abdullah, S. (2015). Hybridizing firefly algorithms with a probabilistic neural network for solving classification problems. Applied Soft Computing 35: 513–524.

[32] Metropolis, N., Rosenbluth, A.W., Rosenbluth, M.N., Teller, A.H. and Teller, E. (1953). Equation of state calculations by fast computing machines. The Journal of Chemical Physics 21: 1087–1092.

[33] Kirkpatrick, S., Gelatt, C.D. and Vecchi, M.P. (1983). Optimization by simulated annealing. Science 220: 671–680.

[34] Barbu, A., She, Y., Ding, L. and Gramajo, G. (2017). Feature selection with annealing for computer vision and big data learning. IEEE Transactions on Pattern Analysis and Machine Intelligence 39: 272–286.

[35] Glover, F., Laguna, M. and Marti, R. (2007). Principles of Tabu search. Approximation Algorithms and Metaheuristics 23: 1–12.

[36] Wang, S., Kong, W., Zeng, W. and Hong, X. (2016). Hybrid binary imperialist competition algorithm and tabu search approach for feature selection using gene expression data. BioMed Research International, Vol. 2016.

[37] Shi, L., Wan, Y., Gao, X. and Wang, M. (2018). Feature selection for object-based classification of high-resolution remote sensing images based on the combination of a genetic algorithm and tabu search. Computational Intelligence and Neuroscience, Vol. 2018.

[38] Giagkiozis, I., Purshouse, R.C. and Fleming, P.J. (2015). An overview of population-based algorithms for multi-objective optimisation. International Journal of Systems Science 46: 1572–1599.

[39] Mohan, B.C. and Baskaran, R. (2012). A survey: Ant colony optimization based recent research and implementation on several engineering domain. Expert Systems with Applications 39: 4618–4627.

[40] Sabeena, S. and Sarojini, B. (2015). Optimal feature subset selection using ant colony optimization. Indian Journal of Science and Technology, Vol. 8.

[41] Sharma, S. (2015). Feature-selection technique using ant colony optimization on keystroke dynamics. International Journal of Scientific & Engineering Research 6: 668.

[42] Wan, Y., Wang, M., Ye, Z. and Lai, X. (2016). A feature-selection method based on modified binary coded ant colony optimization algorithm. Applied Soft Computing 49: 248–258.

[43] Aghdam, M.H. and Kabiri, P. (2016). Feature selection for intrusion detection system using ant colony optimization. IJ Network Security 18: 420–432.

[44] Mehmod, T. and Rais, H.B.M. (2016). Ant colony optimization and feature selection for intrusion detection. pp. 305–312. *In*: Advances in Machine Learning and Signal Processing, Springer.

[45] Neagoe, V.-E. and Neghina, E.-C. (2016). Feature selection with ant Colony optimization and its applications for pattern recognition in space imagery. pp. 101–104. *In*: Communications (COMM), 2016 International Conference on, 2016.

[46] Rashno, A., Nazari, B., Sadri, S. and Saraee, M. (2017). Effective pixel classification of Mars images based on ant colony optimization feature selection and extreme learning machine. Neurocomputing 226: 66–79.

[47] Zawbaa, H.M., Emary, E. and Grosan, C. (2016). Feature selection via chaotic ant lion optimization. PloS One 11: e0150652.

[48] Emary, E., Zawbaa, H.M. and Hassanien, A.E. (2016). Binary ant lion approaches for feature selection. Neurocomputing 213: 54–65.

[49] Mafarja, M., Eleyan, D., Abdullah, S. and Mirjalili, S. (2017). S-shaped vs. V-shaped transfer functions for ant lion optimization algorithm in feature selection problem. p. 14. *In*: Proceedings of the International Conference on Future Networks and Distributed Systems.

[50] Karaboga, D. (2005). An idea based on honey bee swarm for numerical optimization. Technical report-tr06, Erciyes University, Engineering Faculty, Computer Engineering Department.

[51] Ragothaman, B. and Sarojini, B. (2016). A multi-objective non-dominated sorted artificial bee colony feature selection algorithm for medical datasets. Indian Journal of Science and Technology, 9.

[52] Samsani, S. and Suma, G.J. (2016). A Binary Approach of Artificial Bee Colony Optimization Technique for Feature Subset Selection.

[53] Ghanem, W.A.H. and Jantan, A. (2016). Novel multi-objective artificial bee Colony optimization for wrapper based feature selection in intrusion detection. International Journal of Advance Soft Computing Applications, 8.

[54] Hancer, E., Xue, B., Zhang, M., Karaboga, D. and Akay, B. (2015). A multi-objective artificial bee colony approach to feature selection using fuzzy mutual information. pp. 2420–2427. *In*: Evolutionary Computation (CEC), 2015 IEEE Congress on.

[55] Yang, X.-S. (2010). A new metaheuristic bat-inspired algorithm. pp. 65–74. *In*: Nature-inspired Cooperative Strategies for Optimization (NICSO, 2010), (Ed). Springer.

[56] Rozlini, M., Munirah, M., Nawi, N. and Wahid, N. (2015). Bat with Dempster Theory for Feature Selection: A Framework.

[57] Bansal, B. and Sahoo, A. (2015). Bat algorithm for full model selection in classification: a multi-objective approach. International Journal of Innovations & Advancement in Computer Science.

[58] Kumar, D. (2017). Feature selection for face recognition using DCT-PCA and Bat algorithm. International Journal of Information Technology 9: 411–423.

[59] Jeyasingh, S. and Veluchamy, M. (2017). Modified bat algorithm for feature selection with the wisconsin diagnosis breast cancer (wdbc) dataset. Asian-Pacific Journal of Cancer Prevention: APJCP 18: 1257.

[60] Karaboga, D. and Akay, B. (2009). A survey: Algorithms simulating bee swarm intelligence. Artificial Intelligence Review 31: 61.

[61] Pham, D.T., Ghanbarzadeh, A., Koç, E., Otri, S., Rahim, S. and Zaidi, M. (2006). The bees algorithm—a novel tool for complex optimization problems. pp. 454–459. *In*: Intelligent Production Machines and Systems, (Ed.) Elsevier.

[62] Eesa, A.S., Orman, Z. and Brifcani, A.M.A. (2015). A new feature selection model based on ID3 and bees algorithm for intrusion detection system. Turkish Journal of Electrical Engineering & Computer Sciences 23: 615–622.

[63] Boukra, W.G.A. (2017). GAB-BBO: Adaptive Biogeography Based Feature Selection Approach for Intrusion Detection.

[64] Chu, S.-C., Tsai, P.-W. and Pan, J.-S. (2006). Cat Swarm Optimization.

[65] Lin, K.-C., Zhang, K.-Y., Huang, Y.-H., Hung, J.C. and Yen, N. (2016). Feature selection based on an improved cat swarm optimization algorithm for big data classification. The Journal of Supercomputing 72: 3210–3221.

[66] Yang, X.-S. and Deb, S. (2009). Cuckoo search via Lévy flights. pp. 210–214. *In*: Nature & Biologically Inspired Computing, NaBIC, 2009, World Congress on.

[67] Sujana, T.S., Rao, N.M.S. and Reddy, R.S. (2017). An efficient feature selection using parallel cuckoo search and naïve Bayes classifier. pp. 167–172. *In*: Networks & Advances in Computational Technologies (NetACT), 2017 International Conference on.

[68] Askarzadeh, A. (2016). A novel metaheuristic method for solving constrained engineering optimization problems: Crow search algorithm. Computers & Structures 169: 1–12.

[69] Sayed, G.I., Hassanien, A.E. and Azar, A.T. (2017). Feature selection via a novel chaotic crow search algorithm. Neural Computing and Applications, 1–18.

[70] Storn, R. and Price, K. (1997). Differential evolution—A simple and efficient heuristic for global optimization over continuous spaces. Journal of Global Optimization 11: 341–359.

[71] Noghabi, H.S., Mashhadi, H.R. and Shojaei, K. (2015). Differential evolution with generalized mutation operator for parameters optimization in gene selection for cancer classification. arXiv preprint arXiv: 1510.02516.

[72] Wang, J., Xue, B., Gao, X. and Zhang, M. (2016). A differential evolution approach to feature selection and instance selection. pp. 588–602. *In*: Pacific Rim International Conference on Artificial Intelligence.

[73] Shahbeig, S., Sadjad, K. and Sadeghi, M. (2016). Feature Selection from iron direct reduction data based on binary differential evolution optimization. Bulletin de la Société Royale des Sciences de Liège 85: 114–122.

[74] Mafarja, M., Jaber, I., Eleyan, D., Hammouri, A. and Mirjalili, S. (2017). Binary Dragonfly Algorithm for Feature Selection.

[75] Mirjalili, S. (2016). Dragonfly algorithm: A new meta-heuristic optimization technique for solving single-objective, discrete and multi-objective problems. Neural Computing and Applications 27: 1053–1073.

[76] Zhang, Y., Song, X.-F. and Gong, D.-W. (2017). A return-cost-based binary firefly algorithm for feature selection. Information Sciences 418: 561–574.

[77] Alweshah, M., Rashaideh, H., Hammouri, A.I., Tayyeb, H. and Ababneh, M. (2017). Solving time series classification problems using support vector machine and neural network. International Journal of Data Analysis Techniques and Strategies 9: 237–247.

[78] Alweshah, M. (2014). Firefly algorithm with artificial neural network for time series problems. Research Journal of Applied Sciences, Engineering and Technology 7: 3978–3982.

[79] Li, X.-l. (2002). An optimizing method based on autonomous animats: fish-swarm algorithm. Systems Engineering-Theory & Practice 22: 32–38.

[80] Manikandan, R. and Kalpana, A. (2017). Feature selection using fish swarm optimization in big data. Cluster Computing, 1–13.

[81] Saremi, S., Mirjalili, S. and Lewis, A. (2017). Grasshopper optimisation algorithm: Theory and application. Advances in Engineering Software 105: 30–47.

[82] Mafarja, M., Aljarah, I., Heidari, A.A., Hammouri, A.I., Faris, H., Ala'M, A.-Z. and Mirjalili, S. (2017). Evolutionary population dynamics and grasshopper optimization approaches for feature selection problems. Knowledge-Based Systems.

[83] Rashedi, E., Nezamabadi-pour, H. and Saryazdi, S. (2009). GSA: A gravitational search algorithm. Information Sciences 179: 2232–2248.

[84] Nagpal, S., Arora, S. and Dey, S. (2017). Feature selection using gravitational search algorithm for biomedical data. Procedia Computer Science 115: 258–265.

[85] Bostani, H. and Sheikhan, M. (2017). Hybrid of binary gravitational search algorithm and mutual information for feature selection in intrusion detection systems. Soft Computing 21: 2307–2324.

[86] Holland, J.H. (1992). Adaptation in Natural and Artificial Systems: An Introductory Analysis with Applications to Biology, Control and Artificial intelligence, MIT Press.

[87] Bäck, T. and Schwefel, H.-P. (1993). An overview of evolutionary algorithms for parameter optimization. Evolutionary Computation 1: 1–23.

[88] Cerrada, M., Sánchez, R.V., Cabrera, D., Zurita, G. and Li, C. (2015). Multi-stage feature selection by using genetic algorithms for fault diagnosis in gearboxes based on vibration signal. Sensors 15: 23903–23926.

[89] Hong, S.-S., Lee, W. and Han, M.-M. (2015). The feature selection method based on genetic algorithm for efficient of text clustering and text classification. Int. J. Advance Soft Compu. Appl. 7: 2074–8523.

[90] Soufan, O., Kleftogiannis, D., Kalnis, P. and Bajic, V.B. (2015). DWFS: A wrapper feature selection tool based on a parallel genetic algorithm. PloS One 10: e0117988.

[91] Chen, Z., Lin, T., Tang, N. and Xia, X. (2016). A parallel genetic algorithm based feature selection and parameter optimization for support vector machine. Scientific Programming, 2016.

[92] Aalaei, S., Shahraki, H., Rowhanimanesh, A. and Eslami, S. (2016). Feature selection using genetic algorithm for breast cancer diagnosis: Experiment on three different datasets. Iranian Journal of Basic Medical Sciences 19: 476.

[93] Mirjalili, S. and Lewis, A. (2014). Grey wolf optimizer. Advances in Engineering Software 69: 46–61.

[94] Emary, E., Zawbaa, H.M. and Hassanien, A.E. (2016). Binary grey wolf optimization approaches for feature selection. Neurocomputing 172: 371–381.

[95] Geem, Z.W., Kim, J.H. and Loganathan, G.V. (2001). A new heuristic optimization algorithm: Harmony Search Simulation 76: 60–68.

[96] Bagyamathi, M. and Inbarani, H.H. (2015). Feature selection using relative redact hybridized with improved harmony search for protein sequence classification. International Journal of Trend in Research and Development (IJTRD), Vol. ISSN: 2394–9333, www.ijtrd.com.

[97] Mirjalili, S. (2015). Moth-flame optimization algorithm: A novel nature-inspired heuristic paradigm. Knowledge-Based Systems 89: 228–249.

[98] Zawbaa, H.M., Emary, E., Parv, B. and Sharawi, M. (2016). Feature selection approach based on moth-flame optimization algorithm. pp. 4612–4617. *In*: Evolutionary Computation (CEC), 2016 IEEE Congress on.

[99] Eberhart, R. and Kennedy, J. (1995). A new optimizer using particle swarm theory. pp. 39–43. *In*: Micro Machine and Human Science, 1995, MHS'95. Proceedings of the Sixth International Symposium on.

[100] Wu, S. (2015). Comparative Analysis of Partcle Swarm Optimization Algorithms for Text Feature Selection.

[101] Butler-Yeoman, T., Xue, B. and Zhang, M. (2015). Particle swarm optimization for feature selection: a size-controlled approach. pp. 151–159. *In*: Proceedings of the 13th Australasian Data Mining Conference (AusDM 2015).

[102] Aghdam, M.H. and Heidari, S. (2015). Feature selection using particle swarm optimization in text categorization. Journal of Artificial Intelligence and Soft Computing Research 5: 231–238.

[103] Javidi, M.M. and Emami, N. (2016). A hybrid search method of wrapper feature selection by chaos particle swarm optimization and local search. Turkish Journal of Electrical Engineering & Computer Sciences 24: 3852–3861.

[104] Mirjalili, S. and Lewis, A. (2016). The whale optimization algorithm. Advances in Engineering Software 95: 51–67.

[105] Sharawi, M., Zawbaa, H.M. and Emary, E. (2017). Feature selection approach based on whale optimization algorithm. pp. 163–168. *In*: Advanced Computational Intelligence (ICACI), 2017 Ninth International Conference on, 2017.

[106] Mafarja, M.M. and Mirjalili, S. (2017). Hybrid whale optimization algorithm with simulated annealing for feature selection. Neurocomputing.

[107] Mafarja, M. and Mirjalili, S. (2018). Whale optimization approaches for wrapper feature selection. Applied Soft Computing 62: 441–453.

[108] Thuraisingham, B., Khan, L., Masud, M.M. and Hamlen, K.W. (2008). Data mining for security applications. pp. 585–589. *In*: Embedded and Ubiquitous Computing, 2008. EUC'08. IEEE/IFIP International Conference on.

[109] Kshirsagar, V.K., Tidke, S.M. and Vishnu, S. (2012). Intrusion detection system using genetic algorithm and data mining: An overview. International Journal of Computer Science and Informatics ISSN (PRINT) 2231: 5292.

[110] Harshna, N.K. (2014). Fuzzy data mining based intrusion detection system using genetic algorithm. International Journal of Advanced Research in Computer and Communication Engineering 3: 5021–5028.

[111] Al-Qurishi, M., Rahman, S.M.M., Hossain, M.S., Almogren, A., Alrubaian, M., Alamri, A., Al-Rakhami, M. and Gupta, B. (2017). An efficient key agreement protocol for Sybil-precaution in online social networks. Future Generation Computer Systems.

[112] Yu, C., Li, J., Li, X., Ren, X. and Gupta, B. (2018). Four-image encryption scheme based on quaternion Fresnel transform, chaos and computer generated hologram. Multimedia Tools and Applications 77: 4585–4608.

[113] Wu, L., Chen, B., Choo, K.-K.R. and He, D. (2018). Efficient and secure searchable encryption protocol for cloud-based Internet of things. Journal of Parallel and Distributed Computing 111: 152–161.

[114] Gorbach, I., Krishnan, V., Shur, A., Denisov, D., Kuhtz, L., Adabala, S., D'souza, R.P., Entin, M., Clark, M.R. and Saubhasik, G.A. (2018). Trust services for securing data in the cloud. (Ed.). Google Patents.

[115] Tewari, A. and Gupta, B. (2017). A lightweight mutual authentication protocol based on elliptic curve cryptography for IoT devices. International Journal of Advanced Intelligence Paradigms 9: 111–121.

[116] Alsmirat, M.A., Jararweh, Y., Obaidat, I. and Gupta, B.B. (2017). Internet of surveillance: a cloud supported large-scale wireless surveillance system. The Journal of Supercomputing 73: 973–992.

[117] Memos, V.A., Psannis, K.E., Ishibashi, Y., Kim, B.-G. and Gupta, B. (2017). An efficient algorithm for media-based surveillance system (EAMSuS) in IoT smart city framework. Future Generation Computer Systems.

[118] Zkik, K., Orhanou, G. and El Hajji, S. (2017). Secure mobile multi cloud architecture for authentication and data storage. International Journal of Cloud Applications and Computing (IJCAC) 7: 62–76.

[119] Ibtihal, M. and Hassan, N. (2017). Homomorphic encryption as a service for outsourced images in mobile cloud computing environment. International Journal of Cloud Applications and Computing (IJCAC) 7: 27–40.

CHAPTER 9

A Taxonomy of Bitcoin Security
Issues and Defense Mechanisms

Prachi Gulihar and *B.B. Gupta**

1. Introduction

There are many e-payment systems existing in today's world. A financial infrastructure, which works on the internet technology, is the need of the hour. Currently, the most dominant payment system is the credit card system which involves a bank as the trusting authority. Other systems in practice are e-wallets, like Paypal, Paytm which follows an intermediary architecture in which instead of bank, the company stands responsible for fair transactions [1, 2]. The successor to the intermediary architecture was SET architecture, in which the customer did not have to register himself with the company and credit card details also were not necessary to be shared. Cybercoin was based on this architecture, but this architecture did not stay for long. The problem was there was no way to certify which Cybercoin belonged to which real-life user. This problem was solved by bitcoin, which removed the real-life identities from the whole transaction process [3]. In the discussed e-payment systems, the characteristic of anonymity was missing. By making use of clever engineering, bitcoin was able to achieve this. Bitcoin is a famous cryptocurrency which was introduced by Satoshi Nakamoto in 2008 [4]. His name is said to have been derived from subparts of names of four firms, namely Samsung, Toshiba, Nakamichi and Motorola. The first bitcoin was realeased on January 3, 2009. It was a code of 32 thousand lines released online as all bit, no coin. Bitcoin operates in a peer-to-peer (P2P) network. Smallest unit of a bitcoin is called Satoshi, which equals 10^{-8} bitcoins. It is a decentralized cryptocurrency, which works by maintaining a public ledger, like in ethereum [5] called blockchain. Blockchain is a transaction ledger which is maintained in a distributed manner by the anonymous participating entities, called 'miners'. These

National Institute of Technology, Kurukshetra, India; prachigulihar2@gmail.com
* Corresponding author: gupta.brij@gmail.com

miners are responsible for maintaining and extending the blockchain by solving a cryptographic puzzle as a Proof-of-Work (PoW) [6, 7]. The senders broadcast their transaction in the bitcoin network after which it is the responsibility of the miners to include them in the blockchain. They are rewarded by the senders for doing so. This reward comes as a transaction fee. It is an instantaneous method of transferring payments with minimal transaction fees.

2. Overview of Bitcoin Protocol and its Working

The bitcoin system is a distributed system [8] which comprises; five main components as shown in the figure. The first component are the users. Any bitcoin user is strongly related to another component of the bitcoin environment, known as bitcoin wallet. Identity of any bitcoin wallet is its address. The user owning the bitcoin wallet has a single private key which remains secret from other users. The other key is a dynamic public key which is changed every time the user initiates a new transaction. This enhances the security of the transaction. The fixed private key is used by the user to sign his transactions digitally by elliptic curve cryptographic [9] techniques. Bitcoin wallets [10] are mainly of two types—hot and cold. The hot wallets remain connected to the internet continuously, whereas the cold wallets need to be connected to the; Internet only when the transaction is to be performed using the stored bitcoins. Hot wallets are more vulnerable to cyber attacks due to prolonged connectivity; therefore, it is advisable to keep only a small portion of the bitcoins in the hot wallet and keep the rest in the cold wallet. As an analogy, cold wallets are comparable to the safe in our homes. Things of daily use and money are kept in the open, whereas valuables are stored in a separate hidden place. The hidden place in this case becomes the hard disk and other storage devices. Users operate the wallets by installing the bitcoin client on their personal computer or smartphones [11].

2.1 Miners

The next component is the miners, whose function is to club the transactions waiting for confirmation into a single block and then generate the hash of the block. This process is repeatedly done by every miner until the hash value of the block becomes less than the target value. They do so by varying the set of transactions being included in the block. It is the miners who are responsible for verification of authenticity, integrity and correctness of the bitcoin transactions. Miners may work in collaboration with co-miners to form a mining pool [12]. Whenever any miner or mining pool is successful in generating the proof-of-work, then that miner broadcasts the verified block in the bitcoin network. To gain rewards, this block must be verified by the majority of participating miners after which this block is added to the public blockchain. The bitcoins which are rewarded as transaction fees are then distributed among the co-miners in the mining pool in proportion to how much work effort was put in by each individual miner. Some mining pools distribute the rewards equally among the co-miners, but this policy is unfair [13]. The miner who manages the mining pool is rewarded with a higher share for he has the extra work of management of mining pool as well.

Bitcoin exchange is the virtual space in which bitcoins are sold and bought against other functional currencies and developers are a set of programmers, who work to

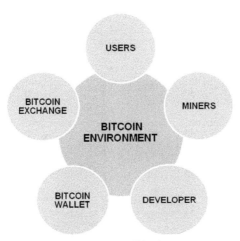

Fig. 1: Components of bitcoin system.

improve the bitcoin cryptocurrency and add new features to it. Their work can be found on Github repository. They mainly focus on developing new services and software which make use of bitcoins.

2.2 Blockchain

The backbone of bitcoin is blockchain, whose basic component is a block, which is a public log of all the transactions taking place in the bitcoin network. This supports the distributed type of computation because it is P2P. These blocks are chained in a form of Merkle tree [14] comprising full transaction nodes. Merkle tree is a tree-like data structure which stores a summary of the transactions in the blocks. Transaction is a script which identifies the owner of the bitcoin. Script is a sequence of instructions which the miners execute. To ensure that the transactions once written in the public log are unchangeable, the bitcoin miners solve a cryptographic puzzle, which is called proof-of-work. The transfer of coin ownership from one person to another is denoted by transactions.

2.3 Transaction Script

A transaction comprises of two components—inputs and outputs. Inputs denote the unspent coins and output denotes the address of the receiver. Inputs are defined by scriptSig which is a signature ensuring that the bitcoins being transferred are unspent and belong to the sender. Outputs are defined by scriptPubKey which specifies the condition for redeeming the bitcoin being transferred. The bitcoin environment comprises of five standard scripts—Public-Key, Pay-to-Public-Key Hash (P2PKH), Multi-Signature, Pay-to-Script Hash [15] and Data Output. Public-Key script defines the key using which the transaction will be signed. P2PKH script is used to find that public key which is hashed to a given value. In Multi-Signature script more than one signature is required for a public key to be able to redeem a transaction.

Pay-to-Script Hash [16] is used to decrease the length of the Multi-Signature script by hashing the non-standard to transaction scripts to standard ones. Data Output

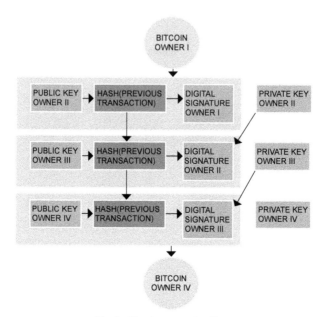

Fig. 2: Bitcoin transaction flow.

script is used to store the transaction messages in a human readable form. It is the task of miners to ensure that the transaction is valid by verifying the signatures and using standard transaction scripts. For new blocks which become a part of the blockchain the miners need to find a random value called nonce satisfying, the condition that it is less than the block header. This block header value is known as target value. This process is called Proof-of-Work (PoW) that depends on the miner's computational power [17]. For validating a transaction, the miners get transaction fees which motivate them to work on extending the blockchain by adding new blocks. The public key of the next owner is the address to which the bitcoins are to be sent. The collection of public addresses of all the e-wallets is saved as a file in distributed database. For a miner to solve one cryptographic hash puzzle, the machine calculates around 650 trillion hashes to get the acceptable one. So, even if the PC is of i7 processing speed, it would still take around 21 years to mine one bitcoin. This brings in the need of formation of mining pools by sharing the computational resources among partner miners. BTC Guide is the largest mining group which has the capacity to mine 25 bitcoins in 20 minutes.

2.4 Byzantine Agreement

Bitcoin relies on the concept of Byzantine agreement [18] to ensure distributed and mutual extension, exchange and acceptance of a final blockchain at any point of time. The Byzantine agreement has to ensure two properties—agreement and validity. Agreement means that all the participating miners return the same blockchain at the end of the validating a transaction. Validity means the blockchain extended by an honest miner is valid and in further used by other participating miners to extend the current blockchain. In the blockchain, the blocks are connected in a chain-like linear structure. The right-most block in the blockchain is the block recently mined and is called 'head'. The left-most block in the blockchain is the first ever block mined and called the 'genesis block'. The length of a blockchain is the number of blocks it has after the genesis block. The bitcoin environment sets a parameter, which determines the difficulty level of mining the next block. It is set by increasing the level of the cryptographic hash puzzle to a level where only one block is generated every 10 minutes, not more or less than that. This difficulty level is recalibrated after every 2,016 blocks.

2.5 Bitcoin Protocol

The bitcoin protocol [19] is the protocol which is implemented by the bitcoin miners. Its two main properties are—common prefix property and chain quality property. The common prefix property means that the blockchain of any two honest miners is different only in the most recent blocks. The chain-quality property ensures that the honest miner mines on a blockchain whose sufficiently long part is absent of any adversarial block. Ideal chain quality is never attained by the bitcoin protocol. Further, it can be divided in three sub protocols-chain validation protocol, chain comparison protocol and PoW protocol.

The chain validation protocol validates the structural properties of the blockchain. For every block in the blockchain, this protocol checks three things. Firstly, whether the PoW is solved correctly; secondly, the timestamp timer has not exceeded; thirdly, the hash of the previous block is present or not. If all the blocks are able to pass the verification test, then the blockchain is valid; otherwise it is rejected. The chain comparison protocol is responsible for finding out the most suitable blockchain from a set of blockchains. The competing chains may be picked up randomly or lexographically and in the output, the blockchain having the longest length becomes the winner. The PoW protocol attempts to extend the winner blockchain by solving the cryptographic hash puzzle by doing brute-force for a predefined number of times. For every new proposed block, this protocol generates the block hash which must be less than the set target value. If the puzzle is solved, then the blockchain is extended by one block; otherwise it remains unaltered. This way the bitcoin protocol is iterated indefinitely.

3. Taxonomy of Attacks against Bitcoin Protocol

Bitcoin environment is uncontrollable and decentralized which makes it more prone to cyber attacks as it becomes easier to generate fraud transactions. Various vulnerabilities are found in bitcoin protocol and bitcoin network. Figure 3 lists the different possible attacks based on various threats.

3.1 Distributed Denial-of-Service Attack

The first category of cyber attack the bitcoin system is prone to the networking infrastructure attacks. They are done by exploiting the vulnerabilities in the communication protocols of the bitcoin peer-to-peer network. First under this classification comes the Distributed Denial of Service (DDoS) attack [20]. In this attack the network resources are exhausted to block access of services to the genuine

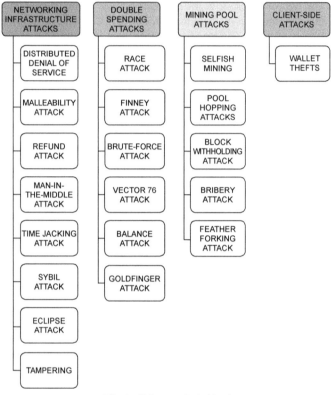

Fig. 3: Cyber attacks in bitcoin system.

users. This attack is performed under the instruction of a master machine, which instructs other compromised machines, also known as slaves machine to overload the victim machine's network by flooding it with malicious request packets. 142 unique kinds of DDoS attacks have been identified to have taken place in the bitcoin network. In this attack, a large number of compromised client nodes send fake transaction requests to an honest miner. After a while that honest miner is burdened and starts discarding all the requests, including those coming from other honest miners, thus causing a denial-of-service. DDoS attacks take place on large mining pools [21] and big bitcoin exchanges due to bigger rewards [22]. Individual miners and small mining pools are safe from such kind of attacks. DDoS attacker's main motive is generation of bulk ransom. DDoS attack discourages the participating miners in the mining pool, leading to their withdrawal from the pool. For example, the malicious miner shows the co-miner that he owns more computing power, enough to snatch his rewards; the honest miner finds it better to step back, thus successfully imposing DDoS in the bitcoin network.

The attacker on gaining the majority of the computing power will be able to launch double-spending attack and Distributed Denial-of-Service (DDoS) attack [43]. Three methods are there to bribe the co-miners—out-band offer, in-band offer and negative fee offer. In out-band offer the attacker offers a direct payment to the co-miners, who are ready to work on the attacker's private blockchain, which comprises of attacker's pre-mined block. In in-band offer, the attacker lures co-miners by offering bitcoins. The attacker creates a blockchain fork and any miner choosing to mine on that fork gets the rewards in the form of bitcoins which are available for free. In negative free offer, the attacker lures the co-miners by paying better reward shares than the current mining pool has to offer.

Although the gains by the co-operating miners are highly rewarding, they are short-lived because in the long run, the honest miners will lose trust in the mining pool, ultimately leading to a crash in bitcoin rates. Another mining pool attack is feather forking. In this attack, the attacker blackmails the user client to pay him some bitcoins or else the attacker shall blacklist the transaction performed by that particular client node. The attacker does this by publicly broadcasting the blacklisted transaction. The attacker creates a blockchain fork and works to extend this fork so that it may outrace the public blockchain. If the attacker fails to do so, then it discards its private blockchain and resumes working on the public blockchain. When the attacker is determined to block and blacklist a particular transaction, then he or she will perform iterative forking to gain trust of the co-miners. The attacker will then be able to block the selected transaction without any inherent cost because then the majority of miners will be entrusting his forking strategy.

3.2 Malleability Attack

Next category of attacks includes malleability attacks [23, 24]. Malleability attacks also lead to denial-of-service. In malleability attacks, the order of the transaction to be processed is disturbed by putting in fake transactions. These fake transaction poses a higher transaction validation fee thus luring honest miners to include this fake transaction with higher priority than the in-queue transactions. The attacker's purpose is to waste miner's resources of time and computing power in validation of a fake transaction which is ultimately in vain. In cryptography, the term 'malleable' is used if the attacker is able to generate an output Y similar to X without knowing the function used to generate X. In the bitcoin network, it is the transaction which is made malleable by varying syntax configuration of two semantically identical transactions. This can be done without knowing the user's private key because there is a bug in the bitcoin protocol which identifies every transaction by its transaction-id. But this transaction id may not be unique which means that transaction of transfer of x bitcoins from Alice to Bob may have two transaction ids – T and T'. This happens because the hash of these transactions is different. Though the amount of bitcoins being transferred is similar, it is a different set of bitcoins which are transferred every time. This loop-hole of the bitcoin protocol is exploited in transaction malleability attack.

Mt. Gox bitcoin exchange was using the custom implementation of the bitcoin protocol which had this bug [25, 26]. The exchange had to block accounts and the transactions were barred due to this attack. The attacker was able to withdraw his or her coins twice from the bitcoin exchange. This was possible because the custom implementation of the bitcoin protocol, which was used by Mt. Gox bitcoin exchange, searched only for matching transaction-id, unlike the recent reference implementations of bitcoin protocol which searches for any transaction which is semantically equivalent. In Mt. Gox [27] case, the attacker had a bitcoin trading account in Mt. Gox. For example, Alice first deposited X bitcoins in her exchange account. She then sent a transaction Tx to Mt. Gox, asking to transfer her coins back to her. In response,

Mt. Gox issued transaction Ty to transfer Alice's coins back to her. Next, Alice generated T' which is semantically same as Tx but has different transaction-id. This transaction T' gets included in the public blockchain and the transaction Tx is rolled back, so Alice now informs Mt. Gox of the unsuccessful deposit transaction Tx. Then, Mt. Gox check its transaction and does not find the transaction Tx which makes him believe that Alice is saying right about the unsuccessful transaction and thus returns X bitcoins back to her. These were the X bitcoins which were never submitted.

3.3 Refund Attacks

Whenever any bitcoin transaction fails, then the refund has to be initiated by the bitcoin exchanges. This is where refund attacks [28] come into place. Border Internet Protocol-70 (BIP-70) is widely accepted by the community of bitcoin developers for performing successful payments, using bitcoins. Almost all major wallet and exchange services use this protocol. There are payment services which provide the exchange and wallet service providers with a platform and infrastructure to operate in a secure environment. The refund attack takes place by exploiting the authentication vulnerability present in BIP-70. In this attack, the user wallet is under the control of a malicious bitcoin node. Whenever the customer begins trading his bitcoins, his wallet address is sent to the malicious node. On completion of trading, the malicious user requests a refund from the node with which the customer traded. In the refund address of the refund request, the malicious user puts in his own wallet address and the bitcoins get refunded to him. The customer is unaware of all these transactions taking place in his name. This happens because the merchant using BIP-70 protocol lacks secure authentication of who is sending the refund requests.

3.4 Man-in-the-Middle Attack

Network infrastructure attacks in bitcoin system also comprise of Man-in-the-Middle (MITM) attack [29]. In this, the attacker traps the victim by attracting him to a Website which poses as a trustable merchant Website helping in making secure bitcoin payments, but, clicking on Webpage reveals the victim's identity which includes the victim's wallet address. Whenever the any customer engages in making any kind of online payment, then he is directed to a legitimate payment gateway. Let us say customer C just bought something from the victim. Then the victim's payment gateway page will be opened in customer's browser. This page is under the attackers control and the details of customer C will be known to the attacker without the victim noticing any of these practices. After customer C pays to the victim, then the attacker sends a refund request in the name of customer C and the refund amount will be sent to the attacker. The customer remains unaware about the refund and the victim loses his coins. Thus, making MITM attack successful.

A revision in BIP-70 protocol is required to prevent this. All the nodes in the bitcoin network have an internal clock maintaining the time count. This is of two kinds —median time and system time. Median time is the time sent to the newly connected node by the adjacent peer nodes at the time of joining the bitcoin network. It is sent in the version message shared by the adjacent nodes. The system time is nodes' own timer clock which is allowed a maximum deviation of 70 minutes from the median time. It is reset every time the maximum deviation limit exceeds.

3.5 Time-jacking Attack

In the time-jacking attack [30, 31], the attacker brings inaccuracy in the timestamps by fake median time to the peer nodes. The attacker may also plant fake nodes for this purpose. This will either lead to abrupt increase or decrease in the timers of the peer nodes. A classic kind of attack involves making the timer of the victim node the slowest by increasing the timers of co-peers. The maximum duration by which the victim can be made to lag is 140 minutes. Acceptance of any blockchain depends on the network time; so by altering the timestamps, the attacker confuses the peer nodes. The victim nodes then start mining on an older blockchain fork which is already discarded by the rest of the network, thus wasting their time and computation power. This leads to a lag in confirmation rates of bitcoin transactions and the nodes may have to wait for more than 6 minutes to entrust that the transaction just validated is not going to roll back.

3.6 Sybil Attack

In Sybil attack [32, 34], the malicious node compromises a part of the network by forming a group of dummy nodes. Their motive is to partition the victim node from the bitcoin network and isolate it so that all the transactions of the victim node are blocked from entering the public blockchain. This is achieved by collaboration in timing attack which leads to higher time in performing the encryption process, thus delaying confirmation of the victim's transactions. The input of the victim's transaction stands waiting in the network for validation which increases the vulnerability of these inputs being used for double spending attack [35].

3.7 Eclipse Attack

In eclipse attack [33], the attacker possesses multiple IP addresses to perform spoofing. The victim is selected by the attacker, after which, all the IP address the victim node tries to contact, are diverted to the attacker's IP address, thus blocking the nodes which the victim wishes to connect to. Eclipse attack is further classified into two kinds—infrastructure attacks and botnet attacks. Infrastructure attack takes place on Internet Service Provider (ISP), which is forced to manipulate the addresses of multiple bitcoin client nodes. In botnet attack, the attacker manipulates the address of a specific range, like private IP addresses affect the peers connected in that private network. This attack is performed by an army of botnets, which work on the instructions of the attacker. After a new block is mined, the information about this newly mined block is broadcasted in the bitcoin network. This information is shared at set intervals of time. The attacker in the tampering attack [36] delays the propagation of this broadcasted information by congesting the network route and by overloading the client node by sending multiple requests.

3.8 Double Spending Attacks

Double spending attack [14] means spending of the same bitcoin for two different transactions. In the bitcoin network, it is the function of the miners at work to verify and process the transactions in the network. They must ensure that only unspent coins are referred as inputs in any transaction. This is achieved by distributed time-stamping and distributed Byzantine consensus protocol. For example, Alice creates a transaction

Tx directed towards Bob at time *t1* using her bitcoins B$_{alice}$. Alice broadcasts this transaction in the bitcoin network. Then at time *t2* which is almost parallel to *t1*, Alice creates another transaction *Ty* using the bitcoins B$_{alice}$ again. This time is directed towards a wallet address, which is already owned by her. Alice will be successful in using the same bitcoins twice if Bob accepts transaction *Tx* and provides the goods or services because Bob is at the risk of being unable to redeem those bitcoins. In the bitcoin network, it is the dynamic responsibility of miners to ensure that only unspent bitcoins are used in any transaction. When any miner *X* will receive two transactions having same bitcoins in the input script, then miner *X* must verify only one of them, hence rejecting the other.

Even after strict ordering in bitcoin transactions and proof-of-work, the bitcoin environment remains vulnerable to double-spending attack. Double-spending attack is successful only when the following four conditions are fulfilled consecutively. Firstly, when a part of the miners in the bitcoin network approves of Alice's transaction *Tx* and Bob receives the approval confirmation after which Bob releases his products and services, later realising the roll back of approved transaction. Secondly, simultaneously the remaining part of the miners in the bitcoin network approve of the transaction *Ty*, thus forking the existing blockchain. Thirdly, Bob receives the approval confirmation of transaction *Ty* after he has shipped the product or services to Alice. Lastly, more that 50 per cent of the bitcoin miners must start mining on the blockchain which accepted *Ty* as the valid transaction. If all these conditions are met, then Alice will be successful in using the same set of bitcoins twice for different goods and services. Double-spending attack takes place with varying levels of difficulties and complexities which can be divided into six sub-categories as shown in the Fig. 3.

3.9 Finney Attack

In finney attack [37], Alice beforehand mines the block having *Ty* transaction. Alice then creates the transaction *Tx* directed towards Bob, using the same bitcoins B$_{alice}$. Alice keeps the pre-mined block private and does not broadcast it to the bitcoin network. Alice waits until the miners accept her transaction *Tx* and declare it valid. Bob, on getting the validity confirmation of transaction *Tx* from the network miners, releases the goods and services. When Alice receives the goods and services from Bob, she then broadcasts her pre-mined block into the network, creating a fork in the blockchain which is of equal length. If the majority of bitcoin miners start working on this newly-created forked blockchain, then Alice will be successful in making the transaction *Ty* valid and *Tx* invalid because, according to bitcoin protocol, the miners will work on the longest blockchain when a fork arises.

This creates a race condition between the two blockchain forks and Alice, on winning the race, will end up getting the bitcoins back to wallet address. To avoid this race condition, Bob should have waited for multiple confirmations from the bitcoin miners before releasing his goods but that will only make the attack harder for Alice and not finish its possibilities.

3.10 Brute-force Attack

In the brute-force attack [38], the attacker is one step ahead. Here the attacker Alice controls *x* nodes in the bitcoin network which work together to pre-mine *n* blocks on the newly forked blockchain. These *n* blocks are not broadcasted in the bitcoin network. To prevent the Finney attack, if Bob waits for *n* confirmations, then Alice will be able to provide him with n confirmations on releasing these pre-mined blocks. On doing this, Alice's forked blockchain becomes the longer one and the miners make a shift to Alice's blockchain, making her successful in double-spending attack.

3.11 Vector-76 Attack

Another form of double-spending attack occurs in the case of bitcoin exchanges. It is known as vector-76 attack [39]. In this attack, Alice pre-mines the block having the transaction of bitcoin deposit in the bitcoin exchange. When the next block confirmation information is flooded in the network, then Alice releases its pre-mined block in the network along with another mined block which comprises the transaction of the withdrawal of the deposited bitcoins. This increases the length of Alice's forked blockchain by one block. Abiding by the bitcoin protocol, the network miners then consider the longest chain as the valid branch and start mining on Alice's forked blockchain. Due to this, the deposit transaction is rolled back and withdrawal transaction is validated, leading to the loss of bitcoins from the bitcoin exchange.

3.12 Balance Attack

Another kind of double-spending attack known as balance attack [40] exploits the proof of work feature of the bitcoin protocol. Its basis is that if the miner is able to generate the proof of work at a faster speed than the fellow miners, then its probability of success in double-spending increases multiplefolds. This will be possible if the miner has additional computing power which helps him to solve the proof-of-work at a faster rate. In balance attack, Alice will try to delay the network communication among the mining groups. The success of balance attack depends not only on the delay generated, but also on how much of the computing power relies with Alice.

3.13 Goldfinger Attack

The last kind of double-spending attack is the goldfinger attack, also renowned as 51 per cent attack. It is a theoretical attack which, if successfully executed, will lead to instability in bitcoin system. This may take place in worst-case scenario when more than 50 per cent of the computing power will reside with a single miner or mining pool. Below is the graph showing the relationship between the number of confirmations, computing power and double spending.

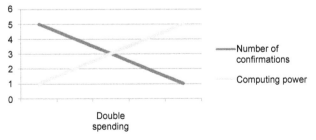

Fig. 4: Graph showing double spending dependencies.

3.14 Mining Pool Attacks

The next category of cyber attacks in bitcoin system is the mining pool attacks [62]. Bitcoin miners clubbed together their resources to increase their computing hash power, which will directly lower the verification time of the set of transactions they choose to put in their proposed block. Lately, the research area of mining strategies has gained a lot of attention. Mining pool attacks can be broadly classified into two kinds—internal and external. In the internal attacks, the miners of the pool become dishonest and try to take more of the fair share of rewards in the transaction fees. In some of the internal attacks, the dishonest miners disrupt the working of the pool by keeping them away from successfully generating the next block in the blockchain, while in the external mining pool attacks, the in-pool miners are not involved. They are performed by external miners, who take undue advantage of their higher hash power to initiate the double-spending attack. Researchers have used a game-theoretic approach for showing instances of selfish mining attacks in which the correct block is discarded by a majority of miners in the bitcoin network.

It is a fact that every participating miner in the bitcoin network is there for the purpose of gaining rewards only, but they do this in an honest and fair manner without harming the co-miners. But in selfish mining attacks, the dishonest miners purposefully hide the information of the block they have already mined successfully. Their purpose is dual-fold. First, they want an unfair share of rewards won by the mining pool. This means that the reward, instead of being distributed according to the mining work done by individual miner, gets unfairly distributed. The dishonest miner gets a share more than the computing power he spent on mining. The second purpose of these dishonest miners is to confuse the co-miners, leading to their resources being expended on the wrong blockchain.

To understand this better, let us assume Alice is the dishonest miner of the mining pool. Alice has kept the information of the pre-mined block with herself. She then keeps on mining on her private blockchain. She releases her private blockchain in the mining pool only when her private blockchain exceeds the length of the public blockchain the rest of the pool is mining on. This creates a fork in the blockchain. Then as a rule of the bitcoin protocol, all of the miners need to switch on the longest chain which is Alice's newly broadcasted blockchain. Doing this, the rewards given to the honest miners on adding block to the previous blockchain are taken back by the mining pool manager, thus leading to wastage of the time, power and money of the honest miners. The attack gets stronger if Alice is able to make her own dishonest sub-pool in the mining pool by collaborating with other dishonest co-miners which may further lead to 51 per cent attack.

3.15 Pool-hopping Attack

The next mining pool attack is the pool-hopping attack. In this attack, the attacker analyzes the mining-share entries the co-miners submit to the mining pool manager. At many instances, what happens is that a large number of co-miners submit their mining shares but the mining pool is unable to find the next block. In this case, even if the block is found at a later stage, then the reward is going to be distributed according to the mining share entries which are very large in number; so the individual share will be negligible. In such cases, the attacker dynamically switches his pool or starts mining independently.

3.16 Block-withholding Attack

Another mining pool attack is called the block-withholding attack [41, 42] in which the attacker never broadcasts the block mined by him. It can be of two types—sabotage attack and lie-in-wait attack. In sabotage attack, the purpose of the attacker is not to gain the rewards, but to make the co-miners lose the rewards. In lie-in-wait attack, the attacker's purpose is to gain the rewards by intentionally concealing the pre-mined block, as in selfish mining. The lie-in-wait attack is not economically viable in short duration but it can do enormous damages in the long run, leading to mining pools losing a lot of money in a few months. The next attack a mining pool is vulnerable to is 'bribery attack'. In this attack, the attacker bribes and lures the co-miner to share their computing resources with the attacker, who might end up gaining control of the majority of resources in the mining pool. Although the duration for which the attacker owns the majority of computing power is less, but it is a threat to the reward gains of the mining pool.

4. Security and Privacy Issues in Bitcoin Protocol

With increasing computational power and advancing cryptanalysis techniques, the cryptographic primitives are at risk of being broken [46]. These primitives weaken over time and rarely undergo abrupt breakage. Although some of the attacks discussed below are theoretical [47], but it is crucial to anticipate their impact so that rigorous contingency plans can be put in place. The flowchart shows the process which bitcoin protocol follows to verify the cryptographic primitives of the next proposed block. Due to variations in mining activities, network delays and presence of dishonest miners, the correct state of the blockchain gets disputed, leading to a fork in the blockchain. To prevent forking, checkpoints are introduced in the blockchain so when a dispute arises, the blockchain is reset to the last checkpoint. These checkpoint entries begin from the first block in the blockchain, called the genesis block. This way consensus is maintained among the miners, who then continue to mine only the longest blockchain [48].

But these kinds of temporary forks increase the potential of the double-spending attack to occur. In simple language, double spending occurs when the attacker firstly credits the bitcoins from his wallet in return of services. Then he reorganizes the transaction ledger in such a way that the initial amount which credited, is debited allowing him to use them in some other transaction. Technically, the adversary attempts to exploit this temporary split in the network. In this attack, the adversary places two different transactions with same input script in two different branches of the blockchain. This is prevented only when the consensus is reached and one of the branches is chosen as the valid branch. For this, the receiver must wait for many confirmation blocks before committing. Satoshi Nakamoto has explained that the bitcoin system has the ability to withstand the double-spending attack if the receiver waits for the sender's transaction to advance ahead in the blockchain by some N blocks. Then, the probability of the sender being able to reorganize the public blockchain drops exponentially by a factor of N.

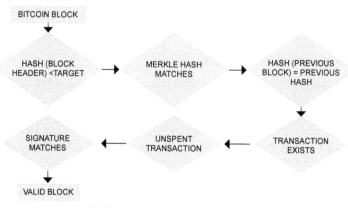

Fig. 5: Bitcoin cryptographic primitives.

Another major security issue in the bitcoin system is when more than half of the computational power resides with a single miner. This led to instability in the system and the chances of this kind of attack increased manifold after the concept of mining pools was introduced. In a mining pool, a group of miners is created to collaborate and work together to find the nonce value. The mining rewards are then distributed in proportion to the work done by each member miner. Once a transaction is verified, the accepted block details flood across the network. There is no need for all the nodes to download the full blockchain at its end; they can just download the block headers of the corresponding Merkle tree and become lightweight clients, which follow Simple Payment Verification (SPV). Over the time new updates and security enhancements are made in the bitcoin protocol by the bitcoin community, by forking the blockchain. When this fork is backward-compatible with the previous version of the protocol, then it is known as a 'soft fork'. It is less strict as not all the nodes need to upgrade to the latest version. When this fork is not backward-compatible, then it is known as 'hard fork'. In case of a hard fork, all the nodes need to upgrade to the latest version in order to participate because the new transactions are rejected by the previous version of the bitcoin protocol.

4.1 Hashing

Hashing in bitcoin is done at two places [49, 50]. Primary hash function has a 256-bit output and the secondary has a 160-bit output. Primary hash function has three functions. Firstly, it is the hash which is used by miners to generate PoW. Secondly, it is used to generate the hash of transactions in a block which is then stored in Merkle tree. Thirdly, it is used while signing transactions with the private key. In primary hash function, the Secure Hash Algorithm is applied twice on the input. Secondary hash function is used in two scripts—P2PKH and P2SH. In secondary hash function, first the input is hashed with SHA-256 and then with RACE Integrity Primitives Evaluation Message Digest (RIPEMD-160). A robust hash function must ensure three properties —pre-image resistance, second pre-image resistance and collision resistance. Pre-image resistance means even if the attacker knows the output, still it is hard to find the input on which hashing is done. Second pre-image resistance means that given an input and a hash equal to its hash, it is impossible to find a different input value. Collision resistance means that it is computationally infeasible to find two different inputs whose hash is same.

For attacking collision and second pre-image resistance property only one of the hashes, either primary or secondary needs are to be revealed whereas for attacking the pre-image resistance property, both the hashes need to be known to the attacker. When the hash breakage leads to compromised collision resistance of the primary hash, then the bitcoins are stolen and destroyed. And when collision resistance of the secondary hash is compromised, then the transaction is repudiated. The compromised second pre-image resistance property leads to double spending and compromised pre-image property uncovers the user addresses and leads to complete failure of the blockchain.

4.2 Digital Signature Scheme

Digital signature scheme used by the bitcoin protocol is Elliptical Curve Digital Signature Algorithm (ECDSA) [51, 52] using secp256k1 parameters. It is used in signing the primary hash value. When a signature scheme is broken, it will lead to transaction malleability attacks. In this attack, the attacker is able to encode the same transaction in multiple ways without validating the signature. The three parameters to measure the robustness of the digital signature scheme are—unforgeability, integrity and non-repudiation. Unforgeability means that without knowing the secret key, it is impossible to generate the signature. When the unforgeability property is compromised, then it leads to four kinds of breakage—total break, universal forgery, selective forgery and existential forgery. Universal forgery enables the attacker to forge all the messages. Selective forgery enables the attacker only to forge some messages of his choice. This risks the bitcoin wallets and the attackers are able to drain them.

Existential forgery is the ability of the attacker to generate a valid signature for a new message, but this is not effective because the new message may not be a valid message. A valid message is the hash of a valid transaction. Integrity ensures that the signature is bound to one transaction and cannot be concatenated to another. When the integrity is compromised, then the claimed payment is not received. It is of two types—collision integrity and second pre-image integrity. Collision integrity is compromised when the attacker knows that which public key is used to generate the transaction, he is able to forge a valid signature for another transaction. Second pre-image integrity is compromised when the attacker knows the public key and the message generated by it, the attacker generates a parallel new transaction for which the same signature is valid.

Non-repudiation is a proof of all the nodes regarding the actual owner of the bitcoins being transferred. Non-repudiation can only be compromised when the secondary hash is broken. These properties are interdependent and breakage in one

Fig 6: Conventional privacy model.

leads to breakage in others. These two figures explain the conventional privacy model and bitcoin privacy model. In any conventional method of exchanging money, like in a bank, the sender does not send it directly to the receiver. There is always an intermediary authority between the sender and the receiver. The details and information, on who is sending how much money to whom, lies only with the trusted authority and is not accessible to public. This ensures pseudo privacy because the sharing of personal information is dependent on the trusted bank.

In the second figure, the privacy model of bitcoin is shown. In the bitcoin system, all the transactions are in the public domain as public blockchain but still full privacy is ensured. This happens because all the entities communicate, using their public keys and mapping of these public keys to real world entities is a computationally infeasible task. This makes the real world entities anonymous. So, although the attacker can gain information like how many bitcoins are transferred at what time, but the linking of transactions to any parties is not possible because the real identities remain in the hidden domain. A new public key is used for every new transaction, making the trace back harder.

4.3 Theft in Bitcoin Exchanges

Bitcoins are prone to theft in bitcoin exchanges. Some mining applications, which miner partners use for collective mining in a mining pool, are malicious as they are compromised by the hacker's distributed corrupt code. An example of such malicious act can be traced back to 2013, when the Skype video call application was held responsible for spreading the Trojan Horse virus. It started with India due to poor cyber security and then it went on to infect Western countries. A malicious program was embedded in the Skype messages exchanged, so that whenever the victim clicked on that message, the linked program was executed, making the machine a bitcoin miner, which led to overloaded sloth CPUs. The attacker was rewarded bitcoins in proportion to the decreased CPU throughput. The processing power of individual PCs was exploited to mine more bitcoins, leading to crash of the user PCs. Bitcoin mining viruses have been accused of stealing bank details by installing Zeus in the backend. They have been accused of converting PCs into botnets by installing Andromeda. Bitcoin wallet malwares [53] have been found attacking the user's browser, which then uncontrollably clicks on the link of the corrupt Website, which generates income for them for the clicks. Almost 2 million computers were infected for three days, each one generating 12,000 clicks a day.

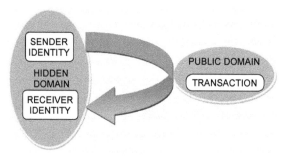

Fig. 7: Bitcoin privacy model.

4.4 Extortion of Money from Bitcoin Wallets

Extortion of money is done by hackers by extorting the bitcoin wallet of the user. Online gaming was one of the first industries to accept payments in the form of digital cash transactions. Game accounts contain points, which are extorted by hackers by spreading malicious files having gaming account-extortion function. Similar file versions have been found executing bitcoin account extortion function to extort bitcoin exchanges in Korea. Malicious code was distributed by trapping the user by downloading an image file that never existed in reality. So, instead of image file that the user is looking for, he becomes a victim and ends up downloading the malicious program.

Bitcoin exchanges have also been the victims of Distributed Denial of Service (DDoS) attacks [54]. They are lucrative attack options because of the amount of currency exchange they cater to. In DDoS attack, the attacker's main motive is to disrupt the services offered by a server by overloading its bandwidth and exhausting its resources. The attacker exploits the vulnerabilities in the system. The compromised systems then are controlled by the attacker himself to expand its army of zombie machines. The compromised machine acts as a master which then further exploits the vulnerabilities in other systems to gain control over. All these form the slave machines, which act on the commands of the master machine, which acts on the commands of the attacker to launch attack against the victim machine. The victim machine is flooded with multiple requests of malicious packets.

BTC China, a famous bitcoin exchange, became a victim of large-scale DDoS attack in 2014. It was a SYN flood attack, which lasted for nine hours and traffic level went up to 100 gigabit. Instead of an army of zombie machines, a network of multiple compromised servers was used to initiate the DDoS attack. Similarly, Bitstamp and BTC-e bitcoin exchanges also became the victims of DDoS attack in 2014 when transactions had to be suspended for 48 hours at the cost of incurring huge losses.

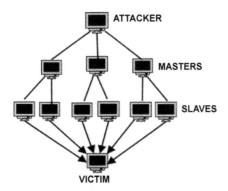

Fig. 8: DDoS attack scenario.

5. Taxonomy of Defense Mechanisms for Bitcoin Security

Increasing popularity of bitcoin cryptocurrency has attracted many users to it. For a user to participate in the bitcoin system, he must have a bitcoin wallet which operates as the bitcoin client node. For secure transactions, proper key-management strategies [44] must be adopted for public-private bitcoin wallet keys. This is of utmost importance because if once the user forgets or misplaces these keys, then the bitcoin associated with them are also lost. A compromise on the confidentiality of these keys is called 'wallet theft' which is caused by using compromised wallet software. Bitcoin protocol uses Elliptical Curve Digital Signature Algorithm (ECDSA) [45] to secure the transactions by digitally signing them. A different private key is used for every new transaction to keep the things very random and dynamic. For stealing bitcoins, the attackers aim at breaking the hash value and the signature scheme. So we can say that unlike traditional currency systems, bitcoin makes use of public-key cryptography to ensure security. This draws attention to safe storage and management of these keys.

Users have a variety of wallet options to choose from, like hot wallets, cold wallets, paper wallets and brain wallets. Nowadays almost all the online wallets have mobile apps which allow easy access to the users. Although hot wallets are more prone to cyber hacks, but the cold wallets are not of much usability. The more advanced wallets are paper wallets in which the private key is written and stored in a physically secure space. Brain wallets are also used which makes use of the user's memory and the key is remembered by the user in the form of a phrase or a sentence.

The biggest bitcoin exchange ever, Mt. Gox was hacked causing a loss of bicoins worth 450 million dollars. After this incident, people lost trust in bitcoin exchanges and started keeping the bitcoins in their wallets. A technique named 'provisions' is adopted by most of the bitcoin exchanges in which the exchanges have to show their input transaction list to ensure that they have enough currency to pay back customers bitcoins. This is similar to the traditional banks which have to maintain a reserve balance amount in their treasure. There is nobody to govern the security of bitcoin exchanges, so the users are themselves responsible for their missing bitcoins. Security loopholes in the user's PC makes the personal information vulnerable to extortion threats; so safeguarding the user's bitcoin wallet is important. Users must ensure that they are using the latest version of the bitcoin wallet software. The recent version comes with the security patches of new malwares. The backup of the bitcoin wallet must be stored frequently as an encrypted file in external storage.

The bitcoin wallets can be divided into two categories—cold wallets and hot wallets. Cold wallets are offline wallets; hot wallets are online wallets. Users have both of them [55]. It's like one keeping the majority of one's money is savings account

Fig. 9: Hardware wallet [63].

and only a limited amount in current account for making purchases. In a similar manner, the majority of one's bitcoins are stored in cold wallets, where the hackers cannot access them. The hot wallets are more vulnerable to security breaches due to their continuous connectivity to the internet. But the bitcoin users are suggested to keep only limited amounts of money in their hot wallet because then the hacker is not keen on wasting resources to get hold of a small amount of money. Bitcoin exchanges fall in the category of hot wallets because in that case, the bitcoins are stored in their servers. Although intensive security practices are ensured in exchanges [56], still it is not advisable to store all your bitcoin collection in there due to the unceasing risks of zero day attacks.

Another kind of hot wallet in use is the desktop wallet. The difference is that in desktop wallets the private key is not stored at the server of the bitcoin exchange; instead. it is stored on the owner's PC. But the bitcoins are still vulnerable if the hacker is able to compromise the owner's system. There are different cold wallets available, the most common being the hardware wallets. These are hacker-proof physical devices that are connected to the PC only at the time of making a transaction. To protect the private key from getting exposed, it is stored in an analog medium, instead of electronic one. If the bitcoin user wants to use only hot wallets, then he must ensure that he uses an application which has been time-tested by the peer users and makes sure that the password set is complex and long enough. It should be a combination of numbers, characters and special symbols. And one must refrain from using generic details like address, phone number and registration numbers as passwords.

While in some countries use of bitcoins is free and legal, in some others, it is a banned currency. If we talk about India, then although not illegal, it is not one of the regularised currencies because of which, users refrain from making transactions using bitcoins. For the transactions to take place from a bitcoin wallet, user authorization is required. Figure 10 shows different authentication methods by which the bitcoin-owner approves of the financial transaction. Each one of them can be used independently as

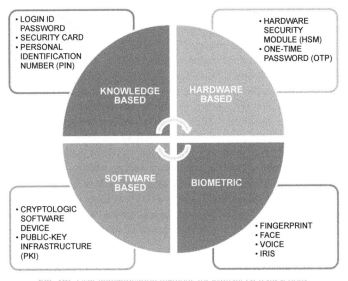

Fig. 10: User authentication methods for e-financial transactions.

well as in combination with one another. When used independently, then it is called 'single-factor authorization'. When a combination of two techniques, like login id password and one-time password, are needed to approve a transaction, then it is called 'two-factor authentication'. Similarly, when more than two factors are required, then it becomes multi-level authentication [57].

5.1 Biometric Technology

Application of biometric technology in the security and surveillance field is increasing day by day. Its increased use can also be seen in internet banking, smartphones, cloud computing and e-commerce. Biometric [57, 58] comprises of two words—bio and metric. 'Bio' means life and 'metric' means to measure. Biometric system takes into account intrinsic behavioral and physiological traits of an individual. This includes DNA, iris recognition, fingerprint, face recognition and heartbeat recognition. A statistical analysis of these features is done by the biometric system and a match is declared only when the stored data and the input data match sufficiently. Voiceprints are a new addition wherein the characteristics of the air expelled while speaking is taken into account. There are many advantages of biometric systems over traditional authentication techniques, like pin, passwords, keys and cards.

The main advantage is that it biologically belongs to one person only, so it cannot be easily compromised. It is user-friendly and takes away the burden of remembering PIN and passwords. Although the initial cost of setting up the biometric system is high, but due to simplified user management, the biometric system helps in cost saving in the long run. Also the speed of finding a positive match from a large collection of data is handled by fast computers running on established algorithms, like image-based algorithms and minutiae-based algorithm. Biometric system can be integrated with existing access control systems to enhance the security levels.

5.2 Hardware Security Module

Hardware Security Module (HSM) is a tamper-resistant device which is used to store, process and manage the cryptographic keys [59]. Its main responsibility is to protect these stored keys. The processing of keys involves encryption, decryption, authentication and generation of digital signatures. They are customizable according to the application's requirement. Nowadays, there is a provision of crypto hypervisors which provide on-demand cryptographic services, using cloud environment.

One-Time Password (OTP) [60] is another hardware-based solution in which the user needs to enter a password, which is newly generated every time the user tries to log in the account. This prevents identity theft as the generated password cannot be used for a second time. Although OTP is a convenient method to ensure strong authentication, it is prone to cyber attacks, like phishing, man-in-the middle attack and keyboard logging. OTP is generated using grid cards and transaction numbers lists. Although this method is cheap, but it is slow as well. A faster way is to generate OTP, using a hardware device. Nowadays microprocessor-based smart cards are used for generating the OTP. They have the advantages of more processing power, portability and data storage.

5.3 Knowledge-based Techniques

A Personal Identification Number (PIN) is a code used to verify the identity of a user. Unlike passwords, it comprises only of numeric characters. In the earlier days, PIN was only used for transactions in Automated Teller Machine (ATM), but now its use has expanded to unlocking doors and smartphones. The user should avoid using PIN in series, like 1234,0000 and should also refrain from using generic information, like date of birth, vehicle number, as PIN. Longer PIN is more robust as the attacker needs to try more permutations and combinations to crack the PIN. Remembering many PINs for different cards is a tiresome job for the user, so here the role of password managers comes in. They store the records of the account and its corresponding PIN. There are many innovative ways to create PIN, like picking up numeric base against alphabets; for example, bitcoin will translate to 2920315914.

5.4 Public-key Infrastructure

Some smart security cards employ Public-Key Infrastructure (PKI) [61] for strong authentication capabilities, so it is a combination of all three-hardware-based, knowledge-based and software-based techniques as shown in Fig. 10. PKI is a software-based authentication method used in e-financial transactions. The management and distribution of the keys used to perform encryption, decryption of bitcoin addresses and digital signatures of the bitcoin client are handled by PKI. It distributes accredited certificates to the participating nodes, which identify them as the bonfires of the bitcoin being transferred. Each issue comprises of an electronic fingerprint.

PKI has two parts—certificate authority and registration authority. Certificate authority has to issue accredited digital certificates in sync with the PKI framework. It checks the background of the applicant to prevent any fraud. The certificate authority signs the certificate with its private key and gives its public key to the entity which wants to be certified. Registration authority works as the subordinate certifying authority. A certificate database is maintained which saves all the certificate requests, issued and cancelled certificates. PKI is widely used in smart card logins, Enterprise Resource Planning (ERP) and for client authentication in Secure Socket Layer (SSL). It is based on the concept of the chain of trust model, which is a centralized model and requires a governing body, but decentralized working is the main characteristic of the bitcoin system. So PKI used in here is based on a decentralized trust model called 'Web of trust model' instead of chain of trust model.

6. Conclusion

Since the advent of bitcoin in 2008, its monetary value has seen many surges, sometimes skyrocketing and at time crashing. The bitcoin system is vulnerable to many threats as proven by multiple instances of security breaches over time. The system has been under DDoS attack, the personal wallets and the exchanges have been hacked. Due to anonymity of users and public availability of the blockchain, it has been widely used in illegal activities, such as drug trafficking and money laundering. However, the importance of digital currencies cannot be ruled out in this era of internet. Till now there is no systematic procedure or security model to ensure the safety of digital currencies, but a minimum of two-level authentication process is suggested.

References

[1] Luisanna Cocco and Michele Marchesi. (Oct. 21, 2016). Modeling and Simulation of the Economics of Mining in the Bitcoin Market, published online.
[2] Szabo, N. (1988). Secure property titles with owner authority. Available: http://nakamotoinstitute.org/secure-property-titles.
[3] Ron, Dorit and Adi Shamir. (2013). Quantitative analysis of the full bitcoin transaction graph. pp. 6–24. *In*: International Conference on Financial Cryptography and Data Security, Springer, Berlin, Heidelberg.
[4] Satoshi Nakamoto. (Nov. 2008).
[5] Wood, G. (2015). Ethereum: A Secure Decentralised Generalised Transaction Ledger. Yellow Paper.
[6] Florian Tschorsch and Scheuermann, B. (2016). Bitcoin and beyond: A technical survey on decentralized digital currencies. IEEE Communications Surveys & Tutorials.
[7] Rosenfeld, M. (2011). Analysis of bitcoin pooled mining reward systems. Cornel University Library, Vol. abs/1112.4980, 2011.
[8] Joseph Bonneau, Andrew Miller, Jeremy Clark, Arvind Narayanan, Joshua A. Kroll and Edward W. Felten. (2015). SoK: Research perspectives and challenges for bitcoin and cryptocurrencies. IEEE Symposium on Security and Privacy.
[9] Miller, V.S. (1986). Use of elliptic curves in cryptography. pp. 147–426. *In*: Lecture Notes in Computer Sciences, 218 on Advances in Cryptology—CRYPTO 85, Springer-Verlag, New York Inc.
[10] Your Wallet, S. (March 2014). The bitcoin wiki. Available: https://en.bitcoin.it/wiki/Securing your Wallet.
[11] Garay, J., Kiayias, A. and Leonardos, N. (2014). The Bitcoin Backbone Protocol: Analysis and Applications: Technical Report.
[12] Kroll, J.A., Davey, I.C. and Felten, E.W. (2013). The Economics of Bitcoin Mining or Bitcoin in the Presence of Adversaries.
[13] Rosenfeld, M. (2013). Mining pools reward methods. Presentation at Bitcoin 2013 Conference.
[14] Narayanan, A., Bonneau, J., Felten, E., Miller, A. and Goldfeder, S. (2016). Bitcoin and Cryptocurrency Technologies: A Comprehensive Introduction. Princeton, NJ, USA: Princeton University Press.
[15] Andresen, G. (Jan. 2012). Bip 16: Pay to script hash. Available:"https://github.com/bitcoin/bips/blob/master/bip-0016".mediawik.
[16] Back, A. (2012). Hashcash—A denial of service countermeasure. http://www.hashcash.org/papers/hashcash.pdf (09.04.2011), Semantic scholar.
[17] Luu, L., Saha, R., Parameshwaran, I., Saxena, P. and Hobor, A. (July 2015). On power splitting games in distributed computation: The case of bitcoin pooled mining. pp. 397–411. *In*: 2015 IEEE 28th Computer Security Foundations Symposium.
[18] Malkhi, D. (Jan. 2012). Byzantine quorum systems. Distributed Computing 4: 203–213.
[19] Garay, Juan A., Aggelos Kiayias and Nikos Leonardos. (2015). The bitcoin backbone protocol: Analysis and applications. pp. 281–310. *In*: EUROCRYPT (2).
[20] Johnson, B., Laszka, A., Grossklags, J., Vasek, M. and Moore, T. (2014). Game-Theoretic Analysis of DDoS Attacks against Bitcoin Mining Pools, Springer, Berlin Heidelberg, pp. 72–86.
[21] Eyal, I. and Sirer, E.G. (2013). Majority is Not Enough: Bitcoin Mining is Vulnerable, Cornel University Library, Vol. abs/1311.0243, 2013.
[22] Dagher, G.G., B˝unz, B., Bonneau, J., Clark, J. and Boneh, D. (2015). Provisions: Privacy-preserving proofs of solvency for bitcoin exchanges. pp. 720–731. *In*: Proceedings of the 22nd ACM SIGSAC Conference on Computer and Communications Security, ser. CCS'15. ACM.

[23] Decker, C. and Wattenhofer, R. (2014). Bitcoin Transaction Malleability and MtGox, Springer International Publishing, pp. 313–326.

[24] Malleability attack a nuisance but bitcoin not broken, pundits say. Available: http://www. financemagnates.com/cryptocurrency/news/malleability-attack-a-nuisance.

[25] (2017). The bitcoin malleability attack: How can it undermine the blockchains credibility? Available: http://www.coinwrite.org/.

[26] Andrychowicz, M., Dziembowski, S., Malinowski, D. and Mazurek, L. (2015). On the Malleability of Bitcoin Transactions. Springer, Berlin, Heidelberg, pp. 1–18.

[27] B.J. (2016). Why buy when you can rent? Financial Cryptography and Data Security' Lecture Notes in Computer Science, Vol. 9604, Springer, Berlin, Heidelberg.

[28] McCorry, P., Shahandashti, S.F. and Hao, F. (2016). Refund attacks on bitcoins payment protocol. http://eprint.iacr.org/2016/024.

[29] Moore, T. and Christin, N. (2013). Beware the Middleman: Empirical Analysis of Bitcoin Exchange Risk, Springer Berlin Heidelberg, pp. 25–33.

[30] Corbixgwelt. (Mar. 2011). Timejacking and Bitcoin. Available: http://culubas.blogspot. de/2011/05/ timejacking-bitcoin 802.html.

[31] Haber, S. and Stornetta, W.S. (1991). How to time-stamp a digital document. Journal of Cryptology 3(2): 99–111.

[32] Douceur, J.R. (2002). The sybil attack. pp. 251–260. *In*: The First International Workshop on Peer-to-Peer Systems, ser. IPTPS '01, London, UK: Springer-Verlag.

[33] Heilman, E., Kendler, A., Zohar, A. and Goldberg, S. (2015). Eclipse attacks on bitcoin's peer-to-peer network. pp. 129–144. *In*: Proceedings of the 24th USENIX Conference on Security Symposium, ser. SEC'15,USENIX Association.

[34] Bissias, G., Ozisik, A.P., Levine, B.N. and Liberatore, M. (2014). Sybil resistant mixing for bitcoin. pp. 149–158. *In*: Proceedings of the 13th Workshop on Privacy in the Electronic Society, ser. WPES'14, ACM.

[35] Karame, G.O., Androulaki, E. and Capkun, S. (2012). Double-spending fast payments in bitcoin. pp. 906–917. *In*: Proceedings of the 2012 ACM Conference on Computer and Communications Security, ser. CCS '12. New York, NY, USA: ACM.

[36] Gervais, A., Ritzdorf, H., Karame, G.O. and Capkun, S. (2134). Tampering with the delivery of blocks and transactions in bitcoin. pp. 692–705. *In*: Proceedings of the 22Nd ACM SIGSAC Conference on Computer and Communications Security, CCS'15. ACM.

[37] Finney, H. (2312). Best practice for fast transaction acceptance: How high is the risk? Available: https://bitcointalk.org/ index.php?topic=3441.msg48384#msg48384.

[38] Heusser, J. (2013). Sat solvingan alternative to brute force bitcoin mining. Available: https:// jheusser.github.io/2013/02/03/ satcoin.html.

[39] Vector 67. (2011). Fake bitcoins? Available: https://bitcointalk.org/ index.php?topic=36788. msg463391#msg463391.

[40] Natoli, C. and Gramoli, V. (2016). The balance attack against proof-of-work blockchains: The R3 test-bed as an example. Cornel University Library, Vol. abs/1612.09426.

[41] Courtois, N.T. and Bahack, L. (2014). On subversive miner strategies and block withholding attack in bitcoin digital currency. CoRR, Vol. abs/1402.1718.

[42] Bag, S., Ruj, S. and Sakurai, K. (2016). Bitcoin block withholding attack: Analysis and mitigation. IEEE Transactions on Information Forensics and Security PP(99): 1–12.

[43] Vasek, M., Thornton, M. and Moore, T. (2014). Empirical Analysis of Denial-of-Service Attacks in the Bitcoin Ecosystem, Springer, Berlin, Heidelberg, pp. 57–71.

[44] Eskandari, S., Barrera, D., Stobert, E. and Clark, J. (2134). A first look at the usability of bitcoin key management [Online]. Available: http://people.inf.ethz.ch/barrerad/files/usec15-eskandari.pdf.

[45] Al Imem Ali. (June 2015). Comparison and evaluation of digital signature schemes employed in NDN network. International Journal of Embedded Systems and Applications (IJESA).

[46] Giechaskiel, Ilias, Cas Cremers and Kasper Bonne Rasmussen. (2016). On Bitcoin Security in the Presence of Broken Crypto Primitives. IACR Cryptology.

[47] Bahack, L. (2013). Theoretical bitcoin attacks with less than half of the computational power (draft). CoRR, Vol. abs/1312.7013.

[48] Mauro Conti, Sandeep Kumar, Chhagan Lal and Sushmita Raj. (July 2017). A Survey on Security and Privacy Issues of Bitcoin, Cornel University Library, Published Online.

[49] Bos, J.W., Halderman, J.A., Heninger, N., Moore, J., Naehrig, M. and Wustrow, E. (2014). Elliptic Curve Cryptography in Practice, Springer, Berlin, Heidelberg, pp. 157–175.

[50] Giechaskiel, I., Cremers, C. and Rasmussen, K.B. (2016). On Bitcoin Security in the Presence of Broken Cryptographic Primitives, Springer International Publishing, pp. 201–222.

[51] Gallagher, P. and Kerry, C. (July 2013). Federal information processing standards (fips) publication 186-4: Digital signature standard (dss). Available: http://nvlpubs.nist.gov/nistpubs/FIPS/NIST.FIPS.186-4.pdf.

[52] Howgrave-Graham, N.A. and Smart, N.P. (2001). Lattice attacks on digital signature schemes. Designs, Codes and Cryptography 23(3): 283–290.

[53] Litke, P. and Stewart, J. (2014). Cryptocurrency-stealing malware landscape. Technical Report, Dell Secure Works Counter Threat Unit, 2014.

[54] Neil Gandal, A.F. Tyler Moore and Hamrick, J. (2016). The impact of DDOS and other security shocks on bitcoin currency exchanges: Evidence from mt.gox. *In*: The 15th Annual Workshop on the Economics of Information Security, Vol. abs/1411.7099, June 13–14.

[55] Chinmay A. Vyas and Munindra Lunagaria. (2014). Security concerns and issues for bitcoin. international journal of computer applications, National Conference-cum-Workshop on Bioinformatics and Computational Biology, NCWBCB.

[56] Hoch, J.J. and Shamir, A. (2008). On the Strength of the Concatenated Hash Combiner when all the Hash Functions are Weak, Springer Berlin Heidelberg, pp. 616–630.

[57] Lim, Il-Kwon, Young-Hyuk Kim, Jae-Gwang Lee, Jae-Pil Lee, Hyun Nam-Gung and Jae-Kwang Lee. (2014). The analysis and countermeasures on security breach of bitcoin. pp. 720–732. *In*: International Conference on Computational Science and Its Applications, Springer, Cham.

[58] Yeom, H.-Y., Jo, H.-J., Lee, D.-H., Jeong, Y.-G., Jang, G.-H. and Lee, S.-R. (2011). Research on security criteria for extension to electronic authentication method usage-based. Final Research Report, Korea Internet & Security Agency.

[59] Bamert, T., Decker, C., Wattenhofer, R. and Welten, S. (2014). Blue Wallet: The Secure Bitcoin Wallet, Springer International Publishing, pp. 65–80.

[60] Wu, Longfei et al. (2017). An Out-of-band Authentication Scheme for Internet of Things Using Blockchain Technology.

[61] Kroll, J., Davey, I. and Felten, E. (June 2013). The economics of Bitcoin mining, or Bitcoin in the presence of adversaries. *In*: Proceedings of the Twelfth Annual Workshop on the Economics of Information Security (WEIS'13), Washington, DC.

[62] Neudecker, T., Andelfinger, P. and Hartenstein, H. (May 2015). A simulation model for analysis of attacks on the bitcoin peer-to-peer network. pp. 1327–1332. *In*: IFIP/IEEE International Symposium on Integrated Network Management (IM).

[63] https://www.amazon.co.uk/ 'Keepkey-Simple-Bitcoin-Hardware-Wallet/dp/B0143M2A5S/ref=sr_1_10?ie=UTF8&keywords=bitcoin%20wallet&qid=1491224892&sr=8-10.

10

Early Detection and Prediction of Lung Cancer using Machine-learning Algorithms Applied on a Secure Healthcare Data-system Architecture

Mohamed Alloghani,[1,]* *Thar Baker,*[1]
Dhiya Al-Jumeily,[1] *Abir Hussain,*[1] *Ahmed J. Aljaaf*[1]
and *Jamila Mustafina*[2]

1. Introduction

Cancer research has evolved over the past decade and most of the advances are attributed to technological innovations and inventions in both medical and biotechnological fields. Of the many challenges of achieving cancer treatment and reducing mortality rate, screening at an early stage and identifying the type of cancer has stood out as a barrier to early prediction and administration of treatment. Besides the technologies, researchers have developed and used different strategies for achieving early cancer prediction. However, these advances have led to the collection, aggregation and storage of a large volume of data and the development has made the concept of big data and the need for better analytics more urgent. Even though an enormous amounts of data are available to the community, predicting the outcome for diseases accurately is one of the most interesting and challenging tasks for both researchers and medical practitioners [1]. It was in the pursuit of accurate health outcome prediction that cancer researchers resorted to machine-learning techniques. Machine-learning algorithms

---

[1] Liverpool John Moores University/3 Byrom St, Liverpool L3 3AF, UK.
 E-mails: T.baker@ljmu.ac.uk; D.Aljumeily@ljmu.ac.uk; A.Hussain@ljmu.ac.uk;
 A.J.Kaky@ljmu.ac.uk
[2] Kazan Federal University, 18 Kremlyovskaya str, Russian Federation; DNMustafina@kpfu.ru
* Corresponding author: M.AlLawghani@2014.ljmu.ac.uk

can detect patterns and relationships hidden in complex data but most importantly, the techniques have proven effective in predicting the outcome of the disease.

This chapter focuses on how machine-learning algorithms can be used in early detection of cancer. The analysis and subsequent inferences are based on survey data. The idea that lung cancer can be detected without exposure to radiation is one that provides a solution to some of the problems that arise from radiation exposure. Nonetheless, the focus is on finding a solution to the persistent problem of detecting cancer at very advanced stages.

According to [2], cancer remains a leading public health problem globally and the second leading cause of mortality in the U.S. The researcher made estimations and computations on the number of deaths averted, including the major risks factors. The analysis relied on mortality data that was collected over a period spanning 1930–2015 and of interest are the projections and conclusions made about the disease in 2018. The study established that over 1,735,350 cases of invasive cancer are expected to affect Americans and the number is much higher on a global scale. The results of the modeling exercise also established that about 609,640 Americans will lose their lives because of cancer in 2018. In their conclusions, the researchers established that cancer vulnerability was dependent on race and age among other critical factors, such as access to high-quality health care. It is, therefore, imperative to develop or adopt alternative factors when considering early cancer detection and its importance in reducing cancer-incidence rates and mortality.

The survey data used in the analysis is part of a continually growing online data and its usage meets the requirement for big data. Big data refers to a large dataset that can analyze, using advanced visual analytics and other programmable languages, and reveal hidden patterns, trends and association. Of great concern are security of the data and the systems that store them.

1.1 Cyber Security and Data-related Security

The advent of industry 4.0 and its concepts introduced technologies, such as cloud storage and other internet-based technologies that have exposed data to innumerable threats, especially given the exposure that comes with the internet of things [3]. The continued adoption and use of big data and all other related cloud technologies and other concepts have increased the risk of sensitive data. Therefore, it is pertinent to consider computer security for systems that collect and store data. Most organizations, including health institutions, are collecting, collating, aggregating and storing digital data and most of which are sensitive since they consist of medical and other socio-economic data.

Of grave concern is the increasing number of advanced and persistent attacks. Organizations, healthcare facility included, have an obligation to revisit their cyber-security threats and move towards the Prevent, Detect and Respond (PRD) paradigm [4]. In the context of the CIA triad, improved detection is the ultimate technique for identifying anomalies or strange patterns in systems. The CID triad is a data-security concept that deals with confidentiality, integrity and availability of data [5]. The confidentiality axiom of the triad focuses on ensuring the privacy of the data that entails all the measures that restrict access to sensitive information. Most organizations use different encryption technologies to protect their data. Most of the commonly used methods include passwords with an overlay of two-factor verification, security tokens and biometric verification among others [5]. The other axiom of the triad—integrity ensures that the data is consistent, accurate and trustworthy and as such, focuses on protection of data during transit. That is, the organization must ensure data cannot be intercepted by man-in-the-middle or man-in-the-browser attackers. The dataset must have a mechanism of verification, including checksums and proper backups may be required to restore any affected files. Finally, availability involves maintenance of data storage hardware and software to ensure that the system does not have unnecessary downtimes. The hardware must perform at set standards and the supporting software and operating systems must function correctly and without any conflicts. To ensure availability of data, it is recommended that organizations adopt extra security, including firewalls and proxies to protect the system against malicious attacks.

The big data technology alongside machine-learning algorithms has proven crucial in attack detection and prevention. However, big data is also posing a challenge to the CIA triad because of the volume of data that needs protection, and not all the data sources or platforms have the required security features provided in their systems. For instance, the survey data used in the study were retrieved from Data World Website. The platform provides datasets for different applications and some of them are meant for financial analysis. In most cases, sensitive information or attributes in the financial data are removed for security purposes and the practices are quite derailing for some specific machine-learning applications. Nonetheless, Data World uses socket security layer that verifies the identity of the remote computer accessing the data. The platform is an example of a system that implements an encipherment to ensure confidentiality of the data [6]. In specific, the system uses a digital signature to verify the identity of the downloaded file. Despite these features, the integrity of the survey data is also dependent on the practices and precautions that the data contributor adheres to during collection. The debate on the security of the data and its system of storage is a sensitive one because information has proven to be a goldmine for attackers. Most importantly, the internal consistency and validity of studies are dependent on the consistency of the data.

As far as health data security is concerned, there are many methods of ensuring protection and hence, the confidentiality and integrity of the data. Some of the emerging techniques for encrypting images include a four-image encryption technique that relies on a quaternion Fresnel transforms [7]. The technique also uses a 2D-based logistic-adjusted-sine map and it is one of the unique image encryption methods.

1.2 Novelty and Research Contributions

It is pertinent to reiterate that not all countries or regions, both political and economic, have access to high-quality cancer care. In fact, most people from regions, considered as either average or poor but opulent, always seek treatment in the U.S. and India and other places with advanced treatments for cancer. Most importantly, a majority of the recent studies have established that there is a growing trend in attempts to assess cancer risks and predict recurrences, using machine-learning techniques [2, 8, 9]. The authors of these papers focused on how to predict the redevelopment of cancer after discharge from hospital using machine-learning algorithms and ascertained that the techniques are more accurate than the alternative statistical techniques. Further, these studies relied on clinical and molecular data to make the prediction, and arguably, they depended on information that would otherwise not help predict cancer in its early stages.

In this study, machine-learning techniques, which have been generalized to be more accurate than alternative statistical analysis, will be used to study cancer-susceptibility factors and their contribution to cancer incidences and rates. Of the available cancer prediction studies, most of them used genetic epidemiology to study susceptibility for different factors and the genetic epidemiology data have been extensively discussed to limitations, when it comes to early detection and prediction of cancer. Furthermore, most of the studies have used artificial neural network, support vector machine, decision tree and Bayes Network to predict susceptibility and recurrence of data. In this study, 12 machine-learning algorithms are implemented and their performance compared to establish the best-performing ones. Additionally, the learning task uses a prognosis survey data with lung cancer predisposing factors, such as smoking, alcohol consumption, age and gender. The approach used in the study blends lung cancer predisposing factors with symptoms of the disease including yellow fingers, anxiety, fatigue and chest pain. The information that the study provides can help identify lung cancer at its early stage because it combines both the symptoms and predisposing factors to develop models that can detect or project the probability of the individual having cancer or not.

2. Literature Review

A number of studies have been conducted to focus on early detection of several types of cancer. However, there are studies that focus on early detection and prediction of lung cancer. For instance, in [10], the probability of predicting lung cancer using MRI images has been explored. The researchers identified pain, breathlessness, cough, fatigue and weight loss as the symptoms of lung cancer and attributed the disease to nicotine, inhalation of toxic gases, genetic mutations and passive smoking. The system that these scientists designed is meant to detect lung cancer at the initial stages, although the authors acknowledged that detecting lung cancer in its first or second stage is quite challenging. The researcher proposed that MRI images can be used as the input data because the effect or influence of radiation is less lethal on MRI scans. The proposed framework consists of preprocessing, segmentation, extraction of the feature and comparison. The comparison is done using neural nets and other data-mining algorithms applicable in image analysis and the model should predict future incidences of cancer-based on the images. Consequently, individuals can go for MRI scans and the images can be used in predictive and personalized medicine. However, with the personalization of medicine and big data, security becomes an issue, especially when dealing with health images.

Regarding health-data security, [11] proposed a new image forgery detection system that verifies healthcare images to ensure their confidentiality and integrity. The system is based on machine-learning classifiers and it uses SVM and extreme-learning-based algorithms as part of its multi-resolution technique. The proposed system applies a regression fitter on a noise map and feeds output to the classifiers. The noise filtering occurs at the edges of the computing resource, while the classification is done at the core of the computing system. The proposed system is an epitome of the influence of machine learning in document verification in the health sector. Imagery plays a crucial role in cancer detection although most techniques are still in laboratory phases. [12] explored the use of machine learning in exploring high-resolution analysis. The researcher proposed variants of single-value decomposition algorithms, arguing that redundancy in images are reducing the amount of information that can be retrieved from the images and postulates that non-local self-similarity and low-rank prior techniques can improve the resolution of the images. The algorithm uses a combination of non-local interpolation and self-similarity to enhance the images. The technique can also be used to detect.

The other notable trends in disease detection are the growing use of quantum information processing. Researchers argue that quantum information processing is more efficient and secure because it is based on the obscure field of quantum mechanics [13]. In image processing, the quantum technique relies on quantum parallelism, hence the efficiency and security. The other notable technologies that have been used to improve image analysis performance include the use of graphical processing units (GPUs) besides parallel programming to implement different algorithms [14]. The concept behind using the GPUs and parallel programming is to improve the computing capability and reduce the duration of processing the medical images. The segmentation and identification of the region of interest as discussed in the article is based on fuzzy c-means algorithm. [15] also used GPU-based enhancement to improve image processing images although the researchers used a hybrid and achieved a performance enhancement of up to 8.9 times without any compromises on the quality of the image segmentation. Despite these advances, early detection of cancer is still a major problem and some of the studies that addressed this matter are discussed below.

The other study of interest is that conducted by [16] because it ventured into early detection and prediction of lung cancer using a combination of urine trace element analysis and Adaptive Boosting analysis. The AdaBoost, as an ensemble machine-learning technique, allowed the researchers to implement an improved decision stump classifier and augmented it with trace element analysis. The researchers partitioned the dataset into testing and train set using Kennard and Stine (KS) algorithm. The findings of the study showed that AdaBoost achieved 100 per cent sensitivity with a specificity of 93.8 per cent and an augmented accuracy of 95.1 per cent. The study concluded that AdaBoost outperforms other algorithms and it has better ways of handling over-fitting. Similar deductions were arrived at in this paper because Ada-Boost algorithm had the highest classification accuracy (92.6 per cent). Hence, deploying adaptive boosting in any algorithm leads to superior performance of the weaker algorithm. However, it is imperative to note that this study was limited to prognosis-survey data and could have easily undermined the influence of factors, such as second-hand smoking on the group characterizations. That is, the algorithm could have classified an incidence as positive for lung cancer despite non-smoking status, suggesting that secondary smoking may have influenced the outcome. However, this study did not pay much attention to such factors.

3. Methodology

The section specifies the data used in the study, including the attributes and features used in learning the data. It also contains a succinct discussion on the algorithms used in the study, as well as the approach used in the study.

3.1 Data Sources and Types

The study used a prognosis survey dataset that was retrieved from a secure online Web system https://data.world. The Web system utilizes a security protocol, such as the Secure Socket Layer (SSL) that is designed to ensure secure communications of data sharing and with it the data encrypted and decrypted as it travels, which ensure a secure connection with sensitive survey data as it is shared. From the system-architecture perspective, the Web system covers all the retrieval and return of patients' data on a Web browser. The Web system is built on a multi-tier architecture that separates different components of the system into several layers according to their functionalities. Each layer operates on a different system, hence, avoids a single point of failure and provides enhanced application and security.

The dataset consists of 16 attributes and 309 instances, which also represent the number of participants in the study. The attributes are summarized in Table 1 and it should be noted that lung cancer attributes serve both as a predisposing factor and the predictand of the models.

The attributes, SOB and SD, are abbreviations for Shortness of Breath and Swallowing Difficulty. The Attribute Lung is both a predictand and a predisposing factor. All the levels of measurements are categorical with two levels, except Age. The Age variable is numeric and it values very high so that using the variable could have led to poor model performance. That is, most machine-learning algorithms, including Decision Tree perform best when applied to categorical variables. In the study, an entropy-based discretization was applied to the Age attribute and it was converted to a

Table 1: Attribute information table and aggregation into predisposing factor and symptom groups.

Predisposing Factors			Symptoms		
Attribute	Value	Label	Attribute	Value	Label
Gender	M	Male	Yellow Fingers	2	Yes
	F	Female		1	No
Smoking	2	Yes	Anxiety	2	Yes
	1	No		1	No
Allergy	2	Yes	Fatigue	2	Yes
	1	No		1	No
Chronic Disease	2	Yes	Wheezing	2	Yes
	1	No		1	No
Peer Pressure	2	Yes	Coughing	2	Yes
	1	No		1	No
Alcohol	2	Yes	SOB	2	Yes
	1	No		1	No
Age	2	Yes	SD	2	Yes
	1	No		1	No
Lung Cancer	2	Yes	Chest Pain	2	Yes
	1	No		1	No

nominal attribute like the others. The entropy equation is shown below and the choice of the method is based on information gained advantages [17].

$$Entropy(D_1) = -\sum_{i=1}^{m} p_i \log_2 p_i \qquad (1)$$

In Eq. 1, the summation iterates from and is the number of discrete classes created from the observations or the classifier values. The is the probability that observation is a member of the class. However, the equation computes entropy for a single split yet the objective of the discretization is to maximize the information obtained from splitting the Age attribute into finite discrete classes. The entropy maximizes the information by attaining the least value. That is, the closure entropy is to zero the better is the binning and discretization of the variable, and the information gained is expressed as sum of the entropies as defined in Eq. 1 as follows [17]:

$$Info_{Att}(D) = \frac{|D_1|}{|D|} Entropy(D_1) + \frac{|D_2|}{|D|} Entropy(D_2) + \cdots + \frac{|D_m|}{|D|} Entropy(D_m) \quad (2)$$

In Eq. 2, the left-hand side represents the information attribute while the right-hand side is a summation of the entropy of each of the bin ($Entropy(D_1)$, $Entropy(D_2),..., Entropy(D_m)$) multiplied by their respective sizes $\left(\frac{|D_1|}{|D|}, \frac{|D_2|}{|D|},..., \frac{|D_m|}{|D|}\right)$. Hence, the information across m bins is proportional to the sizes of the created bins and their entropies [17].

3.2 Study Approach

The preprocessed data was analyzed in SPSS Version 20 although some of the algorithms were implemented in Weka. In general, the study used visual programming, based on the following framework. The planned approach had three phases—preprocessing, model set and learning and evaluation. The preprocessing involved discretization of the Age variable and characterization of Lung Cancer attribute using the other attributes. The attributes were categorized as either predisposing or symptom factors. The group characterization during preprocesses established lung cancer incidences and suggested significant predisposing factors and symptoms in each group. In the study, the results of lung cancer characterization served as a benchmark for evaluating the performance of each algorithm. The group characterization results established that 270 of the incidences represented lung cancer, while only 39 cases did not. In evaluating the respective models, classification accuracy (CA) and test significance (F1) were based on the model's ability to predict the two characterized groups. All the models used a 10-fold cross-validation so that the results of the predicted observations in each class were an average. The rest of the comparative analysis was also based on these averages. The planned approach and the other measures used in the study, including algorithms, are shown in Fig. 1. The study also made deductions based on the results of other studies and augmented its results, based on the reputation of the chosen metrics.

Figure 1 shows the approach used in the study, including data preprocessing, data learning and model evaluation. The preprocess details with the discretization of the Age attribute, using an entropy-based method. The data learners in the visual program

Fig. 1: Overview of Study Approach.

included Decision Tree, SVM, kNN, Neural Network, Naïve Bayes, Stochastic Gradient Descent, Random Forest and AdaBoost. The other algorithms included in the study were ID3, J48, CHAID and Deep Learning. These algorithms are related to Decision Tree regression and neural network (multi-perception layer) algorithms. The models are evaluated based on confusion matrix, ROC analysis, calibration plots and lift curves.

3.3 Algorithms

The study used a total of 12 algorithms and each is discussed in the following subsections. The discussion focuses on the feature of each of the algorithms and primary requirements that make each suitable for the study.

3.3.1 Naïve Bayes

The algorithm is one of the commonly used ones when it comes to predictive classification tasks. The algorithm is based on Bayes' rule of probability [18]. The rule decomposes the computed posterior probability into a prior and a likelihood probability. In application, it is assumed that an experiment, say E, has a subspace S that contains all the possible outcomes. The same applies to learn the data to determine the consequences of the predisposing factors. The assumption is that in each case or in the case of each individual, both predisposing factors and symptoms represent all the possibilities of lung cancer incidence. That is, suppose A and B denote one of the predisposing factors and one of the symptoms, then the conditional probability that a coughing (symptom) individual is smoking (predisposing factor) is expressed as follows [18]:

$$P(B/A) = \frac{P(A \cap B)}{P(A)} \qquad (3)$$

For the likelihood rule to hold, the probability of *A* occurring must be greater than zero and the left-hand side of the equation denotes the probability of *B* given *A*. In the study, the algorithm iterates through the data and determines the probability of lung cancer occurrence given that the individual has been exposed to any of the factors.

3.3.2 Decision Tree, Random Forest and Logistic Regression

The DT classifier is based on the regression algorithms that build branches based on nodes. It represents layers of relationships between the attributes. In the study, a Decision Tree is used to build a predictive model for accomplishing classification and regression modeling tasks [19]. In the interpretation of the predicted lung cancer incidence, the tree is presented as a hierarchical model of choices and their repercussions. For instance, doctors and nurses can use the Decision Tree to identify whether or not the patient or person is susceptible to lung cancer. As a classifier predictive model, the Decision Tree is a recursive partition of the attributes space. The model has nodes in form of a rooted tree so that it is a directed tree with root node without incoming edges, though the rest of the nodes have just one incoming edge as a rule [20]. The node that lacks outgoing edges is called a 'test' or an 'internal node', while the rest of the nodes constitute the leaves and are mainly known as a 'decision' or 'terminal' node. In the study, each of the internal nodes split the space into two or more subspaces based on the discrete function derived from the input attribute values; and this was the basis for discretizing the Age attribute. A random forest is a group of trees and it can also be used as a classifier or as a regression model. Some of the commonly used regression algorithms in both trees and random forests include, but are not limited to, ID3, CHAID and CRT [19]. Nonetheless, it is crucial to assert that a simple Decision Tree is apt and comprehensible in decision making. In conventional methods, the true complexity of a Decision Tree is measured using the total number of nodes, the sum of the number of leaves, the depth of the tree and the number of variables used as inputs.

3.3.3 ID3, CHAID and CRT

The iterative Dichotomiser 3 algorithm is the simplest Decision Tree-constructing algorithm. While relying on the information gained to control splitting, the algorithm terminates tree growth when all cases are assigned a single value of the predictand or when the minimized entropy is zero or almost zero [21]. It is critical to note that the algorithm does not implement any pruning process, nor does it handle non-categorical variables or data with missing information. Regarding the use of information gained to cease growth, ID3 relies on the entropy information formula in Eq. 2.

The Chi-squared-Automatic-Interaction Detection (CHAID) was also initially intended to analyze nominal variables only [22]. The algorithm finds a pair of values for input that is least significantly different with regard to the predictand. The significance of the difference is based on the of the chosen statistical test. The choice of the statistical technical is also dependent on the nature of the predictand. For instance, F test suits continuous predictands; the Pearson Chi test suits nominal predictands, while a likelihood ratio test is suitable for ordinal predictands [21]. After discretization of Age, all the variables used in the study were ordinal and hence, the use of ratio test in the CHAID algorithm.

3.3.4 Neural Network

The NN algorithm is based on multilayer perceptron algorithm. The multi-layer perceptron (MLP) is the neural network used for real-world data mining applications. Generically, it consists of hidden layers of adaptive weights having a complete connection between the layers [21]. The layers are either input layer, hidden layer or an output layer. The algorithm can approximate subjective accuracy and any continuous function from a non-spurious region of input space. For this reason, it is called a universal approximator. The degree of accuracy depends on the number of hidden layers, the weights and biases.

3.3.5 Support Vector Machine (SVM)

The SVM is a high-performing pattern recognition technology used in different fields of research. The algorithm operates by deploying a decision function that tries to divide the training subset into two levels or classes or categories. When selecting the decision function, it is ensured that the distance to the nearest points in the training data on either side of the hyper plane is maximized [19]. If the decision function fails to linearly separate the data points, a kernel function is implemented to transform the data into a different dimensional space with features supported by the SVM algorithm.

3.3.6 Stochastic Gradient Descent

The SGD is one of the gradient descent methods used to penalize linear discriminant analysis. It should be noted that neural net and deep-learning algorithms use SGD to penalize the high cost associated with either the forward or the backward propagation during model learning. The SGD included in these two algorithms lead to faster convergence and the algorithm that updates the parameters θ of the objective operator $J(\theta)$ is as shown below [23]:

$$\theta = \theta - \alpha \nabla_\theta E[J(\theta)] \tag{4}$$

The expectations E […] in the equation is estimated through the evaluation of the cost and gradient associated with the training set. When applied as an algorithm or data learner, SGD omits the expectation function and calculates the gradient of the parameter so that the equation becomes:

$$\theta = \theta - \alpha \nabla_\theta J(\theta; \rho^{(i)}, \beta^{(i)}) \tag{5}$$

In Eq. 5 $\rho^{(i)}, \beta^{(i)}$ a pair of the attributes is drawn from the training and their gradient or slope computed independent of the other observations [23]. The SGD can be thought of as a variant of correlation coefficient but with a focus on using or focusing a single sample of the training dataset. The algorithm minimizes the variances in subsequent updates, leading to more stable convergence. It also uses highly optimized matrix operations that would otherwise necessitate batching.

3.3.7 AdaBoost

Adaptive Boosting or AdaBoost is an algorithm that combines and improves weaker classifiers or learners, that is, it refers to an approach of training an identified classifier and the algorithm only selects features that improve the performance of the model and omits irrelevant one. A conventional boosted classifier can be expressed as follows [24]:

$$F_\tau(x) = \sum_{\tau=1}^{T} f_\tau(x) \qquad (6)$$

where x refers to all model inputs, including the target variable while f_τ refers to the weaker learners or classifiers. During classification, it is assumed that each of the weaker learners yield a hypothetical output $h(x_i)$ in the training subset so that at each interactive step during boosting, a coefficient μ_t is assigned to the weaker classifier. This way, the sum of the error associated with it is minimized [24].

$$E_t = \sum_i E\ [F_{t-1}(x_i) + \mu_t h(x_i)] \qquad (7)$$

In the equation, E_t is the error sum $F_{t-1}(x_i)$ and denotes the classifier that has been boosted. In the research, Decision Tree is the classifier that is boosted and the AdaBoost process relied on SAME algorithm, assuming that the forecasting results were expected to produce strictly two classes [24]. Finally, a linear regression loss function was also specified in the adaptive boosting process.

4. Experiment

The experiment was implemented in different platforms, namely SPSS 20. As discussed in Table 1, the data used in the experiment was divided into predisposing factors and symptoms of the conditions. The specification aided in the characterization of the Yes and No lung cancer groups. The experiment was implemented in several phases as discussed in Section 2.4. The other platform handled algorithms that accepted an attribute and the test results and scoring were based on the average results obtained from the 10-fold cross-validation.

4.1 Experiment Setup

Firstly, the Age attribute was discretized using the discretize function in Weka. The data, in comma separated format, was uploaded and Age discretized; the results written on a data table and the results saved as illustrated are given in the visual diagram below:

Fig. 2: Data Results illustration.

Secondly, the saved data was loaded on the file section in the file tab in Fig. 2 and the analysis was automatically implemented, the scoring results collated and stored in the Test & Score visual element. The element contains the metrics used to compare the different algorithms and a sample of the algorithms and results is as shown below.

Thirdly, the rest of the visual graphs (ROC curve), predictions and performance metrics were retrieved using ROC Analysis, Confusion Matrix and Prediction elements of the visual program in Fig. 3. The CRT, CHAID and the neural net classifiers were

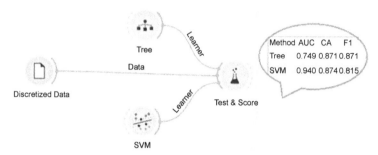

Fig. 3: Discretized data illustration.

constructed in SPSS. The data file was imported and the variables defined to meet SPSS data requirement. The algorithms were implemented in Analyze → Classify → Decision Trees before choosing the different pruning algorithms (CHAID and CRT). The neural net was also implemented using the same path, that is, Analyze → Neural Net → Options.

4.2 Model Evaluations

Both performance metrics and evaluation curves were used to evaluate the models. The metrics used in the study included accuracy, precision, recall, F-1 measure and area under the ROC curve. The formula used in the computation of the metrics is documented in the guide to implement recommender systems in [25]. The curves used in the study include ROC, lift and calibration plots. It is important to note that a 10-fold cross-validation was used in the study so that accuracy and error rates presented in confusion matrix tables for each of the algorithms is an average.

5. Results and Discussion

The output of the group characterization and the results of selected individual algorithms are presented, interpreted and discussed in the following sections:

5.1 Preprocessing and Group Characterization

The characterization of the lung cancer incidence into two groups based on the provided attributes is summarized in Table 2. The aim of the group characterization was to show differences between the groups were based on mean and standard deviation for continuous attributes and recall and precision for discrete attributes. The table shows that 87.4 per cent of the cases were lung cancer positive while only 12.6 per cent were free from lung cancer. The subsequent clustering and learning algorithms should also predict or classify the cases correctly and the best performing ones should achieve an accuracy of about 87 per cent for cancer cases and 13 per cent for non-lung cancer cases. According to the characterization (Table 2), allergy is the leading cause of lung cancer although it can be thought of as the leading predisposing factor. Alcohol consumption follows the leading cause of lung cancer and smoking ranks as the least cause of lung cancer among the lung cancer positive incident group. Conversely, among those who do not have lung cancer, smoking is the leading cause

Table 2: The group characterization of lung cancer using mean and standard deviation.

Description of LUNG_CANCER							
LUNG_CANCER=YES				LUNG_CANCER=NO			
Examples	[87.4%] 270			Examples	[12.6%] 39		
Attribute - Desc	Test value	Group	Overall	Att – Dec	Test value	Group	Overall
Continuous attributes: Mean (STDV)				Continuous attributes: Mean (STDV)			
ALLERGY	5.74	1.62 (0.49)	1.56 (0.50)	SMOKING	−1.02	1.49 (0.51)	1.56 (0.50)
AC	5.06	1.61 (0.49)	1.56 (0.50)	SOB	−1.06	1.56 (0.50)	1.64 (0.48)
SD.	4.55	1.52 (0.50)	1.47 (0.50)	AGE	−1.57	60.74 (9.63)	62.67 (8.21)
WHEEZING	4.37	1.60 (0.49)	1.56 (0.50)	CHRONIC D.	−1.94	1.36 (0.49)	1.50 (0.50)
COUGHING	4.36	1.63 (0.48)	1.58 (0.49)	ANXIETY	−2.54	1.31 (0.47)	1.50 (0.50)
CHEST PAIN	3.34	1.59 (0.49)	1.56 (0.50)	FATIGUE	−2.64	1.49 (0.51)	1.67 (0.47)
PEER_PRE.	3.27	1.54 (0.50)	1.50 (0.50)	YELLOW_ FIN.	−3.18	1.33 (0.48)	1.57 (0.50)
YELLOW_ FIN.	3.18	1.60 (0.49)	1.57 (0.50)	PEER_PRE.	−3.27	1.26 (0.44)	1.50 (0.50)
FATIGUE	2.64	1.70 (0.46)	1.67 (0.47)	CHEST PAIN	−3.34	1.31 (0.47)	1.56 (0.50)
ANXIETY	2.54	1.53 (0.50)	1.50 (0.50)	COUGHING	−4.36	1.26 (0.44)	1.58 (0.49)
CHRONIC D.	1.94	1.53 (0.50)	1.50 (0.50)	WHEEZING	−4.37	1.23 (0.43)	1.56 (0.50)
AGE	1.57	62.95 (7.97)	62.67 (8.21)	SD	−4.55	1.13 (0.34)	1.47 (0.50)
SOB	1.06	1.65 (0.48)	1.64 (0.48)	AC	−5.06	1.18 (0.39)	1.56 (0.50)
SMOKING	1.02	1.57 (0.50)	1.56 (0.50)	ALLERGY	−5.74	1.13 (0.34)	1.56 (0.50)
Discrete attributes: [Recall] Accuracy				Discrete attributes: [Recall] Accuracy			

Some of the additional attribute abbreviations and shortening include AC (Alcohol Consumption), SD (Swallowing Difficulty), PEER_PRE (Peer Pressure), YELLOW_FIN (Yellow Fingers), CHRONIC D (Chronic Disease) and SOB (Shortness of Breath).

of lung cancer, while age, chronic diseases and anxiety are the leading predisposing factors. Regarding symptoms, people without lung cancer are likely to develop it because of the aforementioned exposure conditions and predisposing factors tend to be fatigued and experienced chest pain. As for those with lung cancer, wheezing, coughing and chest pain is the primary symptom of the condition.

5.2 Model Results

The results of the individual algorithms and the comparison of all models are discussed in the following section:

5.2.1 Naïve Bayes

The classification of lung cancer incidences was based on contribution probabilities and absolute importance of each of the predisposing factors and symptoms. Based on the log odds ratios, each of the predisposing factors and symptoms of lung cancer had varying probabilities of influence. In specific, Allergy, Alcohol Consumption, Swallowing Difficulty, Wheezing, and Coughing had log odds ratios of 37 per cent, 40 per cent, 32 per cent, 43 per cent and 46 per cent respectively.

The points in Fig. 4 are all 0.0 although the log odds are computed; and it is critical to note that all the log odds ratio is positive so that 67 per cent, 60 per cent, 68 per cent, 57 per cent and 54 per cent log are more likely for non-occurrence of lung cancer. Under the given probabilities, it is highly likely for individuals predisposed to allergic conditions and consuming alcohol to wheeze, have difficulty in swallowing and cough as precursors to developing lung cancer. In general, the Naïve Bayes algorithm establishes that given an individual has an allergy, drinks alcohol and experiences swallowing difficulty, wheezes and coughs, then there is 87 per cent chance that he has lung cancer.

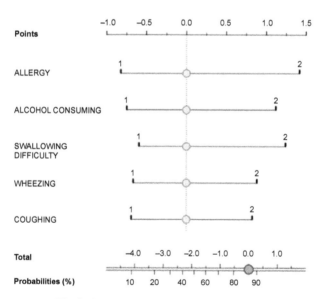

Fig. 4: A nomogram for lung cancer 'Yes Group'.

Regarding the probabilistic performance of the model concerning Lung Cancer 'No' Group, the following nomogram summarizes the results:

The points in Fig. 5 are all 0.0 although the log odds are computed and it is critical to note that all the log odds ratio is positive, so that 67 per cent, 60 per cent, 68 per cent,

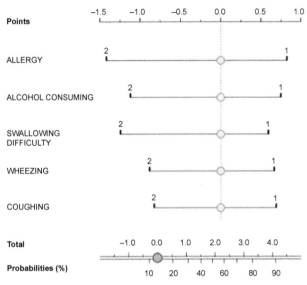

Fig. 5: A nomogram for lung cancer 'No Group'.

57 per cent, and 54 per cent log odds are more likely for the occurrence of lung cancer. Figure 5, unlike Fig. 4, shows that people who do not have lung cancer have higher chances of getting affected by allergies and alcohol consumption. The log odds ratio of a lung cancer-free individual having the disease when allergic, consuming alcohol, having breathing difficulties, wheezing and coughing is 67 per cent, 60 per cent, 68 per cent, 57 per cent, and 54 per cent respectively. Nonetheless, the probabilities of the person being non-allergic, non-alcohol consumer, not having difficulty in breathing and not wheezing or coughing and not having lung cancer are 37 per cent, 40 per cent, 32 per cent, 43 per cent, 46 per cent, and 13 per cent respectively. In general, people who do not have cancer but are exposed to conditions and express these symptoms have 0.13 probability of developing the disease.

5.2.2 Decision Tree and Random Forest

The Decision Tree visualization of the model output is shown in Fig. 6. The tree consists of 29 nodes and 15 leaves. The root node consists of the 'Yes' lung cancer group and it illustrates that 87.4 per cent of the cancer cases were associated with 75.2 per cent of them manifesting as swallowing difficulties (Fig. 6). Of the 103 or 75.2 per cent with swallowing difficulty cases, 92.8 per cent are ascribed to peer-pressure, and 100 per cent or 46 were cancer cases although 78.3 per cent of the non-cancer cases engaged in peer pressure activities that led to wheezing (Fig. 6). More importantly, 57.4 per cent of the swallowing-difficulty cases were associated with wheezing. The model shows that allergy is the root cause of these conditions; that is, of the 309 cases, 167 had allergies, while 137 did not.

The model shows that 270 of the cases were cancerous and the leading predisposing factor was allergy while swallowing and wheezing were the key symptoms. The other

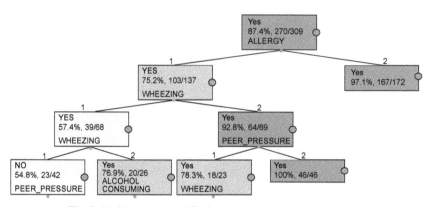

Fig. 6: Decision Tree model for the lung cancer prognosis survey data.

predisposing factors that can point to the symptoms include peer pressure and alcohol consumption.

The random forest, consisting of five trees and three attributes considered at each split, is as shown in Fig. 7.

The model shows that 85.8 per cent of the cases were cancerous and the leading predisposing factors were allergy, smoking and alcohol consumption with symptoms, like swallowing difficulty, wheezing, yellow fingers and fatigue. The model expansion has included the influence of gender on lung cancer although its influence is based on alcohol consumption.

The random forest, unlike the Decision Tree, classifies only 85.8 per cent of the cases as cancerous although it introduces conditions, such as smoking, fatigue and yellow fingers in determining lung cancer. Suppose an individual is not allergic, he or she is 91 per cent likely to develop lung cancer if he or she smokes and 71.1 per cent chance if he or she consumes alcohol. If the person develops lung cancer due to alcohol consumption, that is, there is 62.1 per cent likelihood of him wheezing and 90.2 per cent chance of feeling fatigued. Those who whizz are either coughing (50.9 per cent) or smoking (79.4 per cent). Of those who are fatigued, there is 78.9 per cent chance that they will experience difficulty when swallowing. Regarding smoking, those who smoke stand 100 per cent chance of developing lung cancer, though gender modifies the vulnerability to lung cancer. That is, women have 93.1 per cent chance of developing lung cancer if they smoke and of this proportion, 91.5 per cent of the

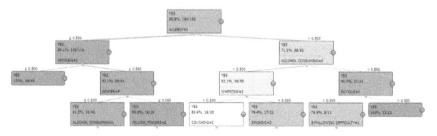

Fig. 7: Random forest model for the lung cancer prognosis data.

susceptibility can be attributed to alcohol consumption and there is a 96.8 per cent chance that the malady will manifest as yellow fingers.

5.2.3 ID3, CHAID and CRT

These are advanced Decision Tree learning algorithms. The three algorithms use different tree growth and pruning approaches. The ID3 model tree is as shown below:

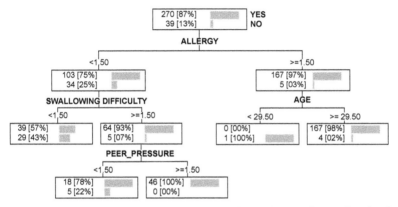

Fig. 8: The ID3 tree model showing lung cancer proportions and the leading predisposing factor (allergy) and symptom (swallowing difficulty).

The model includes Age as a predisposing factor and this marks the difference with a Decision Tree, random forest, and Naïve Bayes models. The ID3 model shows that 87 per cent of the cases are cancerous while 13 per cent are not. This establishes allergy as the leading cause of cancer. In cases of non-allergy cancer, the patient is likely to have swallowing difficulty and peer pressure remains the leading driving factor. However, in cases where lung cancer is associated with age, the model suggests that people who are above 30 years are more vulnerable to lung cancer caused by allergies than those who are less than 30 years old.

The CHAID model produced a similar tree as the ID3 model although it excluded the age factor. The model is shown in Fig. 9:

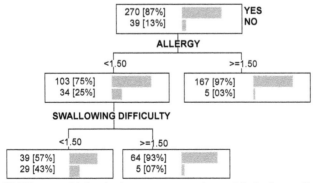

Fig. 9: The CHAID tree model showing lung cancer proportions and the leading predisposing factor (allergy), and symptom (swallowing difficulty)

The performance of the ID3 and CHAID classifiers can be summarized as follows: Table 3 shows that the model accurately predicted 100 per cent of the 'Yes' lung cancer group but predicted the 'No' lung cancer group with an error of 12.6 per cent. Based on the characterization in Table 1, the model ought to have predicted 13 per cent of the 'No' lung cancer correctly, although this was not the case.

Table 3: The performance of the two classifiers (ID3 and CHAID) based on prediction of lung cancer groups.

Error Rate			0.1262			
Values Prediction			**Confusion Matrix**			
Value	**Recall**	**1-Precision**		**YES**	**NO**	**Sum**
YES	1.0000	0.1262	YES	270	0	270
NO	0.0000	1.0000	NO	39	0	39
			Sum	309	0	309

Figures 8 and 9 show that ID3 and CHAID predicted 87 per cent and 13 per cent for lung cancer and non-lung cancer groups. In reference to Table 1, the accuracy of the two models in predicting the cancer cases is 100 per cent, while that of non-cancer cases is 0 per cent. The objective of the models is to detect cancer occurrence among healthy patients and in this regard, it suffices to make an inference that the two algorithms do not suit the machine-learning activity because they ought to predict the lung cancer in 'No' group accurately.

The CRT model also had similar classification performance although the algorithm arrived at a single node. The performance of the model is summarized in the table below:

Table 4: The performance of the CRT based on prediction of lung cancer groups.

Error Rate			0.1262			
Values Prediction			**Confusion Matrix**			
Value	**Recall**	**1-Precision**		**YES**	**NO**	**Sum**
YES	1.0000	0.1262	YES	270	0	270
NO	0.0000	1.0000	NO	39	0	39
			Sum	309	0	309

The error and accuracy rates of the CRT model are similar to those obtained for ID3 and CHAID models. The algorithm partitioned the data into growing and pruning subsets. Based on the table, the CRT had 207 observations predicted in the 'Yes' group and 39 in the 'No' group. Hence, the model had 100 per cent 'Yes' prediction accuracy and 12.6 per cent 'No' group prediction error. The tree growing sequence from the model can be summarized as follows:

The number of leaves and notes that corresponded to the 1-precision value in Table 5 correspond to N = 3 and one is left. The growing error is 0.1208 and the

Table 5: Error associated with CRT growth model. The desired number of trees is achieved when the growing subset error is approximately equal to the 1-precision value in Table 4.

N	# Leaves	Err (growing set)	Err (pruning set)	SE (pruning set)	x
3	1	0.1208	0.1373	0.0341	0.614636
2	6	0.0773	0.1176	0.0319	0.000000
1	8	0.0676	0.1471	0.0351	–

resultant tree has one node and one leaf. The Decision Tree can be summarized as follows:

'LUNG_CANCER = YES (87.92 per cent of 207 examples)'

5.2.4 Neural Network

As stated in the methodology section, both deep learning and neural net algorithms are based on multi-perceptron or propagation association between hidden layers. The multi-perceptron diagram illustrates the hidden layers of the neural net as shown in Fig. 10. The size of the scaling layer is 15 and the number of inputs. The scaling method for this layer is automatic. Table 6 shows the values which are used for scaling the inputs, which include the minimum, maximum, mean and standard deviations.

The number of layers in the neural network is two. The architecture of this neural network can be written as 15:3:1. The number of inputs is 15 and the number of outputs is one. The complexity, represented by the numbers of hidden neurons, is three. The size of the probabilistic layer is one and as the number of outputs. The probabilistic

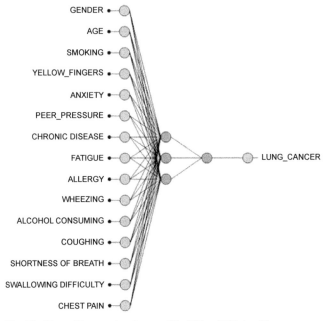

Fig. 10: The multi-perceptron layers of the NN and DP algorithms.

Table 6: The size of scaling layer is 15, the number of inputs. the scaling method for this layer is automatic.

	Min	Max	Mean	Deviation
GENDER	0	1	0	1
AGE	21	87	0	1
SMOKING	1	2	0	1
YELLOW_FINGERS	1	2	0	1
ANXIETY	1	2	0	1
PEER_PRESSURE	1	2	0	1
CHRONIC DISEASE	1	2	0	1
FATIGUE	1	2	0	1
ALLERGY	1	2	0	1
WHEEZING	1	2	0	1
ALCOHOL CONSUMING	1	2	0	1
COUGHING	1	2	0	1
SHORTNESS OF BREATH	1	2	0	1
SWALLOWING DIFFICULTY	1	2	0	1
CHEST PAIN	1	2	0	1

method for this layer is the probability. The graphical representation of the network architecture contains a scaling layer, a neural network and a probabilistic layer. The yellow circles represent scaling neurons, the blue circles, perceptron neurons and the red circles show probabilistic neurons.

The following table shows the values which are used for scaling the inputs, which include the minimum, maximum, mean and standard deviation. It generalizes the minimum, maximum, average and standard deviation. The scaling used in developing the neural network ensures that values exceed the bounds created by the minimum and maximum.

5.2.5 SVM, AdaBoost and SGD

The silhouette plot obtained from the SVM model is shown in Fig. 11. The distances are based on Manhattan method and the figure suggests a possibility of having variations within the two classes.

The variations shown in the figure can be attributed to differences in the degree of influence of each predisposing factor and the symptom to early cancer detection. The results of AdaBoost and SGD model could not be visualized but their performance was compared to the other algorithms and the results are presented in the following section. However, the heat map associated with the SGD suggests the polarity associated with the 'Yes' lung cancer group ranging between –0.29 and 2.15. The elements and chi-square of the sieve diagram deduced from the algorithm suggest increasing risks with increased exposure to the predisposing factors. At the same time, the model result suggests that on observing the symptoms, among patients without historical cancer cases, suggests the very high likelihood of lung cancer incidences.

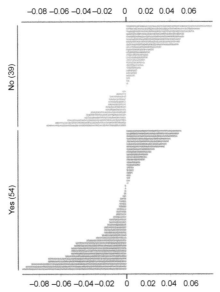

Fig. 11: The SVM Model silhouette plot.

5.3 Model Performance Evaluation

According to Table 2, 'Yes' and 'No' lung cancer groups had 270 and 39 cases respectively. The Decision Tree algorithm predicted 252 and 19 instances correctly. However, random forest predicted 257 and 18 cases correctly. Consequently, the random forest algorithm performed better than the Decision Tree algorithm. The AdaBoost algorithm penalized the errors in the Decision Tree and predicted 258 and 28 instances in the 'Yes' and 'No' lung cancer groups. Based on these predictions, it is evident that AdaBoost performed better than both the Decision Tree and random forest algorithms. However, it is imperative to note that AdaBoost, Decision Tree and random forest are based on regression just as ID3, CRT and CHAID algorithms. When compared, based on regression, the logistic regression predicted 262 and 22 of the 'Yes' and 'No' group incidences correctly.

The metrics summarizing the performance of the models is presented in the table below.

Based on Table 7, SVM is the most efficient algorithm although it was closely followed by logistic regression, Naïve Bayes and Neural Network. The table also shows that AdaBoost was more efficient than the Decision Tree. However, regarding F-1 measure, AdaBoost was the next best performing and it had the highest precision and recall. The visualization of the comparison of the efficiency of the algorithms is shown in Fig. 12.

The accuracy classification score (CA) in Table 7 shows that AdaBoost had 92.6 per cent classification accuracy. Logistic regression, SGD, Neural Network and Naïve Bayes had classification accuracy of 90 per cent, while the rest had classification accuracy of less than 90 per cent. The balanced F-score always show that AdaBoost, Logistic Registration, SGD, NN, and NB had the best scores compared to the rest, although AdaBoost outperformed all of them

Table 7: Average over classes based on the 10-fold cross-validation.

Algorithm	AUC	CA	F1	Precision	Recall	LogLoss
SVM	0.943	0.874	0.815	0.764	0.874	0.239
Tree	0.759	0.877	0.876	0.874	0.877	1.135
Random Forest	0.92	0.89	0.884	0.881	0.89	0.322
kNN	0.9	0.9	0.892	0.89	0.9	0.822
ID3	0.86	0.83	0.876	0.874	0.877	1.135
CHAID	0.88	0.89	0.892	0.89	0.9	0.768
CRT	0.91	0.89	0.92	0.913	0.918	1.906
Naive Bayes	0.934	0.903	0.904	0.905	0.903	0.244
Neural Network	0.931	0.909	0.905	0.903	0.909	0.231
SGD	0.794	0.916	0.915	0.914	0.916	2.906
Logistic Regression	0.941	0.919	0.914	0.913	0.919	0.195
AdaBoost	0.902	0.926	0.926	0.926	0.926	0.392

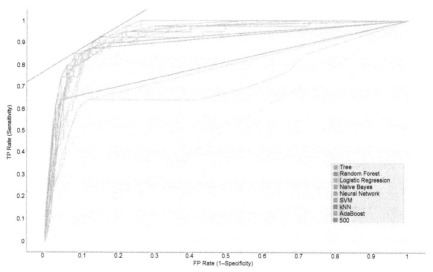

Fig. 12: The AUC's efficiency results of algorithms.

The diagram shows the area under the curve for each of the algorithms. The above figure asserts that SVM, Random Forest, Naïve Bayes, Neural Network, Logistic Regression and AdaBoost are the most efficient algorithms. Even though AdaBoost covers an area of 0.902 while logistic regression ROC curves cover an area of 0.941. As such, logistic regression, NN and NB are more efficient in retrieving information regarding the two groups. However, logistic regression has the least cross entropy value (log loss) although it is non-probabilistic and such can be rated based on this parameter.

6. Conclusion and Future Work

The purpose of the study is to appraise supervised machine-learning algorithm and identify an optimum approach to detect and predict lung cancer incidences at very early stages. The dataset is collected from the online lung cancer Web system. The data treated as a prognosis survey data and implemented on 12 algorithms for the purpose of establishing the most efficient way of making early lung cancer detection. The study relied on prognostic data containing information that was categorized into predisposing factors and symptoms. Of the 12 algorithms, AdaBoost was the most accurate with the highest sensitivity and specificity. The adaptive boosting algorithm was applied the Decision Tree classifier. Among the other 11 algorithms, Decision Tree was the weakest although adaptive boosting rendered it the strongest. The model, AdaBoost, established that allergy, smoking and alcohol consumption are the critical predisposing factors to lung cancer. AdaBoost, like other models, identified allergy as the leading cause of lung cancer and it was deduced that the allergy reported is manifested as an irritation to the lungs and hence the theory that inhalation of toxic substances, including chemical components of tobacco, is the major condition that leads to lung cancer. Other factors, such as alcohol consumption and peer pressure, are contributing to the condition in different ways; that is, alcohol serves as a source of allergens or lethal chemicals as its content and consumption exposes the lungs. The role of peer pressure is more than illustrative in modeling terms. Future studies should consider using peer pressure as an illustrative factor and not a predictor attribute. Moreover, it would be prudent if future studies consider the influence of secondhand smoking on lung cancer detection and prediction. Regarding adaptive boosting, it would be prudent to implant the boosting on some of the best-performing algorithms, such as SVM. The data requirement may prove difficult to handle although a successful implementation may improve the performance of the algorithm.

Acknowledgement

We appreciate the Data World platform for sharing and availing the survey lung cancer dataset (https://data.world/sta427ceyin). Specifically we appreciate Stacey Robert for contributing the dataset to Data World and all those who contributed to the success of the paper.

Key Terminology & Definitions

Machine Learning is one of the branches of artificial intelligence that is interested in the design and development of algorithms as well as technologies that allow computers to have the property of learning. The main task of automated learning is to extract valuable information from the data.

Big Data—is a set of datasets that are so large and complex and often very difficult to address with only one database management tool or traditional data processing applications. Big data challenges include capture, storage and analysis.

Artificial Intelligence—specific characteristics of computer systems that simulate human mental capacities. One of the key features is the ability to learn, infer and react to situations that have not been programmed into the machine.

Cancer—is a group of diseases characterized by its aggressive cells. This disease can affect almost all members of the body. It is also defined as an abnormal growth of tissue from the tissues of the body, so it affects several types of organs and the symptoms usually vary with the different organs or tissues injured.

Secure Socket Layer—is one of the widely used and most powerful security protocol that creates an encrypted link between a server and client, e.g., Web server and the internet.

References

[1] Kourou, K., Exarchos, T., Exarchos, K., Karamouzis, M. and Fotiadis, D. (2015). Machine learning applications in cancer prognosis and prediction. Computational and Structural Biotechnology Journal 13: 8–17.

[2] Siegel, R., Miller, K. and Jemal, A. (2018). Cancer statistics. CA: A Cancer Journal for Clinicians 68(1): 7–30.

[3] Zhang, Z., Sun, R., Zhao, C., Wang, J., Chang, C.K. and Gupta, B.B. (2017). CyVOD: A novel trinity multimedia social network scheme. Multimedia Tools and Applications 76(18): 18513–18529.

[4] Tewari, A. and Gupta, B.B. (2017). A lightweight mutual authentication protocol based on elliptic curve cryptography for IoT devices. International Journal of Advanced Intelligence Paradigms 9(2-3): 111–121.

[5] Gupta, B., Agrawal, D.P. and Yamaguchi, S. (eds.). (2016). Handbook of research on modern cryptographic solutions for computer and cybersecurity. IGI Global.

[6] Memos, V.A., Psannis, K.E., Ishibashi, Y., Kim, B.G. and Gupta, B.B. (2017). An efficient algorithm for media-based surveillance system (EAMSuS) in IoT smart city framework. Future Generation Computer Systems.

[7] Yu, Chuying, Jianzhong Li, Xuan Li, Xuechang Ren and Gupta, B.B. (2017). Four-image encryption scheme based on quaternion Fresnel transform, chaos and computer-generated hologram. Multimedia Tools and Applications 77: 4585–4608.

[8] Kavakiotis, I., Tsave, O., Salifoglou, A., Maglaveras, N., Vlahavas, I. and Chouvarda, I. (2017). Machine learning and data mining methods in diabetes research. Computational and Structural Biotechnology Journal 15: 104–116.

[9] Asri, H., Mousannif, H., Moatassime, H. and Noel, T. (2016). Using machine learning algorithms for breast cancer risk prediction and diagnosis. Procedia Computer Science 83: 1064–1069.

[10] Vishnu, A.V. et al. (2017). Early prediction of lung cancer using MRI images. International Journal of Innovations & Advancement in Computer Science 6: (11).

[11] Ghoneim, A., Muhammad, G., Amin, S. and Gupta, B. (2018). Medical image forgery detection for smart healthcare. IEEE Communications Magazine 56(4): 33–37.

[12] Liu, H., Guo, Q., Wang, G., Gupta, B.B. and Zhang, C. (2017). Medical image resolution enhancement for healthcare using nonlocal self-similarity and low-rank prior. Multimedia Tools and Applications, pp. 1–18.

[13] Abd El-Latif, A., Abd-El-Atty, B., Hossain, M. Rahman, M., Alamri, A. and Gupta, B. (2018). Efficient quantum information hiding for remote medical image sharing. IEEE Access 6: 21075–21083.

[14] Shehab, M., Al-Ayyoub, M., Jararweh, Y. and Jarrah, M. (2017). Accelerating computer-intensive image segmentation algorithms using GPUs. The Journal of Supercomputing 73(5): 1929–1951.

[15] Al-Ayyoub, M., AlZu'bi, S., Jararweh, Y., Shehab, M.A. and Gupta, B.B. (2016). Accelerating 3D medical volume segmentation using GPUs. Multimedia Tools and Applications, 3537–3555.

[16] Tan, C., Chen, H. and Xia, C. (2009). Early prediction of lung cancer based on the combination of trace element analysis in urine and an AdaBoost algorithm. Journal of Pharmaceutical and Biomedical Analysis 49(3): 746–752.

[17] Malla, Y. (2017). A machine learning approach for early prediction of breast cancer. International Journal of Engineering and Computer Science.

[18] Kondratyev, A. and Giorgidze, G. (2017). MVA Optimization with machine learning algorithms. SSRN Electronic Journal.

[19] Wu, C. et al. (2016). Decision Tree induction with a constrained number of leaf nodes. Appl. Intell. 45(3): 673–685.

[20] Koshkarov, A. (2018). Machine learning methods in digital agriculture: Algorithms and cases. International Journal of Advanced Studies 8(1): 11.

[21] Enache-David, N., Sangeorzan, L. and Stella, G. (2017). Data analysis—Between theory and practice. Review of the Air Force Academy (2): 85–92.

[22] Zubek, J. and Plewczynski, D.M. (2016). Complexity curve: A graphical measure of data complexity and classifier performance. Peer J. Computer Science.

[23] Hardt, B.M. and Singer, Y. (2015). Train faster, generalize better: Stability of stochastic gradient descent. arXiv preprint arXiv: 1509.01240.

[24] Rajaguru, H. and Sunil, K.P. (2017). Analysis of AdaBoost classifier from compressed EEG features for epilepsy detection. pp. 981–984. *In*: Computing Methodologies and Communication (ICCMC), 2017, International Conference on, IEEE.

[25] Hahsler, M., Brent, V. and Hahsler, M. (2017). Package 'recommenderlab'.

CHAPTER 11

Preventing Black Hole Attack in AODV Routing Protocol using Dynamic Trust Handshake-based Malicious Behavior Detection

Bhawna Singla,[1,*] *A.K. Verma*[1] *and* *L.R. Raheja*[2]

1. Introduction

Mobile Ad hoc network (MANET) [1] is a collection of nodes in wireless network in which nodes keep on changing position to have a dynamic topology. Topology keeps on changing therefore the path from source node to destination node also keeps on changing, which further is determined by routing protocol. In this work, we are using the reactive routing protocol called Adhoc On Demand Distance Vector Routing Protocol (AODV) [2, 3] where the route is determined on demand, i.e., whenever there is a requirement of route then and only then rhe current route from the source to destination is determined [4–9]. AODV routing protocol has several vulnerabilities, such as

(a) A malicious node can drop any of the control packet or data packets.
(b) A malicious node can modify any field of the control packet and can then forward the packet to its immediate neighbor.
(c) The malicious node can send the faked RREP or route reply acknowledgment (RREP_ACK) in response to the control message or it may send fake response message of its own.
(d) In such a way, the malicious node may cause the route breakage, which may lead to node isolation or flooding of packets, leading to resource consumption. Due to

[1] Thapar University, Patiala, India; akverma@thapar.edu
[2] IIT Kharagpur, India; lrr_2004@yahoo.com
* Corresponding author: bhawna_singla@yahoo.com

its property that malicious node can also modify fields of the control packet, the malicious mode may impersonate any other node or it may leak the confidential information to the unauthorized node.

In AODV routing protocol, the working depends on the genuine cooperation of node. If any of the intermediate nodes is selfish or non-cooperating or malicious, then the working of complete protocol is compromised. Attacks are targeted to damage the basic aspects of security, like integrity, confidentiality and privacy. The nodes performing adverse effects on MANETS are classified into two categories—malicious node [10] and selfish node [11]. Malicious nodes are those nodes that perform an active attack on MANETS and may be active in route establishment or data forwarding phase, while selfish node performs passively by not forwarding the packet just for the sake of saving battery energy.

Due to the above said vulnerabilities, a number of attacks [12–20] are possible in AODV routing protocol. These attacks are broadly classified into two categories, passive or active attacks.

Passive attack: In a passive attack, the attacker's goal is just to obtain information. This means that the attack does not modify data or harm the system. However, the attack may harm the sender or receiver of the message. Main techniques of passive attack are: eavesdropping and timing analysis.

Active attack: Active attack may change the data or harm the system. Attacks that threaten integrity and availability are active attacks. Examples of active attacks on AODV are:

➢ Attacks by dropping the packets, such as Blackhole [21–30] or Grayhole attack [31].
➢ Attacks using modification of protocol message and may include redirection due to modification of Hop-Count or Modified Destination Sequence number. A very common attack in this category is Denial of Service attack [32–34] where the malicious nodes generate unwanted request packets so as to make the resources unavailable to the other nodes.
➢ Attacks using impersonation where malicious node impersonates other node.
➢ Attacks using fabrication where the malicious nodes generate false route error or message or false routing table overflow message.
➢ Other attack, such as Worm Hole attack [35–38] or Byzantine attack [39, 40], etc.

This paper focuses mainly on black hole attack. In blackhole attack, the malicious node intends itself a shaving the shortest path through it. Once it is chose as the intermediate node for the path from source to destination, it drops all the control packets and data packets that are transmitted through it. So, it impacts the performance of the protocol.

2. Implemetnation of Black Hole Attack in an AODV Routing Protocol

2.1 Black Hole Attack

The black hole problem is a type of active attack in which malicious node first claims to have the shortest path. The source node chooses the route containing the malicious node to the destination. Once the traffic is routed through itself, it drops the entire data packet routed through it [41–44]. As shown in Fig. 1, let 1 be the source node and 3 be the destination node and 4 is the malicious node. 4 claims to have the shortest path; that is why the route through 4 (1-4-5-6-3) is selected instead of 1-2-3. But after being selected in the final route, 4 drops the entire data packet. The working of black hole attack is further summarized in Fig. 2. The figure shows that if the packet forwarded is data packet and the node is malicious, then it drops the entire packet; otherwise, if the packet is RREQ control packet and the node is malicious, then it sends the fake RREP so as to claim itself as having the shortest path. Once it is chosen as the intermediate node, it drops the entire data packet routed through it. In all other cases, it behaves normally [88–90].

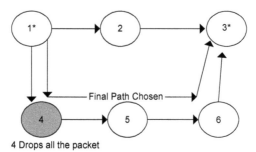

Fig. 1: Example showing the working of black hole attack where 1 is the source node, 3 is the destination node and 4 is the malicious node.

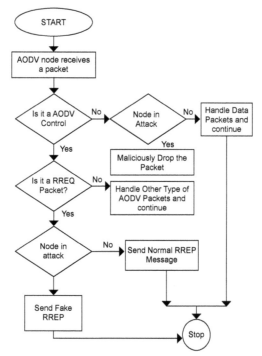

Fig. 2: Pseudo code of AODV routing protocol.

3. Proposed Work

A lot of research work has been done to find the secured AODV routing algorithm [45–54]. The trust [55–58] based on the packet forwarding behavior of neighbor can be used for detecting misbehavior. This model has been previously presented in several literatures [59–63], but, by the same trust-based logic, some of the neighbors who were silent and not actively participated in communications will get wrongly identified as malicious. So, simple trust-based models will mark a lot of non-malicious nodes as malicious nodes. This will initiate a lot of link failures. That is, the link between source to destination will get broken at different locations on their path because of this false identification of malicious nodes.

The dynamic packet forwarding-based trust AODV(DTH-AODV) proposed in this paper will overcome that problem and reduce the possibility of such false marking of non-malicious nodes as malicious nodes. A simple Dynamic Trust Handshake mechanism will help to prevent such false identification.

The main advantage of the proposed detection and prevention scheme is that it will detect and prevent the malicious nodes in the very early stage of AODV route discovery process. So, it will not need any manipulation in routing tables in the route resolving process, because by design, it will avoid including malicious hops in routing table even at the route discovery process itself.

In this work, trust value is associated with each node and initialized as 0. If the node is working genuinely, i.e., forwarding the packet as per the routing protocol instructions, then trust value is incremented; otherwise, it is decremented.

Malicious and faulty nodes are then isolated from the network once they obtain a minimum threshold value.

(a) *Packets Acknowledgment*: Acknowledgment is a method of ensuring that packets sent for forwarding have been forwarded. There are a couple of ways that this is possible but passive acknowledgment is by far the easiest to implement.

```
void TrustNode::increaseTrust()
{
        trustValue++;
}
void TrustNode::decreaseTrust()
{
        trustValue--;
}
```

Fig. 3: Calculation of trust values.

```
bool TrustNode::isNodeTrusted()
{
        if(trustValue <= threshold value)
        {
                return false;
        }
        else
        {
                return true;
        }
}
```

Fig. 4: Calculation of malicious node.

Passive acknowledgment uses promiscuous mode to monitor the channel, allowing the node to detect any transmitted packets irrelevant to the actual destination that they are intended for. With this, the node can ensure that packets it has sent to a neighboring node for forwarding are indeed forwarded. This has been implemented within PTH-AODV using promiscuous mode to monitor the channel.

(b) *Packet Precision*: As defined by Pirzada et al. [64], packet precision ensures the integrity of the data and control packets that are either received or forwarded by other nodes in the network. This type of detection aims to spot packets that have either been corrupted due to a faulty node or have been generated maliciously. This could be done by monitoring the control packets that lead to suitable successful routes. Another possible means is to check that the packet information is within certain tolerances. For example, it may be ensured that the sequence number within a reply is not inconceivably higher than the sequence number within the request, as this suggests that the replying node is trying to ensure it is part of the final route.

(c) *Destination Unreachable Messages*: Although Pirzada [64] mentions that it is possible to use destination unreachable messages, no such messages are returned by Ns2.

4. Implementation of the Proposed Malicious Behavior Detection in AODV

4.1 Implementation of Dynamic Trust Handshake Mechanism

Generally, a trust factor based on the packet-forwarding behavior of the neighbor can be used for detecting misbehavior as previously presented in several literatures. For example, a trust factor of a node can be derived based on the number of forwarded packets at that neighboring node. But, by the same trust-based detection logic, some of the neighbors, who were silent and not actively participated in communications, will get low trust factor and will be wrongly identified as malicious. Because of this, the link between the source to the destination will get broken at different locations on their path due to false identification of malicious nodes. In our proposed dynamic trust handshake-based AODV (DTH-AODV), it will overcome that problem and reduce the possibility of such false marking of non-malicious nodes as malicious nodes by introducing a dynamic trust handshake mechanism. The following flow diagram in Fig. 5 explains the implementation of dynamic trust handshake mechanism in AODV routing agent.

4.2 The Trust Handshake Message-Triggering Mechanism

In this model, the nodes will send a 'trust handshake' in a dynamic fashion based on its local state. This dynamic trust handshake mechanism ensures that at least one handshake packet will be sent just before any new transmission event. But the frequency of such a 'trust handshake' message will be controlled by two variables—the min_trust handshake_interval and max_trust handshake_interval and so, it will not increase the message overhead tremendously.

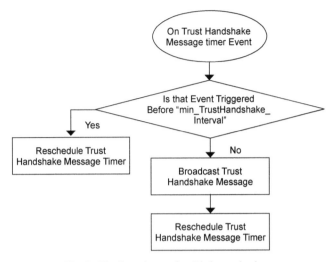

Fig. 5: The dynamic trust handshake mechanism.

The trust handshake message function will be called from different functions of AODV whenever a change in state is expected. For example, after doing a regular route table update, the trust handshake message function will be triggered. But according to the way in which the dynamic trust handshake mechanism works, it will not actually send a handshake message whenever it is triggered. The trigger mechanism may rapidly call the trust handshake messages ending function, but will actually send a new message, if and only if, there was a considerable gap (min_trust handshake_ interval) between two consecutive messages. This will avoid oversending of the trust handshake messages. The following flow diagram explains the implementation of dynamic trust handshake-based malicious node detection and prevention in AODV routing agent.

The process flow and pseudo code of dynamic trust handshake-based malicious node detection and prevention in AODV routing protocol is shown in Figs. 6 and 7.

4.3 The Changes Made in NS2AODV Code for Malicious Node Detection and Prevention

The following two files were modified to incorporate the proposed malicious node detection and prevention mechanism in AODV routing agent.

Changes made in AODV.h: The additional function definitions for detection and prevention of malicious behavior and the variables that will be bound with TCL are declared in AODV.h. By using the variables from a TCL simulation code, we can control the behavior of the routing agent and bring it to detection and prevention mode.

Changes Made in AODV.cc: The actual code of the additional function definitions for detection and prevention of malicious behavior were implemented in AODV.cc and here the new interfaces to the code through the control variables that will be bound with TCL are written here. By setting the variables from a TCL simulation code, we can control the behavior of the routing agent and bring it to detection and prevention mode.

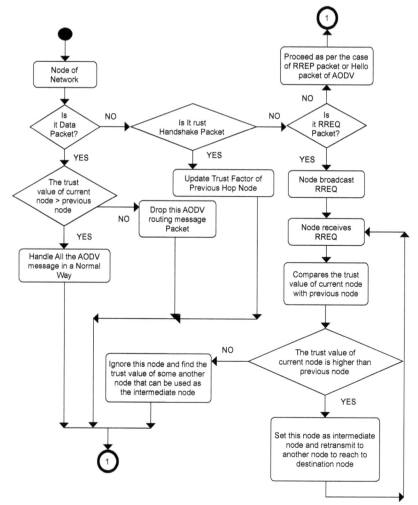

Fig. 6: The process flow of periodic trust handshake-based malicious node detection and prevention in AODV.

4.4 The Functions Modified for Attack Detection and Prevention

The function of trust handshake timer: The dynamic trust handshake mechanism is implemented with the help of a new timer function in AODV.

The function AODV:: Send Trust Handshake Packet: This function will generate a trust handshake packet and transmit it with respect to the conditions explained in Fig. 4.

The function of AODV:: Recv AODV: In this function, the trust-based detection of malicious behavior has been implemented. As shown in the Fig. 4 of previous section, the malicious behavior detection is done based on the trust factor of the previous hop node from which the message was received.

```
Forward(RREQ pkt, delay) {

    // the node receives the RREQ control packet
    // checks whether it is destination node
    if destination{
            // considers the path with highest trust value and sends the route reply along
            that path
            compute_highest_trust_level ()
                    {
                        // the optimal path with highest value value is chosen and route reply
                        is sent along that path
                        highest_trust_value(path)
                        sends_RREP_to_source
                    }
    else (not_destination){
            // if the next intermediate node is not destination then intermediate node
            checks for the packet by computing the trust level
            if RREQ_packet{
            compute_trust_level ()
                    {
                        // compares the trust value of current node with the tryst value of
                        previous node
                        trust_current_node> trust_previous_node
                    }
            if (found not ok)
            // intermediate node drops the packet if its trust level is lesser then previous
            path
            drop(pkt)
            else {
            // if the new path has more trust value then update trust and hop count and
                    rebroadcast it to next neighbour node
            trust--
            // total number of intermediate nodes is incremented by 1
```

```
    // the RREQ packet is rebroadcasted to next neighbour node
            rebroadcast RREQ
                    }
            }
    }
    }

Receive (RREP pkt, delay) {
            // waits for specified period
            if no_duplicate{
                    wait_rrep_wait_time
                    update_trust_metric
                    update next_hop}
                    else
                    {
                    compute trust_path
                    }
            }

Update_Trust_Metric (interval){
    // wait for the minimum trust handshake interval
    wait_trust_handshake_interval();
    broadcast_trust_hanshake;
    trust_value>trust_threshold {
            trust_current_node = trust_current_node - trust_previous_node
                    }
            else {
                    drop (pkt),}
            }
```

Fig 7: The pseudo code of PTH AODV.

5. Results and Discussion

We used network simulator version NS2.35 under Ubuntulinux operating system for obtaining this results [65]. We have implemented the black hole attack as well as attack detection and prevention mechanism on the AODV code of NS2 and did the simulation with the parameters presented in this section and evaluated the performance with respect to the metrics discussed in this section.

5.1 The Simulation Parameters

(a) *Common Parameters*: The following common parameters are used for setting up the network. Moreover the following parameters are also used to set TCP/UDP flows.

Table 1: Parameters values of network in NS2.

Common Parameters	Values	Traffic Parameters for TCP Flows	Values
Topographical Area (m*m)	1800 × 500	Transport Agent	**TCP**
Mobility	20 m/s	No Flows	10
Pause Time	20s	Traffic Type	CBR
Total Simulation Time	100s	Packet Size	1 KB
Routing Protocol	AODV	Interval	100 ms
Mobility Modal	Random Way point	Rate	10 KB
Channel Model	Wireless Channel	**Traffic Parameters for UDP Flows**	**Values**
Propagation Model	Two Ray Ground	Transport Agent	TCP
Phy Model	Wireless Phy	No Flows	10
Mac Model	802_11	Traffic Type	CBR
Antenna Model	Omni Antenna	Packet Size	1 KB
Queue	Drop Tail-Pri Queue	Interval	100 ms
Queue Length	50	Rate	10 KB

(b) *Variable Parameters*: The following parameters are used as variables for analyzing the impact of the attack and detection on different conditions:

Table 2: Total number of nodes, number of malicious node and different attack scenarios.

Parameters	Values
Malicious Nodes	15
Total Nodes	40,50,60
AODV with	(a) No Attack (b) Black Hole Attack (c) PTH Attack Detection

5.2 Analytic Results with Respect to Different Network Size

Here we see the analytic results of comparison of black hole attacks with normal AODV (it means performance without any attack) and it is studied with respect to different network size. In the following analysis, the total number of nodes in the network is as varied as 40, 50 and 60 and among them, the number of malicious nodes kept as 15 and the impact is measured using different metrics.

The following line graph in Fig. 8 shows the impact of attack and detection and prevention mechanism in terms of total data packets sent at the application source. As shown in the line graph, under the presence of a black hole attack, the application source itself is not able to send much, but during detection, the proposed DTH-AODV was able to send as much as normal AODV without any attack.

The following line graph in Fig. 9 shows the impact of attack and detection and prevention mechanism in terms of total data packets received at application destination. As shown in the line graph, under the presence of black hole attack, destination itself is not able to receive anything. But during detection, the proposed DTH-AODV was able to receive as much as normal AODV without any attack.

The following line graph in Fig. 10 shows the impact of attack and detection and prevention mechanism in terms of routing load. As shown in the line graph in Fig. 10, under the presence of black hole, the routing load is very high, but with proposed DTH-AODV-based detection and prevention mechanism, the routing load was almost equal to that of normal AODV. In terms of routing load, the performance of normal AODV and proposed DTH-AODV are almost equal.

The following line graph in Fig. 11 shows the impact of attack and detection and prevention mechanism in terms of MAC load. As shown in the line graph, under the presence of black hole, the MAC load is very high. But with proposed DTH-AODV-based detection and prevention mechanism, the MAC load was almost equal to that of normal AODV. In terms of MAC load, the performance of normal AODV, proposed DTH-AODV are almost equal.

The following line graph in Fig. 12 shows the impact of attack and detection and prevention mechanism in terms of total dropped packets at the application layer. As shown in the line graph, under the presence of black hole attack, a lot of packets were dropped at the application layer. But during detection, the packet dropping of proposed DTH-AODV was very much reduced and almost equal to that of normal AODV without any attack. In terms of application-layer dropped packets, the proposed

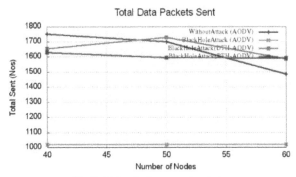

Fig. 8: Network size vs sent packets

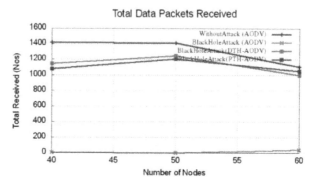

Fig. 9: Network size vs received packets.

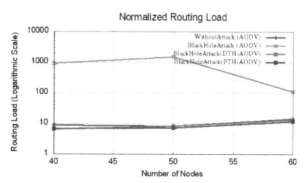

Fig. 10: Network size vs routing load.

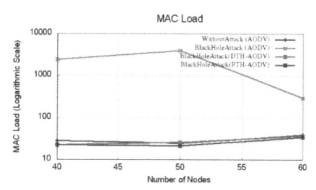

Fig. 11: Network size vs MAC load.

DTH-AODV dropped a little bit high number of packets. This is because the DTH-AODV will try to send more packets than normal AODV.

The following line graph in Fig. 13 shows the impact of attack and detection and prevention mechanism in terms of throughput. As shown in the line graph, under the presence of black hole attack the throughput was almost equal to zero. But with detection, the throughput of the proposed DTH-AODV was very much improved and almost equal to that of normal AODV without any attack.

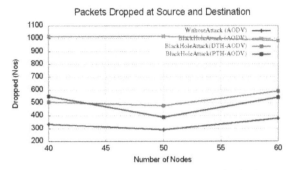

Fig. 12: Network size vs packets dropped at application layer.

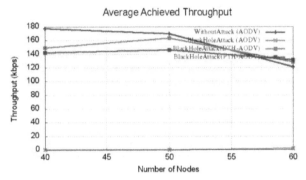

Fig. 13: Network size vs throughput.

The following line graph in Fig. 14 shows the impact of attack and detection and prevention mechanism in terms of PDF. As shown in the line graph, under the presence of black hole attack the PDF was almost equal to zero. And at low network density, PDF is equal to zero. For example, at 40 nodes, it is zero because among the 40 nodes, 15 are malicious and able to break all communication between other nodes. But with detection, the PDF of proposed DTH-AODV was very much improved and almost equal to that of normal AODV without any attack. In terms of PDF, the performance of normal AODV proposed DTH-AODV is almost equal.

The following line graph in Fig. 15 shows the impact of attack and detection and prevention mechanism in terms of End-to-End Delay (EED) of data flows. With respect to the increase of number of nodes in the network, the performance gets decreased. As shown in the line graph, black hole attack seems to be providing lower EED than normal AODV (without attack), but certainly it does not mean that black hole attack is improving the performance of the network. The low end-to-end delay under attack is due to a strange fact that the attack makes disconnection in TCP flows and since the packets are not at all forwarded to any further nodes, indirectly it reduces the message overhead in the network and reduces bandwidth usage, otherwise it will be consumed by the forwarded data packets. So, the flows that were unaffected by black hole attack (the connections where there is no neighboring attack nodes) utilizes that extra bandwidth and gains some performance. Further, keep in mind that the end-to-end delay is only calculated based on the time in which a packet is sent and received.

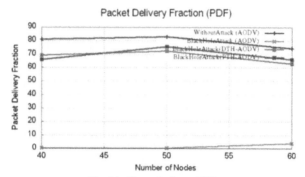

Fig. 14: Network size vs PDF.

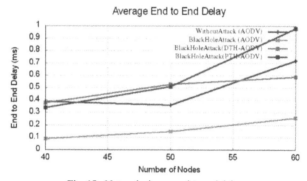

Fig. 15: Network size vs end to end delay.

So if a packet is not received, in that case end-to-end delay cannot be calculated. So this average EED is only the average EED of successfully delivered packets.

The EED of DTH-AODV was a little bit higher than normal AODV because, under attack, detection and prevention, an alternate route will be resolved by avoiding malicious nodes on the path, so that the path length will increase and hence will increase the end-to-end delay.

The following line graph in Fig. 16 shows the impact of the attack and detection and prevention mechanism in terms of consumed battery energy. As shown in the line graph, in the presence of an attack, the battery consumption is lesser than normal AODV (without attack), but certainly it does not mean that these attacks improve the performance in terms of energy consumption. The low energy consumption under attacks is due to the fact that these attacks make disconnection in data flows and since the packets are not at all forwarded to any further nodes, indirectly, it reduces the battery consumption at the other nodes; otherwise it will be consumed for forwarding the data packets. So, the nodes that were unaffected by attacks (where there is no neighboring attack nodes) preserve some battery power. Understanding of this strange fact requires a better visualization of the whole network scenario. It is simple—without any attack. AODV was able to send much and maximum nodes were able to participate in that communication and utilized the energy for transmission/forwarding of packets, so that the energy is consumed in most of the nodes. But in the presence of the attack, the packets get dropped intermediately and the battery powers on other nodes, that are not at all forwarding the packets, get preserved. With respect to the increase of a number of nodes in the network, the performance seems to be decreasing.

But, interestingly, the energy consumption in the case of proposed DTH-AODV is a little bit lesser than normal AODV. This obviously proves a better working of the proposed detection model.

Lots of previous papers say that the attacks will increase energy consumption. Of course, it also may be true, but not in the same sense. For example, if an application will continuously try to send data under attack, then the battery of the sending node and some other nodes between sender and attacker nodes will get reduced rapidly. If the application will vigorously retransmit due to loss, then this will increase the energy consumption. But the transport protocol will handle a loss scenario and just reduce the sending rate to avoid further loss. That is why the average energy consumed in the network seems to be reduced under attack. Understanding this strange fact requires a better visualization of the whole network scenario.

The following line graph in Fig. 17 shows the impact of attack and detection and prevention mechanism in terms of overhead. As shown, under the presence of black hole, the overhead is minimum because the black hole just breaks all the communication. But with the proposed DTH-AODV-based detection and prevention mechanism, the overhead becomes equal to that of normal AODV-signifying that the proposed DTH-AODV works almost equal to normal AODV.

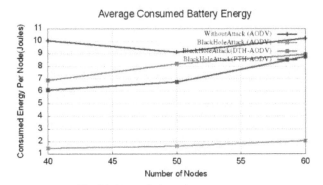

Fig. 16: Network size vs battery energy.

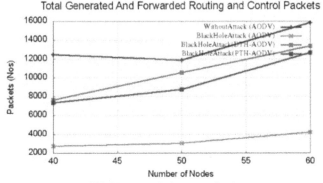

Fig. 17: Network size vs overhead.

6. Comparison of DTH-AODV with Other Trust-based Routing Algorithm

Table 3 compares the characteristics of conventional AODV-based trust routing algorithm with newly developed algorithm DTH-AODV.

Table 3: Comparison of characteristics of existing AODV-based trust routing protocol with PTH-AODV and DTH-AODV.

Secure Routing Algorithm	Characteristics	Advantages	Disadvantages
Collaborative trust-based secure routing against colluding malicious nodes [66]	This protocol design assumes the prior distribution of trust to all the nodes. It also assumes the presence of public key infrastructure because whenever the node transmits the RREQ message containing trust metric to intermediate node, the next node authenticates the previous node by signing with its private key.	This protocol is highly resistant towards attack where a malicious node claims to have genuine identity.	This protocol aims to find the shortest path to the destination node, irrespective of presence of the malicious node; therefore, it is more susceptible to internal attacks. The prior distribution of trust makes the network less dynamic and adaptable to changing situations. The use of public key infrastructure makes the protocol highly expensive to use and it also causes more overhead to maintain all the keys. This protocol fails under the situation of compromised node.
Trust-Embedded AODV (T-AODV) [67]	This protocol is an extension of 131 with the difference that the trust factor is periodically updated by the exchange of routing messages. However, this protocol also assumes the existence of public key infrastructure and it also assumes the radio range of all the nodes are same, which is more theoretical to study.	In this protocol, the following action done by the malicious node is avoided. If the malicious node provides the wrong information in the RREQ packet by changing the hop count field, the destination address, etc. If the malicious node decrypts the sign given by the genuine node with the intention to alter the information given in the header. It is more adaptable to topological changes.	This protocol fails to find the secure end-to-end path from source to destination. More overhead caused on public key infrastructure.

Table 3 contd. ...

...Table 3 contd.

Secure Routing Algorithm	Characteristics	Advantages	Disadvantages
Trust establishment in pure ad-hoc networks [68]	This protocol does not require a trusted third party infrastructure for its operation. All node computer trust the value based on direct feedback.	Malicious nodes are bypassed during route discoveries. This protocol achieves better throughput in the presence of malicious node.	Extra overhead is added due to nature of the protocol. The accuracy of protocol depends on the weight values that are assigned in the calculation of trust values. This protocol is more susceptible to IP spoofing attack and MAC spoofing attack. This protocol fails when the malicious nodes collude.
Opinion-based trusted routing protocol – TAODV [69]	This protocol uses soft encryption technique.	The encrypted parts of the message are forwarded through different routes so that the malicious nodes hardly have access to complete the message.	This protocol is susceptible to internal attack. It takes more time in route selection. It is also possible not to route all the messages securely.
Friendship-based routing algorithm – frAODV [70]	Each node stores the list of friendly nodes and friendship value. The friendship value determines the level of trustworthiness. During control packet transmission, the friendship value gets exchanged between the nodes.	The performance gives better results for the more dynamic network.	The experiment is performed based on five number of nodes so that the performance of protocol for the large number of nodes is undetermined.
DTH-AODV	This protocol does not require public key infrastructure. It also does not use any encryption technique. However, the trust value is exchanged by the nodes only when there is a need of route establishment and the trust value depends on the feedback given by the previous neighbor in the successful communication	Additional overhead caused by periodic exchange of trust metric is also avoided. This protocol detects all the selfish and malicious nodes due to exchange of trust value between the nodes. Moreover, sometimes some genuine nodes that have not been participating in the communication for over a long time, are falsely interpreted as the selfish node. This protocol also avoids false accusation of a genuine node as the selfish node.	More work can be done in case of trust dispersal and trust delay over time. Trust can also be gathered by the malicious scenarios. More work can also be done in case of malicious colluding nodes.

7. Conclusion

In this work we proposed a dynamic trust handshake-based detection of black hole attack. We implemented DTH-AODV under NS2 and compared its performance with the results of standard AODV and standard AODV under attack. The main advantage of the proposed DTH-AODV was that it detected and prevented the malicious nodes in the very early stage of route-discovery process. So, it will not need any manipulation in routing tables in the route-resolving process because by design, it will avoid including malicious hops in the routing table of normal nodes at the route-discovery process itself.

A lot of simulation and analysis is done to arrive at significant and interpretable results. The impact of the attack is measured on the detection and prevention mechanism with suitable metrics and explained the improvements in performance. According to the arrived results, the proposed dynamic trust handshake-based malicious node detection and prevention mechanism worked well and successfully detected the black hole nodes in the network and avoided establishing routes through them. As shown in the results of the previous section, the proposed DTH-AODV improved the throughput and PDF almost equal to that of normal AODV. In this work, we used unencrypted trust handshake messages in the design, but in future works, we may explore the possibility of using a private key or/public key-based encryption mechanism for a more secure operation. It may increase the operational overhead, but one may address issues related with overhead due to encryption-based trust handshake mechanism.

References

[1] Chlamtac, Imrich, Marco Conti and Jennifer J.-N. Liu. (2003). Mobile ad hoc networking: imperatives and challenges. Ad Hoc Networks 1(1): 13–64.
[2] Perkins and Royer, E. (1999). Ad hoc on-demand distance vector routing. 2nd IEEE Workshop Mobile Comp, Sys. and Apps.
[3] Kaur, Dilpreet and Sheetal Kundra. (2016). Comparative analysis and improvement in AODV protocol for path establishment in MANETS. International Journal of Computer Science and Information Security, 213.
[4] Sajid, Ahthasham et al. (2016). Performance evolution of reactive, proactive and hybrid routing protocols in MANET. International Journal of Computer Science and Information Security, 144.
[5] Kodole, Amruta and Agarkar, P.M. (2015). A survey of routing protocols in mobile ad hoc networks. Multi-Disciplinary Journal of Research in Engineering and Tech. 336–41.
[6] Neeli, Jyoti and Dr. Cauvery, N.K. (2015). Comparative study of secured routing protocols in wireless ad hoc networks: A survey. International Journal of Computer Science and Mobile Computing, 225–229.
[7] Belding-Royer, E.M. and Toh, C.K. (1999). A review of current routing protocols for ad-hoc mobile wireless networks. IEEE Personal Communications Magazine, pp. 46–55.
[8] Verma, A.K., Dave, Mayank and Joshi, R.C. (2003). Classification of routing protocols in MANET. National Symposium on Emerging Trends in Networking & Mobile Communication (NSNM-2003), pp. 132–139.
[9] Abolhasan, M., Wysocki, T. and Dutkiewicz, E. (2004). A review of routing protocols for mobile ad hoc networks. Ad Hoc Networks 2(1): 1–22.
[10] Khan, Muhammad Saleem et al. (2017). Isolating misbehaving nodes in MANETS with an adaptive trust threshold strategy. Mobile Networks and Applications, 1–17.

[11] Das, Debjit, Koushik Majumder and Anurag Dasgupta. (2015). Selfish node detection and low cost data transmission in MANET using game theory. Procedia Computer Science 54: 92–101.

[12] Khan, Muhammad Saleem, Qasim Khan Jadoon and Majid I. Khan. (2015). A comparative performance analysis of MANETS routing protocols under security attacks. Mobile and Wireless Technology, Springer, Berlin, Heidelberg, 137–145.

[13] Ferdous, Raihana and Vallipuram Muthukkumarasamy. (2016). A comparative performance analysis of MANETS routing protocols in trust-based models. Computational Science and Computational Intelligence (CSCI), International Conference on IEEE.

[14] Simaremare, Harris et al. (2015). Security and performance enhancement of AODV routing protocol. International Journal of Communication Systems, 2003–2019.

[15] Gharehkoolchian, Mahsa, A.M., Afshin Hemmatyar and Mohammad Izadi. (2015). Improving security issues in MANETS AODV routing protocol. International Conferenceon Ad Hoc Networks, Springer International Publishing.

[16] Ngadi, Md. A., Khokhar, R.H. and Mandala, S. (2008). A review current routing attacks in mobile ad-hoc networks. International Journal of Computer Science and Security 2(3): 18–29.

[17] Ponsam, J. Godwin and Dr Srinivasan, R. (2014). A survey on MANET security challenges, attacks and its countermeasures. International Journal of Emerging Trends & Technology in Computer Science (IJETTCS).

[18] Yih-Chun, Hu and Adrian Perrig. (2004). A survey of secure wireless ad hoc routing. IEEE Security & Privacy, 28–39.

[19] Nguyen, Hoang Lan and Uyen Trang Nguyen. (2008). A study of different types of attacks on multicast in mobile ad hoc networks. Ad Hoc Networks, 32–46.

[20] Wu, Bing, Jianmin Chen, Jie Wu and Mihaela Cardei. (2007). A survey of attacks and countermeasures in mobile ad hoc networks. Wireless Network Security, 103–135.

[21] Bala, Kanchan. (2016). A survey of black hole detection policies in mobile ad hoc networks. International Journal of Future Generation Communication and Networking, 295–304.

[22] Praveen, K.S., Gururaj, H.L. and Ramesh, B. (2016). Comparative analysis of black hole attack in ad hoc network using AODV and OLSR protocols. Procedia Computer Science 85: 325.

[23] Khanna, Nitin. (2016). Avoidance and mitigation of all packet drop attacks in MANETS using enhanced AODV with cryptography. International Journal of Computer Network and Information Security (IJCNIS), 37.

[24] Shahabi, Sina, Mahdieh Ghazvini and Mehdi Bakhtiarian. (2016). A modified algorithm to improve security and performance of AODV protocol against black hole attack. Wireless Networks, 1505–1511.

[25] Ghugar, Umashankar, Jayaram Pradhan and Monalisa Biswal. (2016). A novel intrusion detection system for detecting black hole attacks in wireless sensor network using AODV Protocol. IJCSN—International Journal of Computer Science and Network.

[26] Mjahidi, Mohamedi M. (2015). A survey on security solutions of AODV routing protocol against black hole attack in MANET. International Journal of Computer Applications, 1.

[27] Kumar, Sushil, Deepak Singh Rana and Sushil Chandra Dimri. (2015). Analysis and implementation of AODV routing protocol against black hole attack in MANET. International Journal of Computer Applications, 1.

[28] Kumar, Vimal and Rakesh Kumar. (2015). An adaptive approach for detection of black hole attack in mobile ad hoc network. Procedia Computer Science 48: 472–479.

[29] Jayachandra, S.H. et al. (2016). Analysis of black hole attack in ad hoc network using AODV and AOMDV protocols. Emerging Research in Computing, Information, Communication and Applications, Springer, India, 99–108.

[30] Alem, Yibeltal Fantahun and Zhao Cheng Xuan. (2010). Preventing black hole attack in mobile ad-hoc networks using anomaly detection. Future Computer and Communication (ICFCC), 2nd International Conference on, Vol. 3, IEEE.

[31] Patel, Ankit D. and Kartik Chawda. (2015). Dual security against grey hole attack in MANETS. Intelligent Computing, Communication and Devices, Springer, India, 33–37.

[32] Chhabra, Meghna and Gupta, B.B. (2014). An efficient scheme to prevent flooding attacks in mobile ad hoc network (MANET). Research Journal of Applied Sciences, Engineering and Technology, 2033–2039.

[33] Chhabra, Meghna, Brij Gupta and Ammar Almomani. (2013). A novel solution to handle DDOS attack in MANET. Journal of Information Security, 165.

[34] Gupta, Anurag et al. (2015). Improved AODV performance in DOS and black hole attack environment. Computational Intelligence in Data Mining, Volume 2, Springer India, 541–549.

[35] Patel, Bipin N. and Tushar S. Patel. (2014). A survey on detecting wormhole attack in MANET. Journal of Engineering Research and Applications, 653–656.

[36] Shastri, Ashka and Jignesh Joshi. (2013). A wormhole attack in mobile ad-hoc network: Detection and prevention. Proceedings of the Second International Conference on Information and Communication Technology for Competitive Strategies, ACM.

[37] Sharma, Dhruvi, Vimal Kumar and Rakesh Kumar. (2016). Prevention of wormhole attack using identity-based signature scheme in MANET. Computational Intelligencein Data Mining, Vol. 2, Springer, 475–485.

[38] Shastri, Ashka and Jignesh Joshi. (2016). A wormhole attack in mobile ad hoc network: Detection and prevention. Proceedings of the Second International Conference on Information and Communication Technology for Competitive Strategies, ACM.

[39] Kar, Sumit, Srinivas Sethi and Manmath Kumar Bhuyan. (2016). Security challenges in cognitive radio network and defending against Byzantine attack: A survey. International Journal of Communication Networks and Distributed Systems, 120–146.

[40] Agrawal, Neha, Krishna Kumar Joshi and Neelam Joshi. (2015). Performance evaluation of byzantine rushing attack in ad hoc network. International Journal of Computer Applications.

[41] Patel, Neelam Janak Kumar and Khushboo Tripathi. (2018). Trust Value-based Algorithm to Identify and Defense Gray-Hole and Black-Hole attack present in MANET using Clustering Method.

[42] Tseng, Fan-Hsun, Hua-Pei Chiang and Han-Chieh Chao. (2018). Black hole along with other attacks in MANETS: A survey. Journal of Information Processing Systems.

[43] Nayak, Divyashree and Kiran, Y.C. (2017). Malicious node detection by identification of gray and black hole attacks using control packets in MANETS. Imperial Journal of Interdisciplinary Research.

[44] Yadav, Sakshi et al. (2017). Securing AODV routing protocol against black hole attack in MANET using outlier detection scheme. Electrical, Computer and Electronics (UPCON), 4th IEEE Uttar Pradesh Section International Conference on IEEE.

[45] Khan, Shariq Mahmood, Rajagopal Nilavalan and Abdulhafid F. Sallama. (2015). A novel approach for reliable route discovery in mobile ad hoc network. Wireless Personal Communications, 1519–1529.

[46] Gupta, Brij, Dharma P. Agrawal and Shingo Yamaguchi (eds.). (2016). Handbook of research on modern cryptographic solutions for computer and cyber security. IGI Global.

[47] Yadav, Anita, Yatindra Nath Singh and Singh, R.R. (2015). Improving routing performance in AODV with link prediction in mobile ad hoc networks. Wireless Personal Communications 83: 603–618.

[48] Mathew, Melvin, G. Shine Let and Josemin Bala, G. (2015). Modified AODV routing protocol for multihop cognitive radio ad hoc networks. Artificial Intelligence and Evolutionary Algorithms in Engineering Systems, Springer India, 89–97.

[49] Rathee, Geetanjali and Hemraj Saini. (2017). Secure modified ad hoc on-demand distance vector (MAODV) routing protocol. International Journal of Mobile Computing and Multimedia Communications (IJMCMC), 1–18.

[50] Yang, Hua and Zhiyong Liu. (2016). A genetic algorithm-based optimized AODV routing protocol. International Conference on Geo-Informatics in Resource Management and Sustainable Ecosystems, Springer, Singapore.

[51] Karthikeyan, B.S. Hari Ganesh and Mrs N. Kanimozhi. (2016). Security improved ad-hocon demand distance vector routing protocol (simAODV). International Journal on Information Sciences and Computing 10.2.

[52] Kaur, Pawanjeet and Malkit Singh. (2016). Comparision between AODV and modified AODV in MANET. Global Journal of Computers & Technology 5.1: 239–240.

[53] Yang, Licai and Haiqing Liu. (2016). A data transmitting scheme based on improved AODV and RSU-assisted forwarding for large-scale VANET. Wireless Personal Communications 91.3: 1489–1505.

[54] Choudhury, Debarati Roy, Leena Ragha and Nilesh Marathe. (2015). Implementing and improving the performance of AODV by receive reply method and securing it from Blackhole attack. Procedia Computer Science 45: 564–570.

[55] Cho, Jin-Hee, Ananthram Swami and Ray Chen. (2011). A survey on trust management for mobile ad hoc networks. IEEE Communications Surveys & Tutorials 13.4: 562–583.

[56] Vijayan, R. and Jeyanthi, N. (2016). A survey of trust management in mobile ad hoc networks. International Journal of Applied Engineering Research 11.4: 2833–2838.

[57] Govindan, Kannan and Prasant Mohapatra. (2012). Trust computations and trust dynamics in mobile adhoc networks: A survey. IEEE Communications Surveys & Tutorials 14.2: 279–298.

[58] Yan, Zheng, Peng Zhang and Athanasios V. Vasilakos. (2014). A survey on trust management for Internet of Things. Journal of Network and Computer Applications 42: 120–134.

[59] Subramaniam, Sridhar and Baskaran Ramachandran. (2015). Energy- and trust-based AODV for quality-of-service affirmation in manets. Artificial Intelligence and Evolutionary Algorithms in Engineering Systems. Springer India, 601–607.

[60] Babu, Nelson Kennedy. (2018). Establishing security in manets using friend-based ad hoc routing algorithms. Journal of Computer Science Engineering and Software Testing 4.1.

[61] Janani, V.S. and Manikandan, M.S.K. (2018). Efficient trust management with Bayesian-evidence theorem to secure public key infrastructure-based mobile ad hoc networks. EURASIP Journal on Wireless Communications and Networking 2018.1: 25.

[62] Ahmed, Adnan, Kamalrulnizam Abu Bakar, Muhammad Ibrahim Channa, Khalid Haseeb and Abdul Waheed Khan. (2015). A survey on trust based detection and isolation of malicious nodes in ad-hoc and sensor networks. Frontiers of Computer Science 9(2): 280–296.

[63] Cho, Jin-Hee, Ananthram Swami and Ray Chen. (2011). A survey on trust management for mobile ad hoc networks. IEEE Communications Surveys & Tutorials 13.4: 562–583.

[64] Pirzada, A.A. and Mcdonald, C. (2006). Trust establishment in pure ad-hoc networks. Wireless Personal Communications, Springer, pp. 139–163.

[65] Fall, K. and Varadhan. K. (2002). The ns manual. Notes and documentation on the software NS2-simulator. URL: www. Isi. Edu/nsnam/ns.

[66] Ghosh, T., Pissinou, N. and Makki, K. (2004). Collaborative trust-based secure routing against colluding malicious nodes in multi-hop ad hoc networks. pp. 224–231. *In*: Proc. 29th Annual IEEE International Conference on Local Computer Networks.

[67] Ghosh, T., Pissinou, N. and Makki, K. (2005). Towards designing a trusted routing solution in mobile ad hoc networks. Mobile Networks and Applications, Springer Science 10: 985–995.

[68] Pirzada, A.A., Datta, A. and Mcdonald, C. (2004). Trust-based routing for ad-hoc wireless networks. IEEE, pp. 326–30.

[69] Li, X., Lyu, M.R. and Liu, J. (2004). A trust model based routing protocol for secure ad hoc networks. In Proc. Aerospace Conference, IEEE 2: 1286–1295.

[70] Eissa, Tameem, Shukor Abdul Razak, Rashid Hafeez Khokhar and Normalia Samian. (2013). Trust-based routing mechanism in MANET: Design and implementation. Mobile Networks and Applications 18(5): 666–677.

CHAPTER 12

Detecting Controller Interlock-based Tax Evasion Groups in a Corporate Governance Network

Jianfei Ruan,[1,*] *Qinghua Zheng,*[1] *Bo Dong*[1]
and *Zheng Yan*[2]

1. Introduction

Tax revenue collection is considered a top priority in China. It was reported by the Chinese government that the rate of loss of tax revenue in China is above 22 per cent. At the same time, the number of annual tax-related business records is up to one billion, the daily peak of these records is up to 10 million and the volume of annual data aggregated is 12 TB. How to technically support tax evasion detection based on computer-based case selection, especially identifying suspicious tax evasion corporations/groups (susGroups), from very large scale of business transactions and related data, has become an important and challenging issue. Meanwhile, there is a new tendency for corporations to evade tax via Interest Affiliated Transactions (IATs) that are controlled by a potential 'Guanxi' between the corporations' controllers.

For dealing with these challenges, we propose a Colored and Weighted Network-Based Model (CWNBM) for characterizing economic behaviors, social relationships and the IATs between taxpayers and generating a heterogeneous information network—Corporate Governance Network (CGN). In this, we adopt the shareholder and management-involved role relationships between persons and corporations (such as corporations' executives or managers or legal persons), as well as investment and trading relationships between corporations. In CGN, persons or corporations act as nodes and relationships between persons and/or corporations act as arcs; also the weight of an arc is equal to the Interest Affiliated Degree (IAD) of a direct tie.

[1] Xi'an Jiaotong University; dong.bo@mail.xjtu.edu.cn; qhzheng@mail.xjtu.edu.cn
[2] Aalto University, Xidian University; zheng.yan@aalto.fi
* Corresponding author: xjtu.jfruan@gmail.com

Then, we perform an extensive literature study about potential 'Guanxi' for corporations and find that a phenomenon named 'board interlock' has been researched deeply in Western economics referring to the practice of members of a corporate board of directors serving on the boards of multiple corporations. However, scholars have paid too much attention to the relationship between corporations constituted by the board interlock but neglected the critical role of the interlocking relationship between the corporations' controllers. We believe that the ties between controllers are more important in China or other emerging economies where 'Guanxi' has been rooted in the blood of normal people. Moreover, essentially the economic behavior of a corporation is a concrete embodiment of its controllers' will. Besides, it is verified that in the process of economic transformation in China or other emerging economies, when the corporate governance is not mature, interlocking controllers tend to find loopholes in supervision to maximize their self-interests through legal-like-transactions. So this chapter coins a definition of controller interlock, which characterizes the interlocking relationship between corporations' controllers.

After introducing the definition of CGN and the proposal of controller interlock, this chapter focuses on an important problem that is the detection of Interlock-based Tax Evasion (ITE) in CGN. This problem is then split into three basic questions, (i) how to define the controller interlock according to the economic environment of China (called Type-I question, for short), (ii) how to recognize the controller interlock ties between corporations' controllers (called Type-II question, for short), and (iii) how to detect the suspicious groups of ITE (called Type-III question, for short).

In this chapter, to solve the Type-I question, the concept of controller interlock is coined based on the control relationships in Chinese corporations, drawing on the deeply researched board interlock concept. Moreover, to solve the Type-II question, we adopt a graph projection method to recognize the ties that meet the controller interlock pattern, then propose a component pattern-matching method based on the controller interlock ties to detect suspicious groups of ITE in order to solve Type-III questions. Finally, a number of experiments, based on seven-year period, 2009–2015, of one province in China, were performed and their results demonstrate that our proposed method can greatly improve the efficiency of tax-evasion detection.

2. Related Work

Board interlock analysis is an emerging topic in tax and economic field and has received considerable attention.

Board interlock analysis has been applied to industry dynamic analysis [12, 11, 5, 8] and corporate decision making [1, 3]. Suominen et al. [12] considered board interlock as the tool of inter-organizational flows needed for innovation and growth of the companies, concentrating especially on inter-organizational flows of board networks or interlocks. Robins et al. [11] introduced a bipartite clustering coefficient to compare the global structural properties of the US and Australian interlocking company directors. Ma et al. [8] discovered a structurally autonomous sphere in board interlock network of Chinese non-profits associated with major political and social events in the State-society relationship. Connelly et al. [3] explored the diffusion of an emerging strategy through the interlocking directorate effected by incorporating rational actors, potentially suppressive influences and network structural considerations besides examining a broad social network of interlocking directors in US firms.

Moreover, many researchers pointed out that there exists obviously an imitation effect between interlocked companies which can improve the companies' profitability. For example, Chua et al. [2] used social network analysis to determine the relationship between interlocking directorates and corporate profitability drawn from 2010 Fortune 500 companies and suggested that both interlocks and power asserted a positive linear relationship with companies' profitability. Peltonen et al. [10] indicated the companies that have international revenue interlocking with each other and interlocked boards of directors have the potential to act as important information and resource conduits.

However, negative performance effects associated with board interlocks were also analyzed in some special environment. Liu [7] studied the effect of environmental dynamism on the relation between the interlocking directorates and the corporation's performance in emerging economies, such as China. He pointed out that the output of the firms in the center of social networks constituted by interlocking directorates would be negatively affected. Croci et al. [4] used measures of vertex centrality to examine interlocking directorates and their economic effects and discovered a negative relationship between the firm's value and the degree and eigenvector centrality of board interlock network in Italy.

3. Definition of Corporate Governance Network and Controller Interlock

3.1 Corporate Governance Network

A Corporate Governance Network (CGN) is formed to represent a kind of heterogeneous information network based on the CWNBM. Thus we first coin the definition of CGN used in this research work as follows:

Definition 1: A CGN is formulated as a quintuple:

$$CGN = (\mathbf{V}, \mathbf{E}, \mathbf{W}, \mathbf{VColor}, \mathbf{EColor})$$

where

- $\mathbf{V}$: $\{v_p | p = 1, ..., N_V\}$ denotes a set of nodes;
- $\mathbf{E}$ denotes a set of all existing arcs, and let $\mathbf{E} = \{e_{pq}\} = \{(v_p, v_q) | 0 < p, q \leqslant N_V\}$, where $e_{pq} = (v_p, v_q)$ denotes that there exists an arc from the p-th node to the q-th node;
- $\mathbf{W}$: $\{w_{pq} | 0 < p, q \leqslant N_V\}$ denotes the weight (IAD) of the arc from the p-the node to the q-th node;
- $\mathbf{VColor}$: $\{LC, BC, CC\}$, where LC denotes the color of a legal person or director; CC denotes the color of a corporation; BC denotes the color of a shareholder. Using colors in $\mathbf{VColor}$ to classify $\mathbf{V}$ in CGN, we can draw the following conclusion $\mathbf{V} = \{L \cup C \cup B\}$, where $L = \{v_l | l = 1, ..., N_L, N_L < N_V\}$ denotes all legal person or director nodes marked by the color LC, $C = \{v_c | c = 1, ..., N_C, N_C < N_V\}$ denotes all corporation nodes marked by the color CC, $B = \{v_b | b = 1, ..., N_B, N_B < N_V\}$ denoting all shareholder nodes marked by the color BC, then $N_L + N_C + N_B = N_V$;

- **EColor**: {*CL, HR, IN, TR*}, *CL* denotes the unidirectional actual controller relationship between a legal person/director v_l and a corporation v_c, and if the color of the arc e_{lc} from v_l to v_c is *CL*, then e_{lc} is denoted as e_{lc}^{CL} and its weight is equal to 1, which is denoted as $w(e_{lc}^{CL}) = 1$; *HR* denotes the unidirectional share-holding relationship between a shareholder v_b and a corporation v_c, and $w(e_{lc}^{HR}) \in (0, 1]$; *IN* denotes the unidirectional investment relationship between two corporations v_{c1} and v_{c2}, and $w(e_{c1c2}^{IN}) \in (0, 1]$; *TR* denotes the unidirectional trading relationship between two corporations v_{c1} and v_{c2}, whose formula of weight is omitted because the weight is not considered in the rest of this chapter. Specially, *CL, HR* and *IN* belong to control relationship.

From the view of control relationships and trading relationship, there are two parts in a CGN—the control network and the trading network. The control network covers all relationships (actual control, investment, share holding, etc.), which influence transactions between *corporation* nodes, except for the trading relationship.

3.2 Controller Interlock

A board interlock is defined as sharing a common member on respective boards of directors in which a person affiliated with one corporation sits on the board of directors of another corporation [6, 9].

In this chapter, we extend the connotations of board interlock and develop it into a concept that fits the Chinese economic environment:

(i) The attention is changed from finding the interlocking relationship ties between corporations to mining the potential interlocking relationship ties between the corporations' controllers.

(ii) The persons concerned are extended from simple director to multiple roles of controllers—director, legal person and shareholder. The relationships concerned are extended from simple director relationship between *P* and *C* to the cover of actual control, shareholding relationship between *P* and *C*, as well as investment relationship between *C* and *C* (*C* for a corporation, *P* for a person). Then, the influence from *P* to *C* is accordingly extended from direct control tie (*P-C*) to control trail (*P-C-... -C*), covering both direct and indirect influence.

Based on the above extension, a controller interlock is defined as a tie between two controllers, each of whom sits on the board or executive of or has an indirect influence on a common corporation.

4. Controller Interlock Pattern Recognition

4.1 Interest Community Partition

Considering that the generated CGN is a large-scale graph, the first step is to partition the CGN into a series of small weakly-connected subgraphs by applying divide and conquer strategy. This step is inspired by an intuitive idea that the topology of controller interlock structure is included in one component of a control network as it belongs to a connected graph and is constructed solely by control arcs. Meanwhile, a trading relationship arc that connects two unconnected components of a control network is an unsuspicious trading relationship. Obviously, this means that there is definitely one party (node) not involved in two components at the same time behind the trading relationship arc. Therefore, the *i*-th maximal weakly connected subgraph of a control network and the trading relationship arcs between its corporations' nodes form the *i*-th interest community of a CGN, denoted as subCGN(i). The control part of subCGN is denoted as subCtrlNet, and the trading part of subCGN is denoted as subTraNet.

4.2 Potential Component Pattern Base (PCPB) Construction

Each suspicious group of ITE consists of two potential control relationship trails (see Definition 2) behind an IAT with a controller interlock tie between the start nodes of each trail (controller). Therefore, it is necessary to construct a PCPB to record all potential control relationship trails throughout the corresponding subCtrlNet in the form of *InP-OutC* walk (see Definition 3). To this end, a novel parallel label propagate-based control relationship trail traversal algorithm is presented, which is carried out in each subCtrlNet to obtain its PCPB. The main steps of the algorithm are as follows:

Step 1: Initialization process

First, initialize each node with a unique identification label (*IdLabel*). Meanwhile, suppose that each node in a subCtrlNet carries a local trail set *LTrailSet*, which stores control relationship trails ended by the node itself. Append the unique *IdLabel* of each *Person* node (one node as a trail) to its *LTrailSet* for initialization, and define the *LTrailSet* of each *Corporation* node to a null set. Meanwhile, define a global trail set *GTrailSet* to store all static IRR trails in the subCtrlNet and initialize it to be a null set.

Step 2: Propagation process

Let the *LTrailSet* of each node (*Person* node only for the first loop) propagate to its neighbors by tracing the directions of its adjacent edges. Then, remove the nodes which have not received any *LTrailSet* from neighbors and break the ties to its neighbors. Next, count the number of nodes in the subCtrlNet. If the number is equal to zero, then the algorithm is terminated and *GTrailSet* is the PCPB that records all static IRR trails throughout this subCtrlNet; otherwise, continue.

Step 3: Updating process

Each node updates its *LTrailSet* based on the *LTrailSet* collected from its neighbors. The detailed update operation on each node consists of the following three steps: (i) pop its *LTrailSet* to the *GTrailSet*, (ii) merge the *LTrailSet* collected and remove the component trails containing the *IdLabel* of this node, and (iii) append the *IdLabel* to the rest of the component trails, and update the *LTrailSet* of this node by the set of the new trails obtained.

Steps 2 and 3 are performed iteratively until the termination condition is met. Then, all control relationship trails throughout each subCtrlNet are recorded in the corresponding PCPB, and a copy of each trail is distributed to the *LTrailSet* of the node with *IdLabel* equal to the last element of the trail. The pseudo code of the above three-step approach is shown in Algorithm 1.

Algorithm 1: PCPB Construction

Data: *controlEdges* - represent control relationships between P and C or C and
     C;
     nodes - represent person and corporations;
Param: *thresholdWeight* - parameter used to filter control arc;
     maxLength - parameter used to determine when to stop;
Output: *trailSet* - all trails satisfy weight and length demands
1 *trailSet* = list();
  // filter edges by weight highher than thresholdWeight
2 **foreach** *edge of controlEdges* **do**
3   **if** *edge*[weight] $<$*thresholdWeight* **then**
4   remove *edge* from *controlEdges*;

  // initial node's attribute to a 2D list in which each 1D
    node list regard as a trail
5 **foreach** *node of nodes* **do**
6   **if** *node*[type] $==$ *human* **then**
7   *node*[attr] = list(list(*this*)) ;

  // get all trails along controlEdges
8 pregel (*maxIteration = maxLength,curLength =1*,
9   *sendMsg =*
10   **foreach** *edge of controlEdges* **do**
11   **foreach** *trail of edge*[src] [attr] **do**
12   **if** *trail*[length] $==$ *curLength* **then**
13   *trail.append(edge*[dst]*)*;
14   **if** *trail don't have circle* **then**
15   send *trail to edge*[dst];

16   *mergeMsg = gather Msgs send to same nodes,*
17   *vprog =*
18   append *mergedMsgs to trailSet;*
19   append *mergedMsgs to node*[attr];
20   *curLength += 1;*
21);
22 **return** *trailSet*

Definition 2: *Potential control relationship trail*

A potential control relationship trail is a trail, *T*, meeting the following conditions:
$$T = \{(p, c_1, c_2, ..., c_n)|p \in P, c_1, c_2, ..., c_n \in C, (p, c_1), (p, c_2), ..., (c_n, c) \in E\}$$

Definition 3: *Person-node-start-and-corporation-node-stop walk (InP-OutC walk)*

A *Person-node-start-and Corporation-node-stop walk* is a trail belonging to a set of trails in a control network and does not contain any trading arc, which is started by a *Person* node and stop by a *Corporation* node.

4.3 Controller Interlock Pattern Recognition

A controller interlock is defined as a tie between two controllers, each of whom sits on the board or executive of or has an indirect influence on a common corporation. To address the problem of controller interlock pattern recognition, we first propose a bipartite network-based model for characterizing the control relationship between persons and corporations and generating a Person-Corporation Bipartite Network (PCBN). Then, we accomplish the task of controller interlock tie construction by mapping the PCBN on to a unipartite network of Person called a P-projected graph.

For each potential control relationship trails in a PCPB, in the form of *InP-OutC* walk, the first and last node (p and c) are extracted to form the *Person* node set, P, and the *Corporation* node set, C, and the edge (p, c) forms the edge set E. Then the PCBN can be represented by a bipartite graph $PCBN = \{P, C, E\}$, where P and C are two parts of the nodes in PCBN. E is the set of edges in PCBN. There is no edge between the nodes in the same set of P and C; namely, every edge (p, c) $\in E$ satisfies $p \in P$, and $c \in C$. We use $N(p) = \{c|c \in C, (p, c) \in E\}$ to denote the set of *Corporation* neighbors of *Person* node p in PCBN.

To analyze the controller interlock ties in the PCBN, we map it on to a P-projected graph (see Definition 4).

Definition 4: (P-projected graph). Given a bipartite graph $PCBN = \{P, C, E\}$, its P-projected is defined as a unipartite graph $PCBN_P = \{P, E_P\}$ where the set of edges is $E_P = \{(p_1, p_2)|p_1, p_2 \in P, \exists c \in C, c \in N(p_1) \cap N(p_2)\}$.

From the definition, we can see that if *Person* part nodes p_1 and p_2 in the bipartite network PCBN have at least one common neighbor in the *Corporation* part, then there exists a controller interlock tie (p_1, p_2) in the P-projected graph $PCBN_P$. The detailed process is described in Algorithm 2.

Algorithm 2: Controller Interlock Pattern Recognition

Data: *trailSet* - PCPB returned by Algorithm 1
Output: *ciEdges* - all controller Interlock ties between P and C;

1 *ciEdges* = list();
2 *groupedTrail* = *trailSet*.
3   map((*trail*[head],*trail*[last])).groupBy(*trail*[head]) ;
4 **foreach** *(controller1,controlList1) of groupedTrail* **do**
5 | **foreach** *(controller2,controlList2) of groupedTrail* **do**
6 | | **if** *controlList1 intersect controlList2*
7 | | *and controller1 is not controller2* **then**
8 | | | *ciEdge*.append(Edge(*controller1,controller2*));
9 **return** *ciEdges*;

---

5. Suspicious Tax Evasion Group Identification

According to Definition 5, a series of *FIT-OutC* walks are constructed by carrying out first-interlock-tie join (see Lemma 1) based on the PCPB and controller interlock ties. Then, the topologies pattern of ITE can be redescribed as two *FIT-OutC* walks with the same controller interlock tie that are behind a transaction.

For each trading relationship arc, *tra*, in *subTraNet*(i) ($i = 1, ..., L$), extract the *LIntrlSet* of *tra*'s source vertex *src* and destination vertex *dst* from the *InPCPB* of *subCGN*(i). If there exist *trail$_s$* $\in$ *LIntrlSet*(*src*) and *trail$_d$* $\in$ *LIntrlSet*(*dst*), such that the combination of *trail$_s$*, *trail$_d$* and *tra* meets the topology pattern referred above, then we say that the two trading parties of *tra* are in suspicion of being involved in ITE, that is, *tra* belongs to interest affiliated transaction, and *trail$_s$* and *trail$_d$* are suspicious relationship trails. The detailed process is described in Algorithm 3.

Algorithm 3: Suspicious Tax Evasion Group Identification

    Data: *tradeEdges* - represent trading relationships between C and C;
             trailSet - PCPB returened by Algorithm 1;
             ciEdges - controller interlock ties returned by Algorithm 2;
    Output: *patterns* - nodes which is on suspicion of tax evasion;
1 *trailStartWithCI* = list();
2 *patterns* = list();
   // combine ciEdge with acquired trails
3 *groupedTrail* = *trailSet.*
4   map(($trail$[head],$trail$[last])).groupBy($trail$[head]) ;
5 **foreach** *ciEdge of ciEdges* **do**
6   **foreach** *trail of groupedTrail.get(ciEdge*[dst]*)* **do**
7   *trailStartWithCI*.append(list(*ciEdge*[src])+*trail*)

   // regroup the combined trails
8 *groupedTrail* = *trailStartWithCI*.groupBy(*trail*[last]) ;
9 **foreach** *tradeEdge of tradeEdges* **do**
10   *srcTrails* = *groupedTrail*.get(*tradeEdge*[src]);
11   *dstTrails* = *groupedTrail*.get(*tradeEdge*[dst]);
12   **if** *exist srcTrail in srcTrails, dstTrail in dstTrail:*
13   *srcTrail exchange first two node with dstTrail* **then**
14   *patterns*.append(*tradeEdge*[src]);
15   *patterns*.append(*tradeEdge*[dst]);

16 **return** *patterns*;

Definition 5: *First-controller-interlock-tie-start-and-corporation-node-stop walk* (*FIT-OutC* walk) is a trail produced by a first-interlock-tie join (see Lemma 1), which adds a controller interlock tie to the start of a trail that belongs to the set of *InP-OutC* walks (PCPB). The newly generated local and global *FIT-OutC* walk set is named as *LIntrlSet* and *InPCPB* corresponding to those of *InP-OutC* walk set, respectively.

Lemma 1: If a controller board tie is added to a head in a control relationship tails and it forms a new walk, *nw*, then *nw* is a trail. We can call this controller interlock tie added operation first-interlock-tie join.

6. Experimental Evaluation

6.1 Experimental Design

Data Set and Preprocess. Experiments based on the real tax data of S province in China from 2009 to 2015 are carried out. A series of CGNs are generated from the data monthly, the average of which includes 2,872,469 nodes and 2,488,982 arcs. Meanwhile, audit results of seven years are used as the data set. Each of the results is a conclusion drawn from the inspection of a company's financial affairs, which includes the company's taxpayer ID and inspection identifier (i.e., 1 for tax evasion and 0 for non-tax evasion). In the data set, 19328 ITE companies and 18209 non-ITE companies are included for each month in an average.

Evaluation Technique. For each taxpayer, *t*, in the dataset, identification methods are applied to detect whether it belongs to the trading parties of a suspicious tax evasion group. Then, the obtained conclusion, C, is compared with the actual audit results (i.e., the ground truth) from the following two aspects: (i) when the identifier of t equals to 1, t is correctly identified if C contains any recognized suspicious group; (ii) when the identifier of t equals to 0, t is not falsely alarmed if C is returned to be null.

Evaluation Criteria. To evaluate the performance of our proposed identification methods, two metrics, which are *Identification Precision* and *Hit Rate*, are employed.

Identification Precision (*IP*) is the faction of companies which are predicted to be suspicious, and are really involved in evasion. *Hit Rate* (*HR*) is the fraction of companies which really carry out tax evasion behaviors and are subsequently successfully identified. A higher *Identification Precision* and *Hit Rate* indicates a higher identification capacity.

Method for Comparison. In the experiments, we use traditional board interlock for comparison, since both traditional board interlock and our proposed controller interlock refer to the practice of controllers sharing influence on multiple corporations.

For these two interlock concepts, graph-based models are employed to characterize the patterns, each of which is formed by two potential control relationship trails and one interest affiliated transaction arc. The construction of suspicious relationship trails is affected by two factors: (i) Th: the weight threshold for control arcs, and (ii) L: the max length limit for suspicious potential control relationship trails. Fewer control arcs will be regarded as sufficiently influential if we set a higher Th, and subsequently, fewer trails will be constructed. Likewise, the same condition will occur if we set a smaller L.

The experimental results will show the influence of diversity of suspicious relationship trail construction factors (i.e., Th and L) on evaluation metrics through numerical statistic and analysis.

6.2 Experimental Results

The Effect of Different Methods. Tables 1 and 2 show the comparisons of effects (*Identification Precision* and *Hit Rate*) obtained by controller interlock method and board interlock method. While the effects are different under different evaluation metrics, controller interlock method as a whole achieves a better performance than board interlock method. Specifically, the improvements are significant with respect to *Hit Rate* and much smoother with respect to *Identification Precision*. As shown in Table 1, taking $L = 2$, $Th = 0.3$ as an example, the effects of controller interlock method are described as follows: (i) its *Identification Precision* is approximately equal to board interlock method; (ii) its *Hit Rate* is much higher, with improvements being 237 per cent. In conclusion, compared with board interlock method, controller interlock method greatly improves the *Hit Rate* of tax evasion identification and simultaneously achieves a similar *Identification Precision*.

The Effect of Weight Threshold. Table 1 and Fig. 1(a) show the effects with respect to Th. Similarly, the effects of both methods show substantial differences among various settings of Th. During the increase of Th (from 0 to 0.9), *Hit Rate* decreases sharply. The reason is that as Th increases, a greater number of weak control arcs are filtered

Table 1: Effects with respect to *Th*.

Th	Controller Interlock		Board Interlock	
	IP	*HR*	*IP*	*HR*
0.1	0.7850	0.0331	0.8025	0.0101
0.2	0.7810	0.0316	0.7957	0.0095
0.3	0.7748	0.0283	0.7874	0.0084
0.4	0.7793	0.0258	0.7772	0.0074
0.5	0.7768	0.0225	0.7815	0.0061
0.6	0.7674	0.0188	0.7759	0.0047
0.7	0.7621	0.0162	0.7529	0.0033
0.8	0.7618	0.0151	0.7246	0.0026
0.9	0.7690	0.0131	0.7069	0.0021

Table 2: Effects with respect to *L*.

L	Controller Interlock		Board Interlock	
	IP	*HR*	*IP*	*HR*
1	0.7764	0.0282	0.7874	0.0084
2	0.7748	0.0283	0.7874	0.0084
3	0.7748	0.0283	0.7874	0.0084
4	0.7748	0.0283	0.7874	0.0084
5	0.7748	0.0283	0.7874	0.0084

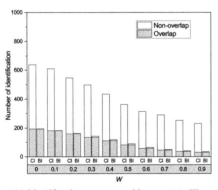

(a) Identification coverage with respect to *W*

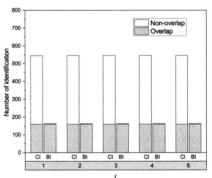

(b) Identification coverage with respect to *L*

Fig. 1: Identification coverage with respect to *W* and *L* (CI for controller interlock, BI for board interlock).

and fewer potential control relationship trails will be involved in PCPBs, which means less evidence can be used to identify tax evasion behaviors. Thus, it will lead to a decrease in *Hit Rate*. Meanwhile, we find that *Identification Precision* remains stable when *Hit Rate* decreases, which means that ITE behaviors are irrelevant to the IAD of control relationship.

The Effect of Max Length. Table 2 and Fig. 1(b) show that *Identification Precision* and *Hit Rate* remain stable with increasing L; the increase is not significant when L changes from 1 to 5. Theoretically, if the max length is larger, there will be more potential control relationship trails involved in PCPBs. We analyze the reason and find that though there are many additional suspicious groups of ITE with larger lengths of control trails when L gets larger, the extra corporations identified are quite few as most of the corporations newly discovered are replicated with prior findings.

7. Conclusion

This chapter focuses on the detection of corporations evading tax via interest affiliated transactions that are controlled by a potential interlocking relationship between the corporations' controllers. A concept named 'controller interlock' is first coined based on the control relationships in Chinese corporations, drawing on the deeply researched board interlock concept. Next, a graph projection method is adopted to recognize the ties that meet the controller interlock pattern between corporations' controllers. Then, a component pattern matching method is proposed to detect suspicious groups of ITE based on the controller interlock ties recognized. Finally, a number of experiments, based on seven-year period (2009–2015) of one province in China are performed and the results demonstrate that our proposed method can greatly improve the efficiency of tax evasion detection.

References

[1] Battiston, S., Bonabeau, E. and Weisbuch, G. (2003). Decision-making dynamics in corporate boards. Physica A: Statistical Mechanics and its Applications 322: 567–582.
[2] Chua, A.Y. and Balkunje, R.S. (2012). Interlocking directorates and profitability: A social network analysis of fortune 500 companies. pp. 1105–1110. *In*: Proceedings of the 2012 International Conference on Advances in Social Networks Analysis and Mining (ASONAM 2012). IEEE Computer Society.
[3] Connelly, B.L., Johnson, J.L., Tihanyi, L. and Ellstrand, A.E. (2011). More than adopters: Competing influences in the interlocking directorate. Organization Science 22(3): 688–703.
[4] Croci, E. and Grassi, R. (2014). The economic effect of interlocking directorates in Italy: New evidence using centrality measures. Computational and Mathematical Organization Theory 20(1): 89–112.
[5] Elouaer, S. (2006). A Social Network Analysis of Interlocking Directorates in French Firms.
[6] Labrinidis, A. and Jagadish, H.V. (2012). Challenges and opportunities with big data. Proceedings of the VLDB Endowment 5(12): 2032–2033.
[7] Liu, T. (2009). An empirical study about the effects of interlocking directorates strategy on the firm's output in the dynamic environment. pp. 818–820. *In*: Artificial Intelligence. JCAI'09. International Joint Conference on IEEE.
[8] Ma, J. and DeDeo, S. (2016). State power and elite autonomy: The board interlock network of Chinese non-profits. arXiv preprint arXiv: 1606.08103.
[9] Mizruchi, M.S. (1996). What do interlocks do? An analysis, critique, and assessment of research on interlocking directorates. Annual Review of Sociology 22(1): 271–298.

[10] Peltonen, J. and Rönkkö, M. (2010). Board interlocks in high technology ventures: The relation to growth, financing, and internationalization. pp. 163–168. *In*: International Conference of Software Business, Springer.

[11] Robins, G. and Alexander, M. (2004). Small worlds among interlocking directors: Network structure and distance in bipartite graphs. Computational & Mathematical Organization Theory 10(1): 69–94.

[12] Suominen, A., Rilla, N., Oksanen, J. and Still, K. (2016). Insights from social network analysis—Case board interlocks in finish game industry. pp. 4515–4524. *In*: System Sciences (HICSS), 49th Hawaii International Conference on IEEE.

CHAPTER 13

Defending Web Applications against JavaScript Worms on Core Network of Cloud Platforms

Shashank Tripathi,[1] *Pranav Saxena,*[1]
Harsh D. Dwivedi[1] *and Shashank Gupta*[2,*]

1. Introduction

Online Social Networks (OSN) are the virtual places that facilitate communication. Today, many OSNs have millions of active users. Most accepted and biggest OSN is Facebook with greater than 1 billion vigorous users. Social networking is not limited to informal use however; it is used for formal purposes. The user on these networks stores the pool of information (personal/professional) and therefore, hackers are paying more attention towards these sites. Hackers can make use of this accessible information for their malicious activities. To offer online users with enhanced services, the OSNs utilize the capabilities of JavaScript or related contemporary platforms of programming language. The support for such language platforms provides fertile platform for JS worm injection vulnerabilities [1]. The injection of such worms gives rise to Cross-Site Scripting (XSS) threats [2]. Actually Cross-Site Scripting abbreviation was CSS but since it matches with Cascading Style Sheet, it was changed and made XSS. Cross-Site Scripting is the name coined by Microsoft Engineers in 2000. Many attacks, which exploited JS vulnerabilities, have been witnessed before; one such instance occurred on the release of My Space Online Social Networking Site on October 4, 2005. The site was attacked by a XSS worm, JS. Space hero (popularly known as Samy) was designed by Samy Kamkar. More than a million users had run the destructive payload, which resulted in the unmatched spreading of Samy. Samy actually carried a malicious script which

[1] Department of Computer Science and Engineering, Jaypee Institute of Information Technology, Sec-128, Noida, UP, India; shashankt865@gmail.com; psaxena130@gmail.com; harshdwivedi124@gmail.com
[2] Department of Computer Science and Information Systems, Birla Institute of Technology and Science, Pilani, VidhyaVihar, Pilani, Rajasthan-333031, India.
* Corresponding author: shashank.gupta@pilani.bits-pilani.ac.in

displayed string, "but most of all, Samy is my hero" on the My Space profile page of the victim. Another XSS attack happened on the multinational e-commerce corporation eBay Inc. in 2016, when their user login page was injected with a pernicious script by the attacker and which would have returned an error message upon entering the user e-mail and eBay password. According to recent statistics, JS worm injection attacks are highly seen vulnerabilities in web applications. Such vulnerabilities are present in web applications because of the incorrectly validated user input. Furthermore, extenuating all possible JS worm injections is impossible because of the scope and complexity of current web applications and the countless behaviors in which browsers call upon their JS engines. Initially, when XSS was exploited due to the injection of JS worm, it was classified in two categories: Stored XSS and Non-Persistent XSS. Later on, a third category of XSS, i.e., DOM-based XSS was defined in 2005 [3].

In the modern era, where cloud computing is very popular, cloud security is a topic of great concern. Nowadays, many commercial IT organizations are using the features of cloud computing by accessing services of Online Social Network (OSN) instead of going in for expensive traditional computing, which necessitates the requirement of a physical computer memory and a high energy cost. In this era of Web 2.0 and HTML5 technologies, OSN is considered to be a most used method for sharing information as it is known that settings used for cloud computing are installed on the internet rather than on a single machine. Therefore, many internet infrastructural web application vulnerabilities exist in the background of cloud computing-based environments. Cross-Site Scripting (XSS) attacks are the most critical security concerns for web applications. According to *Acunetix Web Application Vulnerability Report, 2017*, almost 50 per cent of websites are vulnerable to XSS attack [4]. In 2015, the report stated that only 33 per cent of the web applications were vulnerable to XSS attack. It can be inferred from the report that there has been an increase of 17 per cent in a number of web applications, which are vulnerable to XSS attack. Cross-Site Scripting (XSS) was the most common type of attack used (39.2 per cent) to exploit web applications in 2017. It is also represented that SQL Injection, which used to access sensitive information or run OS commands for further penetration of a system, comes second in being the most-used attack (25.1 per cent). It is expected that Cross-Site Scripting and SQL Injection will continue to be more than half of the attacks used to exploit web applications. In addition to XSS and SQL Injection, Information Leak and XML Injections, which are involved in a disclosure of information, are also among major threats (8.8 per cent). Many researchers have observed that the effect of XSS attacks varies on the type of web browser utilized, so the quality of XSS attacks on the web module depends on the type of browser being used [5–10]. The author's contribution to this paper is to provide a novel approach of using script property function on virtual machines of cloud and on the edge network of Fog device to protect web applications from XSS attacks [11–16]. The following points explains why a large number of web applications are vulnerable to XSS:

Solitary use of Server-side Defense Mechanisms are Insufficient: Server-side content validation is the most commonly adopted practice for XSS defense. Purely server-side mitigation strategies work on the assumption that parsing/rendering on the client browser is uniform with the server-side processing. But in reality, this uniformity has been missing and the client side authentication of input cannot be taken on trust. The flaw has been targeted by several attackers in recent times. A string on server, such as <button onclick:=attack_code> is interpreted by the browser as <button onclick=attack_code>, and attack_code is executed by JavaScript engine.

Content Authentication is a Fallible Mechanism: Validation of untrusted data is also a commonly used mechanism to prevent XSS. Sanitization is one such kind of authentication, which removes possibly malicious elements from untrusted data. Escaping is another kind which converts dangerous elements, preventing them from being interpreted by the JavaScript engine. Special characters have shown that sanitization mechanisms are frequently insufficient to prevent all attacks, especially when the web application developers use custom built-in sanitization functions provided by popular server-side scripting languages, such as PHP. These PHP built-in sanitization functions check the user input and authenticate the data briskly. The filter_var() function of PHP removes the character having ASCII value greater than 127 and also removes the HTML tags; hence it ignores the special characters in the user input, leading to the improper sanitization of the content.

Lack of Developer's Skills: For online Social Networking, the need of the hour is to provide easy mutual receiving and returning of data to the user. This also facilitates direct intercommunication of the attacker with the application's process. Also, developers do not focus on making the Online Social Networking applications more secure against such attacks and hence they increase the vulnerability for the OSN against different attack vectors.

Based on verifiable observations, we have listed out three requirements for XSS protection. The defense should not be solely dependent on server-side sanitization. In addition to it, we should have a client-side sanitization process. The defense should not rely only on detection of common symptoms of malicious activities, like cross-domain sensitive information theft. The defense must restrict unreliable data in a manner which is reliable with both the browser implementation and user configuration.

2. Background of XSS Attack

Cross-Site Scripting is the most prominent attack on OSN-based web applications. These attacks are in use to steal sensitive credentials of the active users or hijack their sessions on the web application by inserting malicious scripts in injection point of the cloud-based OSN application, or can even install a new web browser when the user clicks on to some malicious link which the attacker can deploy into webpage or to our OSN accounts by injection of malicious scripts, using XSS. Web applications, which are having large number of user input fields, like blogging web applications, are more vulnerable to XSS attacks. With XSS, the attacker can take control all input and output data of a website by modifying the content of the website. XSS also plays an important role in bypassing the SOP concept. SOP (Same Origin Policy) concept forbids websites to retrieve content from pages with another origin. For example, Website www.JSsanspf. com/login.html information cannot be accessed by www.cracker.com; instead it can be accessed by www.JSsanspf.com/form.html as it belongs to the same origin. By forbidding access to cross-origin content random websites cannot read or modify data from SOP-enabled web application page while logged into them. XSS attack is tremendously harmful and a prevalent security issue to web applications [8–10]. A successful XSS attack can result in severe security damages for not only the user, but also for the web application. Whenever it is the case that HTML code is generated dynamically and the user input is not sanitized and reflected on the page, it is vulnerable to an XSS attack, as an attacker can insert his own HTML code over there. The web browser's JavaScript engine will run the XSS code as it will consider it as a response sent by the OSN server deployed at the virtual machine of the cloud platform. By doing so, the attacker is able to access other pages on the same domain and can read data or set cookies. XSS has appeared as one of the severe dangers to the World Wide Web. Figure 3 illustrates the abstract view of the procedure used for XSS attack on a vulnerable web application. It is also clearly reflected in the same figure that three parties (i.e., attacker, web browser and victim device) are involved in the XSS attack procedure. For XSS attack to happen, the attacker injects malicious JavaScript in an injection point which can possibly be a comment section or any user input field of the web application. The malicious JS attack vector gets saved into OSN server's database. Now, when the client requests (HREQ) for the web page from the OSN server, which is deployed at a virtual machine of the cloud platform, the OSN server responds with an HTTP response (HRES), which also contains the malicious JS code and the code runs on the machine of the client. As the web application is attacked, all important information of the user can be stolen [11–14].

2.1 Categorization of XSS Attacks

XSS attacks are divided into four categories, namely non-persistent, persistent, and mutation-based and Document Object Model (DOM)-based XSS attacks. All the four categories of attacks differ in the way of exploitation on web applications but all have the same goal of stealing user credentials.

2.1.1 Non-persistent Cross-site Scripting Attack

It is also known as Reflected XSS. In this type of attack, the user sends a user's input via HTTP Request (HREQ) to OSN server, and the server responds with an HRES, which contains whole un-sanitized user input in an error message, or a search result, or any other kind of response. This attack is also known as TYPE 1 XSS because the attack is carried out through single request/response cycle [6–7].
Figure 1 explains one pattern of non-persistent XSS attack.

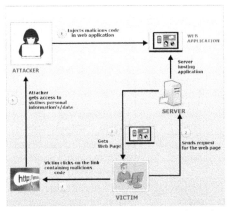

Fig 1: Exploitation of non-persistent XSS attack

2.1.2 Persistent Cross-site Scripting Attack

This attack is also known as Non-Reflected XSS attack. As shown in Fig. 2, in this attack, the attacker injects a malicious code in an injection point and the code gets saved in the database of the OSN server. When the client requests for the page, the OSN server responds by sending the page which also contains malicious code, which is considered a developer-written script by the browser, and the JavaScript engine executes the code. Persistent attacks are less frequent as the injection points are few and are quite difficult in identifying, but at the same time, they are more dangerous than non-persistent attacks, because the latter attack only one user, but persistent attacks affect all those users who requested for the malicious code-containing page [12–14].

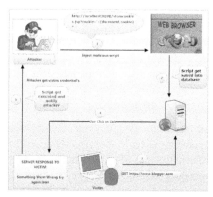

Fig. 2: Exploitation of persistent XSS attack.

2.1.3 DOM-based XSS Attack

DOM XSS is a type of cross-site scripting attack which relies on inappropriate handling of the data from its associated DOM of HTTP response [6–9]. Most of the readers get confused between DOM-based XSS attack and reflected XSS attack. DOM-based XSS modifies the document object model of the web page and the JavaScript code are executed in the application though it look similar to the reflected XSS attack but the difference between them is that when we change the DOM data, the data never goes to the server. This means that the attack takes place at the client side only, that is, the server side attacks are not affected by it. Table 1 highlights the summary of different

Table 1: Summary of types of XSS vulnerabilities.

Type	Name	Vulnerability Location	Description	Exploitation Level
0	DOM-based	Client-side script	Use of document object to write html by client-side script	High
1	Non-persistent	Server-side script	User's supplied data is used by the server to produce results which may not be properly encoded	High
2	Stored, Persistent	Server-side script	User's supplied data stored on server for later display without proper encoding	Low

classes of XSS vulnerabilities. There are several objects in DOM, which an attacker can manipulate in order to create the XSS condition.

2.2 Inefficiency of Existing Solutions

In view of the growing popularity of the XSS-Guard [17] and the limited use of the source code of web applications in their experiments on malicious script detection particularly aiming to look at online users endpoints, non-source code analysis in XSS attack prediction and False Positive Rate (FPR) assessment are gaining momentum, more specifically on the platforms of OSN. In recent years, tremendous progresses has been made in web proxy-based monitoring systems [18, 19] which are constantly omitting the designing and development of precise sanitizers and their appropriate placement in the source code of web applications.

Recently, machine learning algorithms (Naïve Bayes, Support Vector Machine (SVM) and J48 Decision Tree) have emerged as fast and cost-effective alternative techniques for early assessment of malicious script attack vectors [20]. Commercial expert systems (like AppScan, N-Stalker, Zed Attack Proxy, WebInspect and Acunetix) are constantly utilized by web developers in predicting the XSS attack vectors [21]. The drawback of these scanners is that they use regular expressions to detect the presence of dynamic content that results in low fidelity and banning of benign HTML content. Therefore, for many years, FPR of such techniques has been still uncontrollable. Most of the recent XSS defensive techniques are platform-dependent, as different parsers have different ways of interpretation of script content and hence, web browsers have a variety of different parsing quirks. To solve this issue, a customized platform-independent HTML parser (i.e., BIXSAN) generates a parse tree at browser-side to reduce the irregular behavior of web browsers [22]. Later on, it was evaluated on several infrastructures of web browsers (such as Opera, Netscape, Internet Explorer (IE), Firefox Google Chrome) and found to be less adaptive, since it demands re-architecturing of the existing infrastructure of such web browsers. In addition, the infrastructure settings of such existing techniques are not scalable enough to be integrated effectively and evaluated for the web applications hosted on the virtual machines of cloud platforms.

XSS exploitations are practised in two forms—stored and reflected XSS, and malicious script code is usually introduced on the server-side. Innumerable defensive solutions exist for the two classes of XSS worms. Despite the many solutions available, very few are up to the mark against the latest type of XSS attacks; for instance, one factor is introduction of HTML5, though HTML5 is helping to achieve high caliber for the Web applications in terms of facilities, design and evolving user instructiveness at a whole new different level and is also viewed as a substitute for the Adobe's Flash and Microsoft's Silverlight. Also HTML5 hurled us with new security challenges and attack vulnerabilities which need to be taken care of while developing web applications. HTML5 offers Cross Origin Resource Sharing (CORS) which circumvent SOP and hence increases the vulnerabilities against attacks, like XSS, CSRF. HTML5 offers many APIs like web worker, history and properties including <video>, <source>, <autofocus> tags are used for creating new XSS attack vectors like,

<video><source on error= "<script>alert("XSS ATTACK")</script>"></video>

Contemporary web browsers or available XSS attack filters are inefficient against these new HTML5 attack vectors. Furthermore, unlike another language, malicious functions are detected at the runtime [7]. Moreover, most of the web applications allow third-party JavaScript through *<script>*tag or link tag <a> which is the main reason for such attack. Thus, the code is retrieved directly from the third-party source and processed immediately. So this code is not controlled by the application at the host and not even at the web servers. Therefore, there is a requirement for a DOM-based XSS attack-defensive solution, which will ferret out the XSS attack vectors and diagnose them with an effective mechanism of sanitizing/filtering the HTML5 XSS attack vectors. A majority of the available techniques count on methods of string matching, which was not able to recognize the skewed JavaScript attack vector injections. Moreover, the choice of parser also matters a lot as JSOUP parser can sanitize the white list, so that can be a good choice. Sanitization/filtering of dangerous variables of JavaScript code without assessing their context is an ineffective sanitization technique in attenuating the effect of the JavaScript attack vector.

3. Browser-independent Sanitization Framework

The authors suggested a novel script comment estimation and browser-independent sanitization framework that scans for such malicious attack vectors on web applications hosted on the virtual machines of cloud platforms. Existing state-of-art [2–4] based on browser-dependent sanitization mechanism has succeeded in only revealing the external context of such illicit literals/variables of script code and subsequently injects the sanitization primitive routine functions in them. It is undoubtedly true that unless the technique is not able to find such diverse context of script code, the sanitization is not an effective mechanism for alleviating the effect of XSS worms from the web applications. Based on this severe research gap, the authors introduced an enhanced browser-independent run-time dynamic parsing and sanitization of JS worms in the cloud-hosted web applications.

The proposed framework works in two routines, i.e., script comment injection and script comment assessment. Commencing with the script comment injection phase, it includes parsing of the web application module by the web parser to which the DOM generator generates a corresponding DOM tree. DOM generator receives URL links as its input from the parser. The features of the script code extracted from its corresponding DOM tree are estimated and injected in the original source code. Script comment repository reserves the injected features, along with the source code. Further, working with the script comparator routine, the HTTP request is forwarded to the web server to which a corresponding HTTP response is received by the browser. The hidden injection points are managed by the HTML parser and are provided to the script link extractor to excerpt the required script code.

The authors have proposed a novel sanitization mechanism—JavaScript-Sanitizer using Script Property Function (JS-SANSPF), an injection and congregation-based sanitization framework which uses Script Property Function. The framework is proposed to run in two modes, namely training mode and detection mode. Training mode is carried on when the web application is just developed and is ready to be put on OSN server. Detection mode is carried on whenever there is an HTTP request for the web application. It is proposed that the script property functions are calculated in training mode and in detection mode. The calculations done in training mode are saved and are matched with the calculations done in the detection mode. If both the calculations are same, then the HTTP response is free from any XSS attack vector, and if the calculations don't match, it can be inferred that HTTP response contains XSS attack vector and needs to be sanitized.

3.1 Abstract View

Here, the authors briefly discuss the proposed framework, which is as highlighted in Fig. 3. The framework is proposed to work in two modes called Training mode and Detection mode. Training mode is a one-time activity and it runs when the web application is just developed and is ready to be put on OSN server. In the Training mode, it is proposed that the framework will identify and locate possible injection points of the web application, and will calculate property functions of the script found in injection points. Detection mode is executed on the edge network of Fog devices, whenever there is an HTTP Request for the web application. Before executing Detection mode, the original SPF which were calculated by Training mode are appended to the HTTP Response (HRES). In the Detection mode, the framework will calculate Script Property Function (SPF) of HRES and will compare it with the original SPF. If the comparison result outputs that the both SPFs are same, the HRES will be sent to the client. Else, HRES will be sanitized and then sent to the client. The sanitization is a sophisticated process consisting of the congregation of same looking scripts, and then creating templates of the congregation so that scripts of a congregation can be represented by a single template. The templates are then sanitized by using efficient algorithms. Figure 4 also illustrates the flowchart of both the modes of our framework.

3.2 Detailed View

Here, the authors discuss the detailed view of the framework as highlighted in Fig. 6. The JS-SANSPF framework operates in two modes, namely training mode and detection mode.

3.2.1 Training Mode

This mode is a one-time activity, which is executed when a web application is just developed and is ready to be put on OSN server. In this phase, the web application

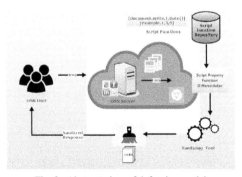

Fig. 3: Abstract view of defensive model,

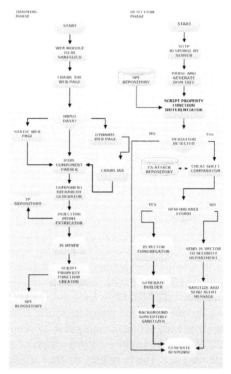

Fig. 4: Flowchart of both the modes of our framework.

module is first crawled to check whether it is a static web page or a dynamic web page on the basis of the input field of the data. The input given to this web module is the extracted web page, retrieved by the crawler. The input is parse to check whether the web page is static or dynamic. If the output of the module is static in nature, then it is directly passed through the parser and generates the Document Object Model (DOM), but if the module is dynamic in nature, then it is crawled by Crawljax crawler. The output of the crawler is then passed through HTML Component Parser, which parses HTML code present in the web page to generate the DOM tree. Output of parser is transferred to Component Hierarchy Generator (CHG), which creates a hierarchy of the components used in the code. The output of CHG is input to tree traverser, which traverses the tree and sends its output to Injection Point Extractor (IPE). IPE also gets input from IP repository, which stores all possible injection points, which can be found in web module. IPE then identifies possible injection point in the tree of the code, by using input of IP repository. The IPE sends its output to JS miner, which identifies JavaScript code used at possible injection points. The output of JS miner is sent to Script Property function creator, which calculates the property function of the script used in possible injection points.

3.2.2 Detection Mode

This mode is executed whenever an HTTP request is made. The mode parses HTTP Response (HRES) and creates DOM tree. The mode then extracts script from DOM tree and calculates Script Property Function (SPFs) of the extracted script. The calculated SPFs are then compared with the SPFs, which were calculated in training mode. If there is no deviation found between the two groups of SPFs, the HRES is sent to the user. If deviation is found, then the scripts whose SPF didn't match with the original SPFs, are compared with a cheat-sheet, which is a list of already used XSS script vector till date.

Case 1: If the script is found in cheat-sheet, the script will be congregated and a template will be created for each congregate to represent all the scripts of the same congregate. The template will be sanitized, using efficient algorithms and the scripts which were represented by this template will be changed accordingly.

Case 2: If the script is not found in cheat-sheet, security department of the web application's owner organization will be alerted about the script. As the framework is not able to find the script in cheat-sheet, the framework will not be confident, whether the script is an XSS attack vector or not. But, it will try its best to prevent any possible XSS attack vector to reach to the client. The steps that the frameworks will take are event handling, removal of doubtful keywords, checking data URI and removal of character escaping. After completing these steps, the framework will send sanitized HRES to the client.

3.3 Key Components of JS-SANSPF

JS-SANSPF works on two modes: training phase and the detection phase. Here, the authors discuss the components of both modes.

3.3.1 Training Phase

HTML Component Parser (HTMLCP): The data of web application is sent to HTMLCP whose job is to parse the HTML markup language into a parse tree. The module HTMLCP parses whole HTML file and stores the parsed data in string S. Since parsing is based on the syntax which is defined by the grammar, so parsing follows the set of rules that are defined in the production rule of the grammar. But unfortunately, in case of HTML, it is not easy to define the grammar for HTML language, so for HTML, we use HTML-DTD (Hyper Text Markup Language—Document Type Definition) grammar.

Component Hierarchy Generator (CHG): This module reads string S and generates a DOM tree of the string in which every component gets separated. DOM tree is the object presentation of the HTML document and acts like the embrasure of the HTML element to the outside world, like JavaScript. The root of tree is the 'Document' object. Generating the DOM tree operation is performed by the syntax analyzer, which generates the parse tree. The generated DOM tree is passed to the semantic analyzer that generates the semantically parse tree that is defined according to the HTML_DTD rule. For example, S="<html><head><title></title><script></script></head></html>".

Tree Traverser: This module traverses whole DOM Tree (data structure D, and outputs all the different used nodes). The Tree Traverser performs tokenization in our case and generates the stream of tokens of the node tag used which we later

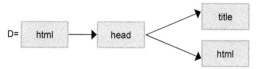

Fig. 5: Structuring of HTML components.

Algorithm 1: DOM tree analysis and its components.

Algorithm: Tree traverse
Input→Array of strucutres returned by Component Hierarchy Generator
Output→Array of nodes

Generate an array S1={x$_i$|x$_i$∈ Structure returned by
 	
                  Component Hierarchy Generator};
Generate an empty array Data[]={x$_i$|x$_i$∈nodes};
int l=S1.length();
inti=0;
WHILE(i<l)
             Data.push(S1[i++].node_name);
END WHILE
RETURN Data;

use to identify the injection point from the injection point repository For example, for above stated data structure D, the module will output an array of nodes, $N=\{$"html","head","title","script"$\}$. Figure 5 exemplifies the detailed structuring of HTML components. Algorithm 1 highlights the DOM tree traversal method.

Injection Point Repository (IPR): This module contains a list of all possible injection points. The repository is globally accepted and is used in finding the vulnerabilities in the web application.

Injection Point Extractor (IPE): This module checks for injection point vulnerabilities in the web application. The Input to this module is the output of tree traversal that is in this case array N, and the list produced by IPR. It contains those injection points which match with the IPE. The output of this module is an array of the injection points, IP.

JS Miner: This module extracts scripts found in injection point extractor. The input to this module is array IPE, and the output is array JS.

Script Property Function Creator (SPF Creator): This module calculates Script Property Function (SPF) of the script based on its properties. The input to this module is array JS. The output of this module is string SPF which is then stored in Script Property Function Repository (SPF Repository).

Script Property Function Repository: This module contains the SPF of the scripts that are present in all possible injection points. This module plays an important role in detection mode too as it appends the SPF in HTTP response.

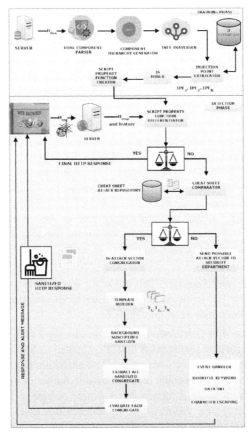

Fig. 6: Proposed defensive framework.

3.3.2 Detection Phase

Script Property Function Differentiator (SPFD): This module calculates SPF of the script present in HRES and then compares it with the SPF calculated in a training phase. The output of comparison is true if SPFs are same and is false if both the SPF are not same. False output implies that an extra piece of script has been inserted into some injection point, which was not originally written in the web module by the developer. Such script can be malicious in nature, so it is needed that such scripts are sent to further components for further investigation before sending them to browser. If the SPFD returns true output, the HRES will be sent to the browser; else the scripts which gave false outputs will be sent to Cheat Sheet Comparator for further investigation.

Cheat Sheet Comparator (CSC): This module compares input scripts with the scripts listed in Cheat-Sheet Attack Repository. The Cheat-Sheet Attack Repository contains all XSS attack vectors, which have been discovered till date. The input to this module are the scripts whose SPFs were not found in the list of SPFs, which was developed in the training phase. If the input scripts were found in the repository, it implies that, the scripts are surely XSS attack vectors, and will be sent JS Attack Vector Congregator for sanitization; else, it is not mandatory that the scripts will be attack vectors, but to be on the safe side, the Cheat Sheet Comparator module will send an alert to the web module-owning organization about the extra script, and will try to sanitize these scripts as per its heuristics.

JS Attack Vector Congregator: This module clusters similar looking scripts. This module compares each script with the other script, and calculates Levenshtein's distance between the two scripts. If the calculated Levenshtein's distance is lesser than or equal to maximum allowable Levenshtein's distance, a congregate is made of the two scripts. Figure 7 highlights an example of grouping of similar scripts.

Template Builder: This module is responsible for creating script templates. The inputs to this module are script congregates. The module finds Longest Common Subsequence (LCS) of the scripts of the given congregate. The module then finds a script which is most similar to the LCS (its Levenshtein's distance with LCS is minimum), let it be string *s*. The module then compares string *s* and LCS. Positions at which character matches, the character at that position is appended in the template string. Else, '-' character is appended in the template string, only if the previous character in template string is not '-' character. Figure 8 highlights an example of template building of a given congregate.

Background Susceptible Sanitizer: This module is responsible for the sanitization of the malicious scripts. The inputs to the module are script templates while the outputs of this module are the sanitized script templates.

Script Property Function Calculation Algorithm: The algorithm accepts script string as an input. The algorithm maintains a linked list called as SPF, which will contain Script Property Functions of the used scripts. The algorithm traverses through script and finds out the number of opening brackets used in the script. If the script contains no opening bracket, then the input script is a simple command like initialization of a variable, or inserting new values in the variable. In this case, whole command is pushed back in SPF Linked List. If the script contains one opening bracket, then the input script is a simple function like document.write(), etc., the algorithm will call simple SPF algorithm and will push back the list returned by simple SPF algorithm

Fig. 7: Grouping of script attack vectors.

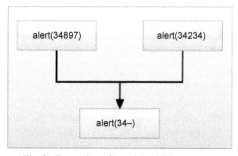

Fig. 8: Formation of template builder process.

Algorithm 2: Estimation of script properties function.

Algorithm: Script Property Function Calculation
Input→Script String
Output→Script Function

Generate an array Script[]={x_i|x_i∈Script String};
 Generate an empty list SPF={x_i|x_i∈Script Property};
 int l=Script.length(),i=0;brackets=0;
 String str='';
 WHILE(i<l)
     IF(Script[i]=='(')
         brackets++;
     END IF
     str+=Script[i++];
 END WHILE
 IF(brackets==0)
     SPF.push_back=str;
     SPF.push_back=0;
     RETURN SPF;
 END IF
 IF(bracket>1)
     SPF=nestedSPF(str);
 END IF
 ELSE
     SPF=simpleSPF(str);
 END ELSE
 RETURN SPF;

into SPF Linked List. If the script contains more than one opening brackets, then the input script is a complex script like document.write(print()), etc. In this case, the algorithm will call Nested SPF, and will push back the list returned by Nested SPF in the SPF Linked List. Algorithm 3 illustrates the process of finding different background of script variables.

Nested SPF algorithm maintains a Linked List SPF, Linked List attr[], variable j, and string str. Variable j tells about how deep are we in the function. The data structure attr is an array of Linked List. For example, attr[i], will contain attributes of ith level function. For example, in script "func(print(1,2))", attr[1] contains attribute of 1st level function, i.e., attributes of print function. String str, contains recent token found between token differentiators like ',', '(',')'. The algorithm traverses through input script.

For example, if the substring is "(print,", str will contain "print". Variable j is initialized to zero. When the algorithm, encounters '(' in its traversal, value of j is checked. If value of j is equal to 0, then string str is name of main function, so string str is pushed back to SPF and value of j is increased. If value of j is more than 1, then the value of j is increased, String str is pushed back to attr [i]. The data stored

Algorithm 3: Nested background determination of script variables.

Algorithm: Nested SPF
Input→Script string
Output→SPF
Generate an array Script[]={x$_i$|x$_i$∈Script String};
Generate an empty list SPF={x$_i$|x$_i$∈Script Property};
Generate an array of list attr[]={x$_i$|x$_i$∈attributes};
inti=0,j=0,l=Script.length();
String str="";
WHILE(i<l)
IF(Script[i]=='(')
IF(j==0)
SPF.push_back(str);j=1
END IF
ELSE
attr[j++].push_back("#"+j+str);
END ELSE
str=""; i++;
CONTINUE;
END IF
IF(Script[i]==',')
attr[j].push_back(str);
str="";
i++;
CONTINUE;
END IF
IF(Script[i]==')')
j--;i++;
CONTINUE;
END IF
str+=Script[i++];
END WHILE
SPF.append(attr[1].length());SPF.append(SPFcreate(attr,1));
RETURN SPF;

in attr *[j]* Linked List when a bracket is encountered, is in the for "#nfunc_name" like "#2print", where '#' is a symbol to denote attribute, 2 is the function level, and "print" is the attribute value. If the algorithm encounters ',', it means that the attributes are being separated. String str is pushed back to attr *[j]* Linked List. If the algorithm encounters ')', it means that level of function is getting decreased. The value of *j* is decreased by one. Figure 9 shows how value of *j* changes while traversing a script. When the algorithm traverses whole script, the size of attr [1] Linked List is appended to SPF. The algorithm then calls SPF create and sends attr Linked List and value 1 to the called function. The list returned by SPFcreate is appended to SPF Linked List.

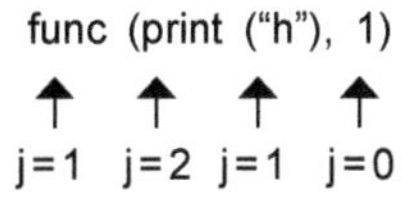

$$\text{func (print ("h"), 1)}$$

$$\uparrow \quad \uparrow \quad \uparrow \quad \uparrow$$

$$j=1 \quad j=2 \quad j=1 \quad j=0$$

Fig. 9: Pattern of different variations observed in values of variables.

Algorithm 4: Estimation of script properties function.

Algorithm: SPF create
Input→Array of list attr[] and index of list(i)
Output→SPF
Generate an empty list SPF={x$_i$|x$_i$∈Script Property};
List iterator itr;
SPF.push_back(attr[i].size());
FOR(itr=attr[i].begin;itr!=attr[i].end();itr++)
String str=*itr;
IF(str.contains('#'))
SPF.push_back('{');
SPF.push_back(str);
SPF.push_back(attr[i+1].length());
SPF.append(SPFcreate(attr,i+1));
SPF.push_back('}');
CONTINUE;
END IF
SPF.push_back(str);
END FOR
RETURN SPF;

SPFcreate algorithm accepts a Linked List attr. The Linked List attr is traversed by the algorithm. If the data contained by the address being traversed contains "#", then the data is name of the function. The function name is pushed back into SPF Linked List. Size of one level deeper function is pushed back into the data structure. Then the algorithm calls itself, so that the list of deeper functions could be pushed back into list. If the data doesn't contain '#', then the data is pushed back into SPF Linked List.

Simple SPF algorithms take script string as input. It traverses the string and separates the text before the bracket and text after the bracket. Text before the bracket is the function name and is the first entry in SPF. The text after brackets contains attributes separated by a comma. So, the algorithm stores attributes. SPF is appended by the numbers of attributes and finally gets appended by the attributes. Algorithm 5 illustrates the process of deletion of useless symbol by Script Property Functions (SPF).

Congregation algorithm takes all JS vectors as input. The algorithm matches each JS vector with every other vector and if the Levenshtein's distance between the two is less than Levenshtein_distance$_{max}$, their congregate is made. Algorithm 6 illustrates the process of estimation of Levenshtein distance between the two attack vectors.

Algorithm 5: Deletion of useless symbol by Script Property Functions (SPF).

Algorithm: Simple SPF
Input→Script string
Output→SPF
Generate an array Script[]={x_i|x_i∈Script String};

```
Algorithm: Simple SPF
Input→Script string
Output→SPF
Generate an array Script[]={xᵢ|xᵢ∈Script String};
Generate an empty list SPF={xᵢ|xᵢ∈Script Property};
Generate an empty list attr={xᵢ|xᵢ∈attributes};
inti=0,j=0,l=Script.length(),bracket=0;
String str="";
WHILE(i<l)
    if(Script[i]=='(')
        i++;
        SPF.push_back(str);
        str="";
        WHILE(i<l&&Script[i]!=')')
                        IF(Script[i]==',')
                            attr.push_back(str);
                            str=""; i++;
                        END IF
                        ELSE
                            str+=Script[i++];
                        END IF
        END WHILE
        attr.push_back(str);
    END IF
    ELSE
        str+=Script[i++];
    END ELSE
END WHILE
int n=attr.size();
SPF.push_back(n);
FOR(i=0;i<n;i++)
    SPF.push_back=attr[i];
END FOR
RETURN SPF;
```

Algorithm 6: Calculation of Levenshtein distance between the two attack vectors.

Algorithm: Congregation
Input→JS Attack vectors
Output→Congregated JS attack vectors

Generate an array JSAV[]={x_i|x_i∈JS attack vector};
 	
Generate an empty array CongregatedJSAV[][]={x_i|x_i∈JS attack vector};
Levenshtein_distance$_{max}$=d;
 	
FOR EACH JSAV as attack_vector1
 	
FOR EACH JSAV as attack_vector2 FROM (index$_{attack\_vector1}$+1)→ n
 	
d_{calc}=Levenshtein_distance(attack_vector1,attack_vector2);
 	
IF(d_{calc}<=d)
 	
CongregatedJSAV[index$_{attack\ vector1}$].push(attack_vector2);
 	
END IF
          END FOR
END FOR
RETURN CongregatedJSAV;

4. Implementation and Experimental Assessment

This section discusses the implementation and experimental assessment outcomes of XSS attack on online social networking framework for the cloud-hosted web applications.

4.1 Implementation

The authors utilized the capabilities of ICAN cloud simulator for the designing of their prototype model and integrated their infrastructural settings by creating different virtual machines of this simulator. Different tested freeware platforms of web applications were also hosted on such virtual machines of ICAN cloud simulator. An impenetrable perspective was taken to assess our approach towards shielding against XSS worm attacks. Efficient and stable after-effects were detected as a consequence of exhaustive testing of the system. The liable XSS detection as well as prevention methodology of our framework was tested on five virtual platforms of web applications, namely Humhub, Wordpress, Joomla, Drupal and Elgg. Installation of all these platforms was done using local host server, XAMPP. After the installation, the next step involved searching the vulnerabilities in our websites with the help of XSS attack vector repositories [14–18]. These repositories comprise of a number of script or payloads accountable for XSS worm injection vulnerabilities.

Table 2: Categories of different HTML attributes and related events.

HTML Attribute	List of Event
Windows	onload, onlocation, onunload, onstorage
Mouse	onmouseover, onmousedown, onmouseout, onmouseup
Form	onfocus, oninput, onsubmit, onblur, onformchange, onforminput
Keyboard	onkeyup, onkeydown, onkeypress

In addition, we refined a template of our framework in Java via software development environment, Java Development Kit (JDK) and included its settings on the simulated machines of Eclipse IDE for Java Runtime Environment (JRE). The Eclipse IDE is deployed at a 2.24 GHz Dual Core Processor with 4 GB RAM running Window 10 single home and Google Chrome. We used JSOUP which is a Java library used for working with Hyper Text Markup Language. It is used to parse HTML document. It offers Application Program Interface to remove and handle data from Uniform Resource Locator or HTML file. It uses Document Object Model, CSS and Jquery-like methods for removing and handling a file. Initially, we verified the performance of our proposed framework against three open-source XSS attack repositories, which include the list of past and recent XSS attack vectors. Very uncommon XSS attack vectors were able to evade our server-side design. The parsing of the removed H5 web submission modules was performed by utilizing the competencies of JSOUP. This parser aids the data centers with the application program interfaces that remove document object model tree nodes which provide information about a vulnerable point in the source code and the malicious strings.

The competencies of JSOUP were also used while manipulating the comments of malicious JavaScript code. The injection of comments was done on the consistent javascript nodes of the document object model tree and addresses of such nodes were saved in the existing frame of web applications. Table 2 shows the list of precluded attributes.

4.2 Experimental Assessment

This segment discusses the assessment outcomes of the proposed framework on different platforms of Online Social Network-based cloud-hosted web applications and evaluates the execution and accuracy of the system. We have tested our proposed system on three open-source real-world OSN platforms, i.e., Elgg, WordPress, Drupal, etc. This has been done for estimating the XSS attack vector drop by the deployed tools on these open-source Online Social Network cloud-hosted web applications. In terms of correctness, we approximate what percentage of detection of XSS attack vectors were improved by our system. We have also performed session hijacking on these OSN web applications by injecting XSS attack script to steal cookies. To provide a deep view on how to exploit XSS vulnerability present in the web application, we have provided the screenshot through figures.

These entire steps are as follows:

- Initially, attacker imitates as a legal user by providing user's credentials.
- The attacker tries to perform maliciously by finding out the vulnerable points like comment box and injects some malicious string.

- Web application processes it as a comment from the valid user and stores it in a database without validation.
- Henceforth, a genuine user logs in into the system and starts reading the comments posted by another user.
- As victim reads the malicious comment posted by an attacker, the injected malicious script gets executed by the browser and the victim's login credentials are sent back to the attacker.

Figures 10 and 11 illustrate the process of exploiting the XSS vulnerability on real OSN-based web application, i.e., WordPress and Elgg. It shows what are the steps involved to induct an H5-XSS attack by the attacker. We have combined the framework of our proposed model into the OSN cloud-hosted web application. The stimulus to select these cloud-based web applications is that we simply need to mark that websites can utilize the competencies of the proposed model, regardless of the input testing. This will help in improving H5-XSS malicious code injection vulnerability concerns. The ease with which we are capable of syndicating our cloud-based framework in these popular OSN cloud-based web applications determines the malleable compatibility of our proposed framework. We deploy these OSN cloud-based web applications on a XAMPP web server with MySQL server as the back-end database. We have tested and evaluated the detection capability of our proposed solution on three real-world OSN web applications, namely Elgg, WordPress and Drupal. We have injected the different H5-XSS attack vector on the injection points of these cloud-based web applications and find that the injection point is vulnerable to XSS attack and that our model is capable of sanitizing such H5-XSS attack vector efficiently and effectively.

In Fig. 11, we were able to perform XSS attack in WordPress framework. We found an injection point in the framework. If a script was written in a new page post section, the script would execute on loading. So, we inserted a script belonging to the jQuery library of JavaScript. The script would send the cookies of the victim to the database of the attacker site, without redirecting the browser towards the attacker's site. So, in this way, the victim can never know that his credentials have been stolen. We have also estimated the comments of script code. The comment-estimation process receives retrieved script strings as an input. It analyzes every sequence of the script for uniquely estimating the features. Every sequence of the script owns unique remark statements for its recognition. Such statements are then processed for policy generation. To inject malicious HTML5 script code, an attacker amends the script method explanation for injection suspicious-sequence of scripts. The sequence of script is as follows:

```
<input type="text" name="username" value="<%request.getParameter("U_name")%>">
```

At this point, no sanitization methodology is processed on the user's credentials before it gets interpreted in the response page. Presume that an opponent throws a malicious function, as: `<script>alert("document.cookie");</script>`. Therefore, the original code becomes `<input type= "text" name= "username" value="<script>alert("document.cookie");</script>">`. Subsequently, when the browser perceives the response, then cookies are re-directed to an alert box. Hence, method-call and method-definition arrangements are retrieved via an effective script as its distinctive remark statement. Table 3 presents examples corresponding to remarks of the suspicious scripts.

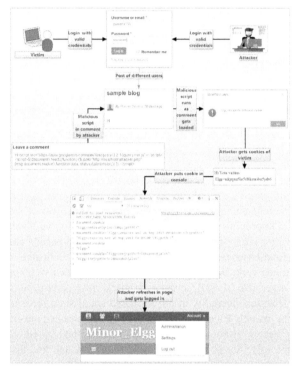

Fig. 10: Session hijacking in Elgg.

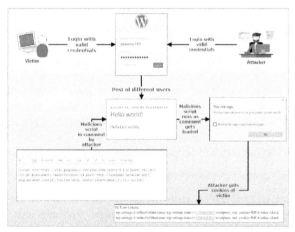

Fig. 11: XSS attack exploitation in WordPress.

In Fig. 10, we were able to perform the hijacking session in Elgg framework. We performed the same tasks that we did in the WordPress framework. Here, we were also able to hijack the session of the victim. The cookies were with the attacker, who opened the browser, went to Elgg site and opened the console. He removed his already made cookies and created a new cookie with the value that he got from the cookie of

Table 3: Comment estimation process of scripts.

Type	Example	SPFC
Integral procedure call	Math.sqrt (x,5)	{sqrt,2,3,5}
User defined procedure call	function check (a,b,c){...};	{check,3,a,b}
Mysterious procedure call	var data=info (x,y){...};	{data,2,x,y}
Document object procedure call	var Data=document. Get Element byTag ("value"); Data.innerHTML="XSS ATTACK";	{document. getElementByTag,1,value}
User defined nested procedure call	check(3,check(4,7))	{check,2,3,{check,2,4,7}}
Integral nested procedure call	Math.sqrt(4,Math.max(3,4))	{sqrt,2,4,{max,2,3,4}}

Table 4: Different infrastructures of web applications.

Web Module	Version	Lines of Code
WordPress	4.8.2	41,281
Drupal	8.4.0	43,035
Elgg	2.3.3	23,702

the victim and then refreshed the page. He was now in victim's session and could do any malicious activity that he wanted. Table 4 highlights the explicit configuration of web applications.

4.2.1 Response Time Calculation

To analyze the response time, we used the Google Chrome web browser along with the XAMP server. The network tool of chrome is used to calculate the response time. For each XSS attack prevention, we have used the JS-SANSPF-based filter. Around 10 times the page has been loaded and response time have been recorded. An average response time has been taken as we can see in Fig. 10 and Fig. 11 respectively that the response ratio for both the WordPress and Drupal is almost constant.

For a better analysis of the response time, we executed the settings of our framework in two different scenarios. In the first scenario, we executed our framework on the core of network, i.e., we executed all the modules of offline mode. The static unambiguous features of all the variables of script code were determined and accordingly our proposed sanitization method was executed on them. The second scenario is our default proposed online distributed intelligence network of cloud infrastructure. Here, the feature of those variables will be determined which could not be found statically on offline mode.

Hereafter, initially, we calculate the response time of execution of the feature-based context-familiar sanitization on offline mode. Likewise, we computed the response time of context-aware dynamic parsing, feature estimation and run-time sanitization on cloud computing architecture. The performance of our framework is entirely dependent on how quickly our framework will respond with the different contexts of illicit variables of script code and accordingly executes the nested context-aware sanitization on them. We have calculated the different features of extracted scripts in virtual machines of cloud and subsequently executed the sanitization on them. On the other hand, we injected the set of vulnerable script code on the cloud server and executed the run-time feature-based nested context-familiar sanitization on these servers. In both the scenarios, we computed the response time of both the modes. Tables 5 and 6 show the response time of our work on both the infrastructures of web applications. In addition, Figs. 12 and 13 show the graphical representations of response ratios on both the platforms (i.e., Drupal and WordPress) of web applications.

Response ratio = Response time with sanitization – Response time without sanitization/ Response time without sanitization

Table 5: Response time data on Wordpress.

Size in KB	Response Time without Sanitization (sec)	Response Time with Sanitization (sec)	Difference in Response Time	Response Ratio
3.0	0.841	2.532	1.691	2.01
56.6	1.34	4.06	2.72	2.03
60.6	1.74	5.55	3.81	2.19
76.9	1.94	5.96	4.02	2.07
110	2.02	6.06	4.04	2.00
119	2.23	6.72	4.49	2.02
$\sum$ Response time No. of record	1.69	5.15	3.46	2.05

Table 6: Response time statics on Drupal.

Size in KB	Response Time without Sanitization (sec)	Response Time with Sanitization (sec)	Difference in Response Time	Response Factor
10	1.40	4.25	2.85	2.04
14	1.50	4.52	3.02	2.01
27	2.37	8.00	5.63	2.37
152	2.55	8.65	6.10	2.39
174	2.60	8.80	6.20	3.60
226	2.70	9.10	6.40	3.70
250	3.50	11.50	9.97	2.50
$\sum$ Response time No. of record	2.77	9.13	6.69	3.10

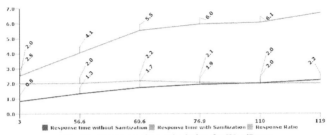

Fig. 12: Response time calculation for WordPress.

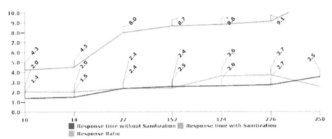

Fig. 13: Response time calculation for Drupal.

We have also compared the performance analysis of our work (i.e., NoScript) with our work. No Script is primarily supposed to be a tool, which interrupts execution of malicious scripts from maximum web pages, though obstruction of the execution of scripts is not a genuine choice for web applications. For resolving this issue, NoScript also acts as a client-side XSS filter, which sanitizes the HTTP request transmitted via web browser. Here, we have verified the malicious XSS attack payload-detection capability of NoScript by injecting 135 XSS worms on our two web applications. Table 7 highlights the performance comparison of our framework with NoScript.

Table 7: Performance comparison of JS-SANSPF with other existing XSS defence models.

Web Module	True Positive (TP)	True Negative (TN)	False Positive (FP)	False Negative (FN)	Positive P = TP + FN	N = TP + TN + FP + FN
JS-SANSPF						
WordPress	42	3	3	8	50	56
Drupal	41	3	3	9	50	56
No Script XSS						
WordPress	40	2	4	10	50	56
Drupal	39	3	3	11	50	56

Evaluation with F-test: To verify that detected H5-XSS count (True Positives (TP)) is lower than injected H5-XSS attack vectors count; we have utilized the concept of hypothesis. We have assumed two hypotheses defined as follows:

Null Hypothesis (H_0): JS-SANSPF detects more than 78 per cent of attack vectors correctly ($p > 0.78$).

Alternate Hypothesis (H_A): JS-SANSPF detects not more than 78 per cent of attack vectors correctly (p <= 0.78).

The assumptions for the P-test are as follows:

1. Randomized data: The sample was randomly collected.
2. 10 per cent condition: Sample size is less than 10 per cent of attack vectors population.
3. There must be at least 10 successes and 10 failures (np > 10 and nq > 10): np = 50*0.78 = 39 > 10, and nq = 50*0.22 = 11 > 10.

As all the three assumptions are fulfilled, we can proceed with the P-test, considering
Significance level (α = 0.05).

$$T - Value = \frac{p' - p}{\sqrt{\frac{p * q}{n}}}$$

where,

p' = Probability of true positives in our evaluation
p = Probability that has to be checked
q = 1–p
n = Number of JS vectors used

$$T - Value = \frac{0.82 - 0.78}{\sqrt{\frac{0.78 * 0.22}{50}}} = 0.683$$

$$P - value = normalCdf\,(T - value) = 0.2483$$

As the P-value is more significance level, so we failed to reject null hypothesis. There is enough evidence to support the claim that JS-SANSPF was able to detect more than 78 per cent of JS attack vectors. The performance evaluation results also revealed that the HTTP response time was very low on the online core network of cloud infrastructure in comparison to our recent work [13–15]. The proposed framework optimizes the usage of bandwidth and reduces the communication distance to half by deploying multiple virtualized cloud-like servers in the core network of cloud infrastructure.

5. Concluding Remarks

In order to better assist HTTP response time of web servers, the virtual machines of cloud computing spread themselves at the core of network and facilitate the online users (accessing OSN-based web applications) with location-aware facilities. Recent existing frameworks suffer from runtime performance overheads as all excessive run-time processing was executed on the online web servers. In this article, the authors present a cloud-hosted framework that senses the JS attack vector injection vulnerabilities on the core web servers of web applications hosted on virtual desktop systems of cloud infrastructures. Initially, the framework computes the comments of such script code as a part of static pre-processing on virtual desktop systems of cloud-data centers. Such machines re-execute the comment computation process on the script code present in the HTTP response. Any abnormality observed between these features and those computed statically would alarm the exploitation of malicious JS worms on web servers hosted on the core network of cloud-computing infrastructure. Very low and acceptable rate of FPs and FNs were found during JS comment assessment of the attack vectors on the web servers incorporated on virtual desktop systems of ICAN cloud simulator. For our future work, we would prefer to analyze the response time of our framework by completely integrating its settings on the online edge servers of Fog computing infrastructure without the intervention of cloud infrastructure.

Key Terminology and Definitions

Malwares: Malware is a software or piece of software that is used to attack on computer machine by interrupting the normal behavior of the system or by gaining important private information from the victim machine. Malware is short for 'malicious software' and can penetrate into the victim's system in form of executables. These are downloaded from the websites or from attractive ads, intentionally or unintentionally by the victim. Malwares include computer viruses, worms, spy ware and Trojan.

File Virus: File virus is attached with the binary executable files or .com files. Once this virus stays in memory, it tries to infect all programs that load on to memory.

Macro Virus: Some of the applications, such as Microsoft Word, Excel or PowerPoint allow macro to be added in the document, so that when document is opened, that macro code is automatically executed. Macro virus is written in macro code and is spread when such infected document is opened. Once infected, repairing of these files is very difficult.

Multipartite Virus: It is the combination of MBR virus and file virus. It is attached to the binary executable file at initial stage. When the user runs this file, virus gets inserted inside the boot record and when next time the computer is booted, virus is loaded into memory and infects other files and programs on computer.

Polymorphic Viruses: These types of viruses change their forms each time they attach on the system. This property of these viruses makes them very difficult to detect and find cure for them.

Stealth Viruses: These viruses affect the files, directory entries of file systems, disk head movement, etc. Their effect is hidden from the user's perspective, but affects the memory storage or processing speed.

Backdoor: Backdoor attack is a malware which bypasses the security authentication in the computer system and offers unauthorized remote access to the attacker.

Rootkit: Rootkit is another type of malware that enables privileged access to the computer, while hiding the existence of certain processed and programs. Once the attacker has administrator access or root access to the computer, he can obtain the confidential information which was otherwise not possible to access with simple user access.

Logic Bombs: Logic bomb starts infecting the system when certain conditions are satisfied. It is a piece of code that remains hidden until desired circumstances are not met. For example, on some particular day, it will start showing messages or start deleting the files, etc.

Denial-of-Service: Denial-of-service attack compromises service availability by sending voluminous data to the server that is supposed to work for other legitimate requests. Victim server becomes unavailable to work for other users and gets busy in servicing the attacker's packets. Also, some attacker sends malicious requests, so that system's resources overflow with the data and then tries to drop the packets belonging to the legitimate users.

Phishing: Phishing is type of cyber-attack in which attacker creates an entity, such as bank website, social media website or online payment processing website that is exactly similar in look as legitimate website. Then attacker makes the user to enter the confidential information, such as bank account password, credit card information and uses it for his financial gain or any other motive. Phishing is carried out with the help of spam e-mails or instant messaging services and are sent in bulk to thousands of users. Users are made attracted to such phishing e-mails, like they may get rewarding money if they login to their bank website and then webpage of phishing website gets opened.

Cross-Site Scripting (XSS): Cross-site scripting (XSS) is a type of computer security vulnerability typically found in web applications. XSS refers to that hacking technique that makes use of vulnerabilities in the code of a web application to allow the hacker to send malicious content from an end-user and collect some type of data from the victim. Web applications that take input from the user side and then generate output without performing any validation are vulnerable to such attacks. The attacker injects client side script into web pages that are accessed by other users. When the browser of that user executes the script, it may perform malicious activities, like access cookies, session tokens or other sensitive information that is supposed to remain with the user.

Spoofing: In spoofing attack, one person or program pretends to be another user or another device and then tries to perform some malicious activities, like launching denial-of-service attacks, stealing data, spreading virus or worms, bypassing access controls, etc. There are various types of spoofing techniques, like IP spoofing, ARP spoofing, DNS server spoofing, e-mail address spoofing.

IP Spoofing: IP address spoofing is most common and widely used spoofing among other spoofing attack types. Attacker spoofs the source address of the attack packet with other IP address. So, when the attack is detected, the true source of the attack packet can not be found out. For example, in case of DDoS attack, a large number of attack packets is sent to the victim server with spoofed source address. Another way of IP spoofing is to spoof the target address. When service packets are replied back to target address, they go to the spoofed address, i.e., the address of victim.

ARP Spoofing: Address resolution protocol (ARP) is used to resolve IP addresses into link layer addresses when packet is received from outside the network. In ARP spoofing, attacker sends spoofed ARP message across LAN, so that the link layer address of the attacker gets linked with IP address of some machine in the network.

This results in redirection of the messages from the legitimate source to the attacker instead of the legitimate destination. The attacker then, is able to steal information or alter the data that is in transit or prohibit the all the traffic that is coming in the LAN. ARP spoofing is also used to perform other types of attacks, like DoS, Man-in-the-Middle attack, session hijacking, etc.

DNS Server Spoofing: Domain Name Server (DNS) maps domain name with IP addresses. When the user enters the website address into the browser, it is the responsibility of the DNS to resolve that domain name address into IP address of the server. In DNS spoofing attack, the attacker changes entries in the DNS server, so that when user enters the particular website address, it gets redirected to another IP address that generally belongs to attacker or it also can be related to the victim to which the attacker may want to flood the traffic. Attacker may reply to these messages with malicious programs that are sent to harm the victim's computer.

E-mail Address Spoofing: E-mail address spoofing uses different e-mail addresses instead of the true source of the e-mail. The recipient cannot tell the actual sender of the e-mail, since they can only see the spoofed e-mail address. Since, Simple Main Transfer Protocol (SMTP) does not provide any proper authentication, the attacker is able to perform the e-mail address spoofing. Generally, prevention is assured by making the SMTP client to negotiate with the mail server, but when this prevention is not taken, anyone who is familiar with the mail server can connect to the server and use it to send messages. E-mail information that is stored and forwarded to the recipient can be changed and modified as the attacker wants it to be. Therefore, it is possible for the attacker to send the e-mail with spoofed source address.

References

[1] Faghani, Mohammad Reza and UyenTrang Nguyen. (2013). A study of XSS worm propagation and detection mechanisms in online social networks. IEEE Transactions on Information Forensics and Security 8(11): 1815–1826.

[2] Hydara, Isatou, Abu BakarMd Sultan, HazuraZulzalil and Novia Admodisastro. (2015). Current state of research on cross-site scripting (XSS)—A systematic literature review. Information and Software Technology 58: 170–186.

[3] Chaudhary, Pooja, B. B. Gupta and Shashank Gupta. (2019). A framework for preserving the privacy of online users against XSS worms on online social network. International Journal of Information Technology and Web Engineering (IJITWE) 14(1): 85–111.

[4] Acunetix Web Application Vulnerability Report. (2017). Available at https://www.acunetix. com/blog/articles/acunetix-vulnerability-testing-report-2017/. Last Accesses on January 17th, 2018.

[5] Shar, LwinKhin and HeeBengKuan Tan. (2012). Predicting common web application vulnerabilities from input validation and sanitization code patterns. pp. 310–313. *In:* Proceedings of the 27th IEEE/ACM International Conference on Automated Software Engineering, ACM.

[6] Gupta, Shashank and Gupta, B.B. (2016). XSS-secure as a service for the platforms of online social network-based multimedia web applications in cloud. Multimedia Tools and Applications, pp. 1–33.

[7] Gupta, S., Gupta, B.B. and Chaudhary, P. (2017, Jun. 12). Hunting for DOM-based XSS vulnerabilities in mobile cloud-based online social network. Future Generation Computer Systems.

[8] Chaudhary, P., Gupta, S., Gupta, B.B., Chandra, V.S., Selvakumar, S., Fire, M., Goldschmidt, R., Elovici, Y, Gupta, B.B., Gupta, S. and Gangwar, S. (May 2016). Auditing defense against

XSS worms in online social network-based web applications. Handbook of Research on Modern Cryptographic Solutions for Computer and Cyber Security 36(5): 216–45.

[9] Chaudhary, P., Gupta, B.B. and Gupta, S. (2018). Defending the OSN-based web applications from XSS attacks using dynamic JavaScript code and content isolation. pp. 107–119. *In*: Quality, IT and Business Operations, Springer, Singapore.

[10] Gupta, S. and Gupta, B.B. (Oct. 2017). Smart XSS attack surveillance system for OSN in virtualized intelligence network of nodes of fog computing. International Journal of Web Services Research (IJWSR) 14(4): 1–32.

[11] Gupta, B.B., Gupta, S. and Chaudhary, P. (Jan. 2017). Enhancing the browser-side context-aware sanitization of suspicious HTML5 code for halting the DOM-based XSS vulnerabilities in cloud. International Journal of Cloud Applications and Computing (IJCAC) 7(1): 1–31.

[12] Gupta, S. and Gupta, B.B. (Sept. 2016). An infrastructure-based framework for the alleviation of JavaScript worms from OSN in mobile cloud platforms. pp. 98–109. *In*: International Conference on Network and System Security, Springer International Publishing.

[13] Gupta, S., and Gupta, B.B. (Nov. 2016). XSS-immune: A google chrome extension-based XSS defensive framework for contemporary platforms of web applications. Security and Communication Networks 9(17): 3966–86.

[14] Gupta, S. and Gupta, B.B. (2016). Alleviating the proliferation of JavaScript worms from online social network in cloud platforms. pp. 246–251. *In*: Information and Communication Systems (ICICS), 7th International Conference on 2016 Apr. 5, IEEE.

[15] Gupta, Shashank and Gupta, B.B. (2016). JS-SAN: Defense mechanism for HTML5-based web applications against javascript code injection vulnerabilities. Security and Communication Networks.

[16] Bates, D., Barth, A. and Jackson, C. (2010). Regular expressions considered harmful in client-side XSS filters. pp. 91–100. *In*: Proceedings of the Conference on the World Wide Web.

[17] Bisht, Prithvi and Venkatakrishnan, V.N. (2008). XSS-Guard: Precise dynamic prevention of cross-site scripting attacks. pp. 23–43. *In*: International Conference on Detection of Intrusions and Malware, and Vulnerability Assessment, Springer, Berlin, Heidelberg.

[18] Parameshwaran, Inian, Enrico Budianto, Shweta Shinde, Hung Dang, Atul Sadhu, Prateek Saxena and Dexter, J.S. (2015). Robust testing platform for DOM-based XSS vulnerabilities. pp. 946–949. *In*: Proceedings of the 2015 10th Joint Meeting on Foundations of Software Engineering, ACM.

[19] Khan, Nayeem, Johari, Abdullah and Adnan Shahid Khan. (2015). Towards vulnerability prevention model for web browser using interceptor approach. IT in Asia (CITA), 9th International Conference on. IEEE.

[20] Vishnu, B.A. and Jevitha, K.P. (2014). Prediction of cross-site scripting attack using machine learning algorithms. p. 55. *In*: Proceedings of the 2014 International Conference on Interdisciplinary Advances in Applied Computing, ACM.

[21] Bau, Jason, Elie Bursztein, Divij Gupta and John Mitchell. (2010). State of the art: Automated black-box web application vulnerability testing. pp. 332–345. *In*: Security and Privacy (SP), 2010 IEEE Symposium on, IEEE.

[22] Chandra, V. Sharath and Selvakumar, S. (2011). Bixsan: Browser independent XSS sanitizer for prevention of XSS attacks. ACM SIGSOFT Software Engineering Notes 36(5): 1–7.

Additional References

ALmomani, A., Gupta, B.B., Atawneh, S., Meulenberg, A. and Lmomani, E.A. (2013). A survey of phishing email filtering techniques. IEEE Communications Surveys & Tutorials 15(4): 2070–2090.

Advances in Information Security, Springer. Retrieved from http://link.springer.com/chapter/10.1007/978-0-387-74390-5_4.

Esraa, A., Selvakumar, M., Gupta, B.B. et al. (2014). Design, deployment and use of HTTP-based Botnet (HBB) Testbed 16th IEEE International Conference on Advanced Communication Technology (ICACT), pp. 1265–1269.

Guamán, Daniel, Franco Guamán, Danilo Jaramillo and Manuel Sucunuta. (2017). 'Implementation of techniques and OWASP security recommendations to avoid SQL and XSS attacks using J2EE and WS-Security. pp. 1–7. *In*: Information Systems and Technologies (CISTI), 12th Iberian Conference on, IEEE.

Gupta, B.B., Joshi, R.C. and Misra, M. (2009). Defending against distributed denial of service attacks: issues and challenges. Information Security Journal: A Global Perspective, Taylor & Francis Group 18(5): 224–247.

Heiderich, Mario, Christopher Späth and Jörg Schwenk. (2017). DOMPurify: Client-side protection against XSS and markup injection. pp. 116–134. *In*: European Symposium on Research in Computer Security, Springer, Cham.

Meyerovich, L.A. and Livshits, B. (2010). Conscript: Specifying and enforcing fine-grained security policies for JavaScript in the browser. *In*: Proceedings of the 2010 IEEE Symposium on Security and Privacy, sp, 10, Washington, DC, USA. doi: doi>10.1109/SP.2010.36.

Nagpal, Bharti, Naresh Chauhan and Nanhay Singh. (2017). SECSIX: Security engine for CSRF, SQL injection and XSS attacks. International Journal of System Assurance Engineering and Management 8(2): 631–644.

Rodríguez, Germán Eduardo, Diego Eduardo Benavides, Jenny Torres, Pamela Flores and Walter Fuertes. (2018). Cookie scout: An analytic model for prevention of cross-site scripting (XSS) using a cookie classifier. pp. 497–507. *In*: International Conference on Information Theoretic Security, Springer, Cham.

Singh, A., Singh, B. and Joseph, H. (2008). Vulnerability Analysis for http. 37: 79–110. New York, USA.

Srivastava, A., Gupta, B.B., Tyagi, A. et al. (2011). A Recent Survey on DDoS Attacks and Defense Mechanisms. Book on Advances in Parallel, Distributed Computing, Communications in Computer and Information Science (CCIS), LNCS, Springer-Verlag, Berlin, Heidelberg, CCIS 203, pp. 570–580.

Teto, Joel Kamdem, Ruth Bearden and Dan Chia-Tien Lo. (2017). The impact of defensive programming on I/O Cybersecurity attacks. pp. 102–111. *In*: Proceedings of the Southeast Conference, ACM.

CHAPTER 14

Importance of Providing Incentives and Economic Solutions in IT Security

Amrita Dahiya and B.B. Gupta*

1. Introduction

Information security management is evolving as complex, prompting and sensitive domain because of dynamic and unpredictable structures of attack, industries' and firm's reliability on internet for doing business, less expertise of security professionals, lack of coordinated solution that can work in heterogeneous network environment, lack of sharing information regarding computer's security, lack of outsourcing information security through contract [1] and improper regulated incentives resulting into conflict of interests between user and security mechanism over his system. Proper embedded firewalls, huge investment in detecting intruders and attack traffic, well-formulated access control policies, application of highly complex cryptographic protocols on data do not make our information secure. We have to think differently and start viewing information security more than this. Nowadays security professionals have to legitimize the protection of information assets within the security investment of the concerned organization with a little knowledge of the incentivization concept. A much more convinced and understandable information security management can be grasped through argument based on economic principles and well-formed economic incentive's chain.

Over the last few years, people have realized that security failure is caused at least as often by bad incentives as by bad design [2]. A large part of research community is concerned with providing technical solutions in order to protect data while research in economic perspective is less; only a handful of work is present yet in this particular domain [3–5]. Big organizations and firms lack incentives to cooperate in defending

National Institute of Technology, Kurukshetra, India.
* Corresponding author: gupta.brij@gmail.com

attacks and threats. Solution to any attack could be either technological or economical. There is no denying the fact that co-operative technological solutions like [6, 7] are effective solutions but organizations lack incentives to implement them; therefore, more preference must be given to economic solutions to provide stronger incentives to the right party.

Generally, parties defending against threats and attacks are not suffering parties, while suffering parties are not defending parties.

This is the main reason why after so much advancement in software world and IT industry, we lack economic solutions. For implementing economic solutions, we need to understand what is the incentive chain and to which party incentives should be provided on internet. The internet world is collaboration of multiple entities like server, ISP, router, content provider and end user. Each party requires strong motivation be it in the form of money or in any kind of profit to co-operate with each other in order to provide quality service to the end user. This motivation is called incentive lacking in technological solutions. For example, a gateway router at border shields a particular network from attacks. Due to this action, the end user and content provider are the most benefited parties. Gateway router is hardly expiated of any loss by the end user or content provider. We need to give a strong incentive (motivation) to the gateway router so that it can perform better in future. It does not mean that the end user and content provider give money directly to the gateway router. In fact, we need to set the incentive chain right and make the money transfer from one party to another and finally to the gateway router.

Every organization has a specified budget for investing in security against cyber attacks. Decided by the security professionals, a fixed part of the budget is used to spend over methods and tools, like installing firewalls, intrusion detection systems and anti-virus on the systems. The remaining amount is used for hiring a third party to handle all security-related breaches. Now, hiring a third party totally depends on the organization's size and available resources. Small-scale firms cannot afford large premiums of third party. This is the main disadvantage of economic solutions; it is not friendly for all scales of organizations. Technological solutions do have their own limitations and economic solutions have their own, but to make an effective solution, we need to blend technological solutions with economic incentive-based solutions. In this paper, we focus on economic solutions and strong need of economic incentive-based solutions.

2. Economic Solution-based Literature Survey

As already mentioned above, information security management is a well-established domain focusing largely on technological solutions (firewalls, intrusion detection systems, cryptography, access rights) like [8–14], whereas very less work is available for economic solutions in the same domain. In 2001, Anderson [15] and Gordon and Loeb [16] raised interest in this topic. In [15], many research gaps and problems, that security researchers had discovered earlier, are resolved with the concepts of micro economics. He also condensed many problems of information security through economics' concepts. In spite of having resources, organizations lack a conceptual framework that tells how to invest resources in information security. In [16], an economic model is presented from the business perspective to calculate n optimum amount to be invested in information security by an organization. This model considers two main factors—vulnerability of information and potential loss associated with the vulnerability, to calculate the optimum level investment to secure information. The main aim of this paper is to maximize profit from investments in security.

In [17], a budgeting process of capital investment using Net Present Value (NPV) model is presented to calculate expenditure on security based on cost-benefit analysis. The main driving force behind using the NPV model is that it allows firms to calculate their benefits and profits from investment in information security. However, in his survey, he also showed that some respondents do not use the NPV model to estimate expenditures on security but most of the respondents consider potential loss due to a security breach and the probability of occurring such breach in future as important factors to estimate the amount to be spent on security. But anyhow, every organization is using economic analysis and principles to processing a budget for ensuring security of information.

In [18], researchers showed evidence that if a security breach hinders confidentiality of the firm, then the market has a considerable negative reaction towards the suffering firm and it definitely affects the economic performance of the firm in future. But this survey holds many limitations too. Another author Riordan, in [19], studied providing economic incentives solutions in the presence of network externalities. Similarly, in the same direction, authors [20] studied that investment in security depends on other agents present in the network. They first developed an interdependent security (IDS) model for airlines system and then applied the same model for cyber security, fire protection and many other applications. Every node on one network makes investment in security against external attacks, but not from internal viruses that could easily compromise the system and make it a source of the attack for other networks. In [21], authors have shown how user's incentives affect network security when the patching cost is high and negative externalities influence network security. They have shown that for a proprietary software patching rebate (users are compensated by vendors for actually patching the vulnerability) and self-patching are dominating while for freeware software's self-patching is strictly needed. In [22], the authors proposed a model for examining the optimal decision for vulnerability disclosure under negative market externalities.

In a case study by Moore et al. [23] on the attack by 'Code Red' worm on July 19, 2001 though the attack was an international event and major attention is drawn on the losses by big enterprises, but a large section of victims were home users and small firms. It became clear from this fact that home users have less incentives as compared to big organizations.

In [24], authors presented a game theoretic model for patch management. They showed that vendor's patch release cycle and firm's patch update cycle are not necessarily proportionate but can be made in sync with the help of assigning liability and cost sharing. In [25], Moore et al. said that Return on Investment (RoI) could be used in some situations optimally but not in all scenarios.

In the case where firms have interconnected IT relations, there exists a vast literature survey. In [26], interdependency among firms becomes negative externality and results in underinvestment from an optimum level by firms. Many works present in the past decades when researchers proposed models to enable firms to take optimal decision for security investment, optimal decision for disclosure of vulnerabilities under positive and negative market externalities, for various works showing the importance of incentives at lower level or for home user and importance of outsourcing security.

Further sections discuss the importance of incentives and cyber insurance and applications of providing economic incentive solutions.

3. Cyber Insurance

Till now, we mentioned the importance of properly aligned incentives. On the same track, we studied about insurance, which is a powerful tool for risk management in the internet world. Generally, risk management on internet involves investment in techniques and methods to mitigate risks to reduce the intensity of attack and consequences and thus has been the matter of attraction for large research community for decades [27–33]. Retaining risk and analyzing it for future betterment is also an option along with mitigation of risk option. After deploying the above-mentioned options, we are still left with some residual risks. Here cyber insurance plays the role. Cyber insurance makes an individual or a firm capable of recovering the loss done by uncertain events, using predefined periodic payments. In further sections, we discuss the need of cyber insurance and its importance.

In spite of having advancement in techniques and methods to detect and mitigate cyber-attacks, we are no closer to perfect security due to many reasons, like (i) lack of technological solutions which provide protection against all types of attack, (ii) weak incentive chain between end users, security providing vendors and cyber laws running authorities, (iii) free riding problem, i.e., entities escape from investing in security measures by taking advantage of consequences initiated by investment by other parties, (iv) 'lemon market' effect where security-providing vendors lack good incentives to launch a robust product in market [34], (v) no internet entity is ready to take liability of the damage caused to the other party without strong incentive. Keeping all these points in mind, we need to shift the paradigm to more effective solutions for risk management. Cyber-insurance is emerging as a full-fledged tool in the same direction.

In cyber insurance, an organization hires a third party for handling all the risks of attacks in return for an insurance premium (economic incentive). Cyber insurance is a kind of outsourcing security of a firm. Many factors, like prevailing competition in market, budget of the concerned firm, quality and type of security the firm wants affects the decision made by a particular organization. Main advantages we get from cyber-insurance are firstly, it puts some responsibility on organizations to secure their systems through enabling firewalls or installing anti-viruses in order to reduce premiums [35, 36]; secondly, this practice of cyber insurance sets benchmarks and standards for insurers to seek and provide best service to users [30, 37]; thirdly, it improves the quality of protection; and last but not the least, it improves societal welfare [36].

We need to mention one important fact that some people perceive cyber insurance as an alternative for 100 per cent robust security. This is not right. In fact, cyber insurance is an option to manage and handle cyber risks on internet.

Now let's discuss some recent trends in this domain. According to a survey by Insurance Information Institute [38] breaches in 2016, strike up to 1093 (in million) from 780 in 2015, but at the same time reported braches are only 37 million in 2016 which are very less than 169 million in 2015 and which is not good. The sector which is hugely affected by these breaches is business sector with 45.2 per cent of total breaches in 2016. In 2017, there were 1,202 breaches till November 29. The biggest data breach was reported by Yahoo on October 2017 affecting nearly 3 billion user accounts. According to a survey by Health Informatics, 21 per cent of the enterprises admit that monetary loss from cyber attacks have increased year by year. From the report by Kaspersky [39], on an average, a single cyber-security incident now costs large businesses \$861,000. Meanwhile, small and medium businesses (SMBs) ends up paying \$86,500. Most alarmingly, the cost of recovery significantly increases, depending on the time of discovery. McAfee and CSIS calculated that every year annual loss to global economy falls in the range between \$375 billion to \$575 billion. Cyber insurance has emerged from U.S. The US cyber insurance market is estimated to reach \$5 billion by 2018 and \$7.5 billion by 2020.

Let's understand how firms take decision by not assuming market externalities, demand and competition externalities by example [4].

For example, if an organization has a budget for investing in security totaling 5 million dollars and the company spends 2 million on devices like firewalls, intrusion detection systems, anti-virus, etc., and uses the remaining 3 million for hiring a third party to monitor and analyze all security related activities. Suppose the monthly expenses of the company towards security breaches is 1,00,000 or 2,00,000 with equal probability. Then we assume that third party will easily manage these monthly breaches and (12*1,00,000) 12,00,000 or (12*2,00,000) 24,00,000 are savings from outsourcing information security. There is equal probability for both outcomes. So, the net value for extra investment is (0.5*12,00,000 + 0.5*24,00,000) 18,00,000, which is handled by the third party. Now, (30,00,000–18,00,000) 12,00,000 is the extra amount invested by the organization on IT outsourcing. Here, it is not a good option to invest 3 million dollars in cyber insurance.

At the same time, if we reduces the budget to 2 million dollars, then the organization can think of outsourcing security as (20,00,000–18,00,000) 2,00,000 and which is very much less than 12,00,000. But if we further lessen the investment to one million dollars, then (18,00,000–10,00,000) 8,00,000 is the net expected value on this extra investment. So, the expected return on this value is (8,00,000/10,00,000) or 80 per cent which is a very good number. If 80 per cent is the expected return, then the company can invest one million dollars in security without giving it a second thought. This is how decisions are taken in industry.

In further sections, we discuss about evolution of cyber insurance and risk assessment options.

3.1 Development of Cyber Insurance

In this section, we highlight some remarkable events which contributed to evolution of cyber insurance in the past years.

Earlier, insurance policies did not consider data loss on internet as physical damage, neither did they cover it. Insurers consider cyber property as intangible property and thus it is impossible to do any damage to it; this became a matter of conflict between the insurers and firms. Despite having commercial general liability (CGL) coverage, firms started feeling unsafe against threats. Then in the early 1990s, some rules and laws have favored the fact that loss of data on internet is considered as physical damage. [40] is the first case in which conflict between insurer and client was highlighted and in this case the court announced that computer tape and data are tangible property since the data has permanent value. Another case [41], court had favored the insured where data is lost due to system failure. Similarly, in another case [42], the court had announced that loss of programming in a computer RAM constituted physical damage or loss. Likewise, there were many cases which continuously raised the questions and court had shown affirmation to the insured ones. Furthermore, another problem is that while most CGL policies do not have worldwide coverage, many cyber-torts are international [43]. In 1998 primitive forms of hacker insurance policies were introduced by ICSA (International Computer Security Association). There was not a sudden transformation of cyber risks and vulnerabilities to demand for cyber insurance. Two major events that drastically changed the concept of cyber risk were namely, Y2K and 9/11 attacks. Y2K is the infamous virus called 'Millennium Bug'. It formatted all calendar data worldwide of indistinguishability of 2000 and 1900. It was the first incident where companies realized how cyber risk can cause a potential harm to e-business.

Second event was the 9/11 attacks on the World Trade Center which completely changed the perception of cyber risk to security professionals. Several US government agencies, like NASA, FBI, US Senate and Department of Defense (DoD) were attacked by hackers during that year nearly to 9/11. It was a series of DDoS attacks launched against major US corporations. 2011 was marked as a year of serious association of IT insurance and cyber risks.

There is one more important event which is biggest in history—TJX hack [44]. In 2007, 45 million customer's credit and debit card information was stolen over a period of 18 months. Hackers had planted illegitimate software in the company's network and stolen all users' personal data. It is said that the attack took place in May 2006 and was discovered after seven months in mid-December 2006. This event cost nearly $5 million to the company in addition to the many other costs. Moreover, this event has completely ruined the reputation of the company. Along with the financial losses, the company has a huge reputational loss and brand damage in the market. After court hearings, it was found that actually 94 million customers were affected, which was almost double of 45. This data breach is the biggest ever in history and after this event, companies were forced to think about alternate solutions against cyber risks. They started realizing that despite having huge investment in 100 per cent robust security solutions, they are not safe and definitely this is the case with big brand companies and big enterprises. If this is the case with big names, then what about small organizations and firms?

Apart from major events which contributed to growth of cyber insurance, there are some laws in various countries which are also the driving force behind the appreciable growth of cyber insurance. In 2003, the first data breach notification law was enforced where companies are accountable to both customers and governing authorities in black and white [45]. In EU, E-Privacy Regulation was enacted in 2009 for telecommunication and ISPs, as these companies have huge databases of personal data of customers. So the regulatory authorities had thought of imposing a fine on banks and insurers and in case of any data breach event, the company had to inform the customer as well as the higher authority. Cyber insurance companies were continually modifying their policies according to the changing laws in order to cover every single possibility of cyber risk and vulnerability with affordable premiums by the customers. Although there are many factors which make cyber insurance this big-like market status, competition from other insurance providers, desirous of making profit from polices which are so well designed that they cover every possible cyber risk from the insured's point of view but at the same time, policies support the insurer in case of any data breach or conflict. However, the major events in the last decade from 2000 contribute a lot to the immense growth of cyber insurance from traditional, basic and primitive to a comprehensive one.

3.2 Risk Assessment

Cyber insurance is an umbrella term which covers first party and third party losses due to a cyber-attack or data breach event. There are many challenges faced by cyber insurance providers, like design of policies that cover every probable risk but at the same time, they tried to ignore some precarious events with high probability by applying some conditions, exceptions or limits. The process involving risk assessment is complex and not defined in such a constantly changing landscape; the translation of identified risks and designed policies into well-formulated premiums which are profitable (not too low as the company faces loss and not too high as then it is not in demand by clients); the moral hazard problem. In this section, we only discuss about the risk assessment factor.

Risk assessment is an important part of underwriting cyber insurance policies as cyber attacks have an economic impact on the victim and any kind of miscalculation or not taking small factor into formula for calculating premiums can leave the company in dire straits (financial as well as reputational loss). From the beginning, risk assessment methodology is flawed because policy writers had faith in factors that are insufficient, like the firm's past data breach events, standards or benchmarks set by local governing authorities and IT market (competitor's policy framework and state of market). But changes in internet and ubiquity of cyber attacks are affected by more factors than mentioned above and which are really impactful to be considered. In the last decade, there has been a tremendous change in the paradigm from traditional risk-assessment process of only checking properly installed and deployed firewalls, IDSs and anti-virus to new holistic pre-agreement with thorough analysis of the company's profile. Now-a-days risk-assessment procedures not only provide static data summary of the company's security profile but also provide much more meaningful statistics of the company's state.

Traditional risk-assessment process insufficiently found the impact of company's security investment in mitigating only significant risks, but now risk-assessment companies generate questionnaires or checklists that have questions ranging from physical security to cyber vulnerabilities; they have questions about company's IT infrastructure, security investment budget, practices followed by authorities and employees to ensure security, type of data managed by company, services that are outsourced by the company, the company's past and present relations with other insurers, the company authorities must cite state laws stating that providing false information is a crime. They divide the checklist into four main domains, namely organizational, data management, cyber-culture and degree of exposure to risk. Table 1 presents the description about these domains [46].

Table 1: Description about various domain names.

Sr. No.	Domain Name	Description
1.	Organizational	• General • Infrastructure • Crisis management • Senior management • Data breach history • IT security budget
2.	Data management	• Type of data to be secured (Client's personal data like credit/debit card info. or trade secrets or proprietary knowledge) • Access management • Deployment of security devices • Recovery measures
3.	Cyber culture	• Staff skills to detect threat • Responsive measures against threats • Threat visibility • Usage policies of internet
4.	Degree of exposure to risk	• Type of business • Vulnerability assessment • Dependency on outsourcing services • Relation with outsourced vendors
5.	Legal and compliance	• Implementation of state laws is strictly followed or not • Standards related to a particular industry are followed?

This new holistic approach of checking and dividing the company's profile into different domains thoroughly quantifies the cyber risks differently. This new approach checks the company's outsourcing dependencies (taking third party help for service providing) to check what kind of vulnerabilities or risks might have been added by the third parties or sub parties into the applicant's network. Apart from monitoring and analyzing other externalities that could contribute to a data loss event, this approach also checks the cyber culture implemented by the organization within its premises. In post-agreement approach, cyber insurers do risk assessment periodically to optimize decisions about prioritizing risks.

Factors like advancement in IT market, diversity in attack methods, emergence of new groups of cyber terrorists with faultless plans of performing cyber-attacks, competitors in market and lack of robust security make this approach suitable for risk assessment.

4. Need of Incentives

To make the internet a safe place to store data and to make safe transactions, all government, non-government (profit and non-profit public or private firms) need to come together in defending cyber attacks and viewing cyber security as societal welfare. Organizations stop viewing their investment in cyber security as a money-making process. This is the first thing we should do to make internet safe.

Secondly, until we make the persons guarding the systems, the persons who suffer loss, we cannot tackle this problem. We need to create incentives at the points where persons guarding the systems face the loss of cyber attacks. Incentive is the tool that can encourage various internet entities to invest in security and make their systems secure and avoid free-riding problems. The need for incentives arises from the uncovered losses. Let's take the case of a naive home user, who uses his computer for light-weight applications like surfing internet, send or receive mail or use social networking sites. He does not have any confidential data or trade secrets that he wants to secure. He does not protect his system until some malfunction occurs and interferes with the normal usage of computer. This home user does not care if his system is changed to a bot and is used to carry out an attack on some third party. We have to provide incentives to this naive home user so that he wants to make his system safe from viruses or trojans.

Next, if we talked about a big enterprise, then this enterprise has a much broader view of IT security than the home user. This enterprise not only wants to keep its systems safe and unhampered from viruses and attacks, but also wants to make its services available to clients in time along while assuring that the integrity and confidentiality of data remains intact. So, here the enterprise's owner has much stronger incentives than the home user to make his systems secure. Since, security on internet is interdependent, if one entity on internet is not sufficiently safe, then other entities taking services from the former entity are also not safe in either way.

The first thing we must do to avoid fraud and misuse of resources is to generate incentives at the right time and for the right party. A security researcher named Ross Anderson at Cambridge University wrote a paper [47], where he discussed the importance of liability and creation of incentives. He found out the reason behind the failure of ATM machines in Britain while in USA in spite of investing less in security, ATM is a huge success. In Britain, liability is on the user and in case of any dispute between the user and the bank, the user has to prove that the bank has committed a mistake which is a difficult task for the user. While in the USA, the liability is on the bank and in a case of dispute between the user and the bank, the bank has to prove whether the user is at fault or not. So, in simple terms, in the USA, the bank has proper incentive (motivation) to invest in information security and risk management techniques, like installing cameras and deploying security guards so that no user can try to cheat the bank, but in Britain, banks lack incentive (motivation) and the complete burden is on the user. This real life example shows a simple fact that liability and incentives should be associated with the party which is in best position to protect the information or is able to manage risks. Banks are in a better position to secure money and data, so more liability and incentives must be associated with the bank. This example shows how the incentive works.

We must use this concept in defending data and information available on internet. For example, *A* is an attacker and *C* is the victim. *A* attacks *C* using *B*'s system. *B* has nothing to do with *C*'s loss. But if *B* is responsible towards *C*, then *B* has stronger incentive to make his system more secure.

4.1 Disproportionate Incentives

Sensitive issues, like usage of computers by naive users and generation of incentives at right place to avoid security breaches, have always been neglected by security IT professionals and researchers. Misaligned incentives in the banking sector, as stated above, have created initial interest in economics of information security, as we have already discussed that ATM machines are a success in the USA because of proper incentives at right party and well-allocated liability. This example shows that economics of liability and game theory in cyberspace are as important as cryptographic protocols and firewalls. Liability must be assigned to the party that can defend systems and manage risks efficiently among all the parties, but this does not mean that the user can escape all responsibilities. A conventional approach of liability allocation must be needed to assign liability to the right party.

A DDoS attack was witnessed in early 2000 when a series of DDoS attacks was launched against big companies, like Yahoo, Amazon. This DDoS attack involves flooding the target with huge packet traffic. Varian [48] explained this as the case of misaligned incentives where users are willing to spend on antivirus software but cannot spend on software that can prevent their systems from being used to launch attack against the third party.

Cyber attacks are a result of lack of incentives not only at network level but also at organizational level. Hackers and cyber attackers have their own hidden market economy which facilitates easy entry of those who are just computer literate, have well-distributed incentives, new ideas of carrying out attacks. This hidden market economy has completely different features than visible defensive market economy. Attackers always take advantage of imbalanced incentives between attackers and defenders. According to a survey report [49] on 800 IT security professionals, there existed three major incentive mismatches that put defenders in disadvantage. Following are the points:

1. First incentive mismatch is between the attacker and defender. Security professional's incentives are outlined by many factors, like hierarchical (top-down) approach of decision making and restricted by protocols set by a particular firm and governing authorities of that particular state while the attacker's incentives are influenced by personal benefit and hidden decentralized market economy.

2. Imbalanced incentives between planned approach and its implementation. Every organization has a well-defined approach to fight against cyber attack but due to certain reasons, many organizations fail to implement them practically. Many strategies remain only in documentation.

3. Incentive misalignment between policy makers and policy runners. There is a lot of difference in perception of incentives between higher authorities that make policies and operators that put those policies into practice.

Cyber attackers have direct incentives like money, fame but cyber data defenders have indirect incentives. One more fact from the report is that higher authorities, who design policies, have more faith in effects of incentives than in the policy runners at the lower level. It could be due to unawareness of lower staff about existing incentives. If the things learnt from the attacker's hidden market economy are implemented in the organization's working environment and security-providing market, then many misalignments could be corrected. We need to grow more incentives at the operator level and make a uniform distribution of incentives hierarchical. The hacker community is leveraging market forces very efficiently than the defenders. They know how to take advantage of competition and market prices to reduce levels of difficulty for new entries and help them in exploiting exposed vulnerabilities and implementing new ideas for carrying out attacks. Secondly, public disclosure of vulnerabilities has a different impact on the attacker and defender. Public disclosure of vulnerabilities reduces the attacker's effort as well as cost in vulnerability research while it takes some time for the defenders to fix these loopholes.

So, assigning liabilities to the right party, growth of incentives at the right point and at policy runner level, experimentations to give the right metrics for risk assessment and incentive creation process, more use of outsourcing security and proliferating information sharing, can help in correcting misalignments of incentives.

4.2 External Factors Affecting Security

The internet world is affected by many pervasive, loose and huge externalities. Externality refers to cost (or benefit) a company has to suffer because of the decision of the other party. The impact of externality does not reflect in market prices. It is the loss or gain of welfare of a company due to decisions made by others. For example, if a factory has skilled labor, then the factory's yield is definitely good; so this is positive externality. But if a user does not secure his system, then it can be a negative externality towards other users as this insecure system can be exploited as a part of botnet to launch attacks against other parties.

Software vendors have competition to grab the user's wallet first without caring much about the most important feature—security. We pay for IT security breaches in many ways: in the form of financial loss, reputational loss. Installing security features needs extra money, effort, time, decreased performance of the device, annoyed users and delay in product launch in market. Consequences of not paying attention towards security are less detrimental to the software vendors as they have to face intermittent bad press, unavailability of services for some time and switching of some users to other service providers. So, every big enterprise does exactly what every organization does in terms of security. These are not the only factors influencing network security externalities.

Author Hirshleifer [50] talked about the weakest link rule. Traditionally, we assume that protection of a public good relies on the sum of efforts by all members of that protection group. But in this paper, the author explains by an example that protection of a public good depends on the effort by the weakest link of the specified group. He illustrates this by an example where an island named Anarchia can be protected from flood when every family residing on the island constructs defenses and now the safety of the island depends only upon the defense created by weakest (laziest) family. He further compares this example with the ICBM attack. Varian [51] extended this work by showing dependability of system reliability on three cases: total effort, weakest link and best shot. In the case of sum of efforts, system reliability depends mainly on the link possessing the highest cost benefit ratio and all other units' free ride on the measures taken by this link. Another author [52] defines positive externalities as the reason for free ride problem. He states that a person taking defensive measures can generate positive externalities for others but at the same time, it can encourage other users not to take sufficient defensive measures on their part. This theory has real life implications also. If two neighbors shares one garden and one family cleans it regularly, then it has positive externality on another family but at the same time it discourages the other family not to clean the garden. Generally, positive externality and free ride problems are two sides of the same coin.

In [53], authors have shown the fact that positive consumption externalities influence the adoption of a new technology, either directly or indirectly, like the number of users currently using that technology affects the user's decision for adopting the technology. Utility drawn from the technology by the user is directly proportional to the number of users consuming the technology. Similarly, another network externality is compatibility. If the new technology is able to run on the hardware already present with the user, then it can help the user to adopt new technology. Likewise, deployment of an efficient defensive mechanism depends on the number of users using it and its compatibility with the already available technology with the user. But everyone waits for the other user to use the technology first to give his reviews and as a result, an efficient technology could never have been implemented. [54] tried to give solutions to these problems.

Moreover, software vendors take security as an extra feature and desist from performing their best unless the gain some profit for companies. Vendors try to balance the cost of secure software against benefit of insecure software. Because of their inappropriate approach of considering only the cost of secure software incurred to them, they ignore the cost incurred on the whole internet space and by other firms and users. Due to this software, buyers have to spend money on their security. This is also a kind of externality to network security. Many factors, like limitation of available quality product by software monopolies, 'lock-in effect' where companies provide minimal security initially and once they get established in market or achieve a certain number of users for doing a good business, they start adding security slowly to lock its customers while compatibility issues make very hard for customers to switch the company that again acts as externality to network security.

It is very sad that no company makes security as a distinguishing feature. Companies choose occasional bad press rather than embedding security features from the very start of the software development cycle. It's not like that we lack secure solutions or protocols, but we lack positive externalities to implement them.

5. Applications

In this section, we discuss applications of economic solutions to security data breaches.

5.1 Trade of Vulnerabilities

Whether public disclosure of vulnerabilities leads to hygienic cyberspace or not has been a matter of discussion between software vendors and security researchers since a very long time. There have been many articles about the market where software vulnerabilities are sold and purchased at a very huge amount. Not only software companies and cyber attack market which are involved in this business, but also there are many government organizations that are also doing the same business. This market of selling and buying vulnerabilities has grown substantially since the last decade. Some researchers have presented papers about the economics related to vulnerabilities. Further, we discuss some of the important work done by the authors.

In [55], authors stated that software with many vulnerabilities and fixing one does not make a large difference in the probability of finding another by the attacker. Unless the inter-dependent vulnerabilities are discovered, there should be no revelation or persistent patching of the vulnerabilities.

Open source and free software vendors believed that making source code available to all users is a good step for security as users, professionals and researchers find the vulnerabilities while the opponent community believes that if the software code is made publicly available, then it would take fewer efforts to find vulnerabilities by the attacker. Now we have question: which public disclosure of vulnerabilities helps—the attackers or defenders?

Anderson [56] in 2002 showed that under certain assumptions, openness of source code helped attackers and defenders equally. According to his research, he found that open software and proprietary software have the same level of security. Another author [57] Ozment talked about immediate disclosure (where a vulnerability report is released before the patch for the same is ready) and responsible disclosure (where a vulnerability report is released after the patch was ready). The community favoring immediate disclosure said that immediate disclosure of vulnerabilities puts pressure on software vendors to provide patches immediately and at the same time encourage users to mitigate the risk of attack. The community favoring responsible disclosure said that immediate disclosure would lead attackers to find vulnerabilities more easily and thus result into an attack that could incur huge cost. Next in this paper, he investigates the vulnerabilities in FreeBSD and found that vulnerabilities are correlated and indeed likely to be rediscovered [57]. Further, Ozment and Schechter also found that the rate at which unique vulnerabilities were disclosed for the core and unchanged FreeBSD operating system has decreased over a six-year period [58]. These researchers claim that disclosure of vulnerabilities has helped in increasing the lifetime of software by improving security. Disclosures of vulnerabilities provide necessary motivation to software vendors to provide patches [59]. In [60], Arora et al. extended the work done in [59] and found through empirical studies that private disclosure makes vendors less responsive towards vulnerabilities as compared to open disclosure. He also stated that though disclosure leads to an increased number of attacks, at the same time it leads to a decreased number of reported vulnerabilities.

So, we can clearly see that disclosure of vulnerabilities has a mixed view by researchers and software vendors. Software vendors need to understand the cost of design secure software is less than the cost of patching after vulnerability is announced. Apart from the cost, vendors have to face reputational loss and how can we ignore the cost of writing and deploying the patch? As discussed earlier in [16], Gordan and Loeb presented a model where they show that securing highly vulnerable information is expensive, so companies focus on securing information with average vulnerabilities. They also analyzed that in order to optimize the expected profit from security investment, a firm should spend only a fraction of expected loss due to a security breach event.

A social-minded white hat hacker, on finding a vulnerability, will responsibly report to the respective firm in expectation of some monetary reward. In this case, no party faces loss as the respective company now can fix the patch, while the white hat hacker receives his reward and there is no cyber attack related to that vulnerability. But it is an irony that reporting the same vulnerability to the hidden cyber attack market by the black hat hacker is more profitable comparatively. We need to encourage white hat trades as many companies have started many programs to provide vulnerability-reporting incentives. For example, Google's 'Project Zero' is such an initiative where full-time researchers are employed to find vulnerabilities and report to the related firm. There is one more effective provision in this scheme—if the related company does not fix the vulnerability within 90 days, then the Website publishes its name publicly. There is no denying the fact that one party from all internet entities has to pay for these vulnerabilities either way in the form of fee to the consultant, or fee to the third party for outsourcing security or many other direct and indirect losses. So, it would be better to invest in fixing bugs and vulnerabilities from the beginning of the software development.

5.2 Economics of Privacy

Internet is the world where it offers the chances of unparalleled privacy but at the same time offers loss of privacy. There are many laws and regulations enforced to protect privacy and they have a huge support from users and professionals. There are also many technologies that can offer good security to maintain privacy [61]. We already discussed the fact that firms do not have stronger incentives to implement good technical solutions.

Everyone is concerned about privacy but few do anything about protecting privacy. People make intense concerns regarding privacy, but do not take a minute to give their personal details on internet with very little or no incentive. This gap of statements and actions on internet has motivated many researchers to invest their time in this direction. Like in [62], authors proposed a model to study the effect of various regulation laws on security investment. They stated that firms that have been compromised have to report to authorities regarding the security investment, which leads to over-investment in breach detection and under-investment in prevention techniques. In [63], a utility approach is proposed for firms to decide whether they need to invest in internal security measures or outsourcing security. In a paper [64], authors have shown that user's decision about securing privacy is affected by many factors, like incomplete information and even with complete information, they get attracted towards short-term benefits at the cost of long-term privacy as the user is unable to act rationally on huge dataset.

In another kind of behavioral research [65], where authors surveyed students to measure their privacy preferences in different scenarios and showed that a group of students read all privacy policies to guarantee that their answers could never be linked to them in future and answered only a few questions out of the total questions. Another group, which was made to give answers on some website having no privacy feature (not linked to their college's site), revealed all their personal information. In [66], authors had investigated the impact of disclosure of privacy flaws and data breaches related to privacy on consumer's purchases in future. If [18], companies reporting security data breach lose confidential information, then stock prices of such companies are more likely to fall. So, economics of privacy mainly works on the actions of companies and users related to personal data. Here also, incentives play a major role for users and firms to enhance security feature.

6. Conclusion

Finally, we conclude that economic solutions for security breaches are as important as technological solutions. In this chapter, we tried to give some useful insights about the importance of economical solutions, cyber insurance, liability assignment, growth of proper incentives at right party, externalities faced by security and trade related to vulnerabilities and privacy. If we succeed in providing incentives as well as in correcting economic problems, then vendors can implement efficient technological solutions. Users at any level, whether it is the home user or the security managers of big enterprises, must understand their liabilities towards other entities on network and come up with the best efforts in securing their systems as cyber security is interdependent. Externalities also play a major role in regulating network security.

References

[1] Wu, Yong et al. (2017). Decisions making in information security outsourcing: impact of complementary and substitutable firms. Computers & Industrial Engineering.
[2] Anderson, Ross and Tyler Moore. (2006). The economics of information security. Science 314.5799: 610–613.
[3] Anderson, R. (2001). Why information security is hard—An economic perspective. *In*: Proceedings of the 17th Annual Computer Security Applications Conference, New Orleans, LA.
[4] Gordon, L., Loeb, M. and Lucyshyn, W. (2003). Information security expenditures and real options: A wait-and-see approach. Computer. Sec. J. 19(2): 1–7.
[5] Gordon, Lawrence, A., Martin P. Loeb and William Lucyshyn. (2003). Sharing information on computer systems security: An economic analysis. Journal of Accounting and Public Policy 461–485.
[6] Gemg, X. and Whinston, A.B. (2000). Defeating distributed denial of service attacks. IEEE IT Professional 2: 36–41.
[7] Naraine, R. (2002). Massive DDoS Attack Hit DNS Root Servers.
[8] Simmons, Gustavus, J. (1994). Cryptanalysis and protocol failures. Communications of the ACM 37qq: 56–65.
[9] Denning, Dorothy, E. and Dennis K. Branstad. (1996). A taxonomy for key escrow encryption systems. Communications of the ACM 39.3: 34–40.
[10] Sandhu, Ravi, S. (1998). Role-based access control. Advances in Computers, Vol. 46, Elsevier, 237–286.
[11] Schneier, Bruce. (2007). Applied Cryptography: Protocols, Algorithms, and Source Code in C., John Wiley& Sons.
[12] Goyal, Vipul et al. (2006). Attribute-based encryption for fine-grained access control of encrypted data. Proceedings of the 13th ACM Conference on Computer and Communications Security, Acm.
[13] Vigna, Giovanni and Richard A. Kemmerer. (1999). NetSTAT: A network-based intrusion detection system. Journal of Computer Security 7.1: 37–71.
[14] Axelsson, Stefan. (2000). The base-rate fallacy and the difficulty of intrusion detection. ACM Transactions on Information and System Security (TISSEC) 33: 186–205.
[15] Anderson, Ross. (2001). Why information security is hard-an economic perspective. Computer Security Applications Conference, 2001. Acsac 2001, Proceedings of 17th Annual. IEEE.
[16] Gordon, Lawrence, A. and Martin P. Loeb. (2002). The economics of information security investment. ACM Transactions on Information and System Security (TISSEC) 5(4): 438–457.
[17] Gordon, Lawrence, A. and Martin P. Loeb. (2006). Budgeting process for information security expenditures. Communications of the ACM 49.1: 121–125.
[18] Campbell, Katherine et al. (2003). The economic cost of publicly announced information security breaches: empirical evidence from the stock market. Journal of Computer Security 11.3: 431–448.
[19] Riordan, Michael. (2014). Economic incentives for security. Powerpoint Slides Presented at Cybercriminality Seminar at Toulouse School of Economics on, Vol. 4.
[20] Kunreuther, Howard and Geoffrey Heal. (2003). Interdependent security. Journal of Risk and Uncertainty 26: 231–249.

[21] August, Terrence and Tunay I. Tunca. (2006). Network software security and user incentives. Management Science 52: 1703–1720.

[22] Choi, Jay Pil and Chaim Fershtman. (2005). Internet Security, Vulnerability Disclosure and Software Provision.

[23] Moore, David and Colleen Shannon. (2002). Code-red: A case study on the spread and victims of an Internet worm. Proceedings of the 2nd ACM SIGCOMM Workshop on Internet Measurement, ACM.

[24] Cavusoglu, Hasan, Huseyin Cavusoglu and Jun Zhang. (2008). Security patch management: Share the burden or share the damage? Management Science 54: 657–670.

[25] Moore, T., Dynes, S. and Chang, F.R. (2015). Identifying How Firms Manage Cybersecurity Investment, pp. 32.

[26] Ogut, H., Menon, N. and Raghunathan, S. (2005). Cyber insurance and IT security investment: Impact of interdependent risk. Proceedings of the Workshop on the Economics of Information Security (WEIS05), Kennedy School of Government, Harvard University, Cambridge, Mass.

[27] Ferguson, P. and Senie, D. (2000). Network ingress filtering: defeating denial of service attacks that employ IP source address spoofing. Internet RFC 2827.

[28] Mirkovic, J., Prier, G. and Reihe, P. (April 2003). Source-end DDoS defense. *In*: Proc. of 2nd IEEE International Symposium on Network Computing and Applications.

[29] John, A. and Sivakumar, T. (May 2009). DDoS: Survey of traceback methods. International Journal of Recent Trends in Engineering, ACEEE (Association of Computer Electronics & Electrical Engineers) 1(2).

[30] Joao, B., Cabrera, D. et al. (2001). Proactive detection of distributed denial of service attacks using MIB traffic variables—a feasibility study. Integrated Network Management Proceedings, pp. 609–622.

[31] Park, K., Lee, H. (2001). On the effectiveness of probabilistic packet marking for IP traceback under denial of service attack. pp. 338–347. *In*: Proc. of IEEE INFO COM.

[32] Mahajan, R., Bellovin, S.M., Floyd, S., Ioannidis, J., Paxson, V. and Shenker, S. (2002). Controlling high bandwidth aggregates in the network. Computer Communication Review pp. 62–73.

[33] Mirkovic, J., Reiher, P. and Robinson, M. (2003). Forming alliance for DDoS defense. *In*: Proc. New Security Paradigms Workshop, Centro Stefano Francini, Ascona, Switzerland.

[34] Akerlof, G.A. (1970). The market for lemons—quality uncertainty and the market mechanism. Quarterly Journal of Economics 84(3).

[35] Anderson, R., Böhme, R., Clayton, R. and Moore, T. (2008). Security economics and the internal market. Study Commissioned by ENISA.

[36] Majuca, Rupertom P., William Yurcik and Jay P. Kesan. (2006). The evolution of cyber insurance. arXiv preprint cs/0601020.

[37] Mansfield-Devine, Steve. (2016). Security guarantees: building credibility for security vendors. Network Security 2: 14–18.

[38] Wilkinson, C. (2015). Cyber risk: Threat and opportunity. Insurance Information Institute, Link: https://www.iii.org/sites/default/files/docs/pdf/cyber_risk_wp_final_102015.pdf.

[39] Link: https://www.kaspersky.co.in/about/press-releases/2017_kaspersky-lab-survey-shows-real-business-loss-from-cyberattacks-now-861k-per-security-incident.

[40] Retails Systems, Inc. v. CNA Insurance Companies, 469 N.W.2d 735 [Minn. App. 1991].

[41] Centennial Insurance Co. v. Applied Health Care Systems, Inc. (710 F.2d 1288) [7th Cir. 1983].

[42] American Guarantee & Liability Insurance Co. v. Ingram Micro, Inc., Civ. 99-185 TUC ACM, 2000 WL 726789 (D. Ariz. April. 18, 2000).

[43] Crane, Matthew. (2001). International liability in cyberspace. Duke Law & Technology Review 1: 23.

[44] Biggest ever; It eclipses the compromise in June 2005 at Card Systems Solutions by Jaikumar Vijayan. Last accessed on 18 Dec, 2017. Link: https://www.computerworld.com/article/2544306/security0/tjx-data-breach--at-45-6m-card-numbers--it-s-the-biggest-ever.html.

[45] Mark Camillo. (2017). Cyber risk and the changing role of insurance. Journal of Cyber Policy 2: 53_63. DOI: 10.1080/23738871.2017.1296878.

[46] Romanosky, S., Ablon, L., Kuehn, A. and Jones, T. (2017). Content Analysis of Cyber Insurance Policies: How Do Carriers Write Policies and Price Cyber Risk?

[47] Anderson, Ross. (1993). Why cryptosystems fail. Proceedings of the 1st ACM Conference on Computer and Communications Security, ACM.

[48] Varian, H. (June 1, 2000). Managing Online Security Risks. Economic Science Column, The New York Times, June 1, 2000. http://www.nytimes.com/library/financial/columns/060100econ-scene.html.

[49] Tilting the Playing Field: How Misaligned Incentives Work Against Cybersecurity by Mcafee. Link: https://www.csis.org/events/tilting-playing-field-how-misaligned-incentives-work-against-cybersecurity.

[50] Jack Hirshleifer. (1983). From weakest-link to best-shot: The voluntary provision of public goods. pp. 371–386. *In*: Public Choice, Vol. 41.

[51] Hal Varian. (2004). System reliability and free riding. pp. 1–15. *In*: Economics of Information Security, Kluwer.

[52] Howard Kunreuther and Geoffrey Heal. (2003). Interdependent security. *In*: Journal of Risk and Uncertainty 26(2-3): 231–249.

[53] Michael Katz and Carl Shapiro. (June 1985). Network externalities, competition and compatibility. *In*: The American Economic Review 75(3): 424–440.

[54] Ozment, Andy and Stuart E. Schechter. (2006). Bootstrapping the Adoption of Internet Security Protocols. Fifth Workshop on the Economics of Information Security. June 26–28, Cambridge, UK.

[55] Eric Rescorla. (May 2004). Is Finding Security Holes a Good Idea? Third Workshop on the Economics of Information Security, Minneapolis, Mn.

[56] Anderson, R. (2002). Security in Open Versus Closed Systems: The Dance of Boltzmann, Coase and Moore, Technical report, Cambridge University, England.

[57] Andy Ozment. (2005). The likelihood of vulnerability rediscovery and the social utility of vulnerability hunting. Fourth Workshop on the Economics of Information Security Cambridge, Ma.

[58] Ozment, A. and Schechter, S.E. (2006). Milk or wine: does software security improve with age? pp. 93–104. *In*: USENIX Security Symposium.

[59] Arora, Ashish, Rahul Telang and Hao Xu. (2008). Optimal policy for software vulnerability disclosure. Management Science 54: 642–656.

[60] Arora, Ashish, Ramayya Krishnan, Anand Nandkumar, Rahul Telang and Yubao Yang. (2004). Impact of vulnerability disclosure and patch availability—an empirical analysis. *In*: Third Workshop on the Economics of Information Security 24: 1268–1287.

[61] Miaoui, Yosra, Noureddine Boudriga and Ezzeddine Abaoub. (2015). Economics of privacy: A model for protecting against cyber data disclosure attacks. Procedia Computer Science 72: 569–579.

[62] Laube, Stefan and Rainer Böhme. (2015). Mandatory security information sharing with authorities: Implications on investments in internal controls. Proceedings of the 2nd ACM Workshop on Information Sharing and Collaborative Security, ACM.

[63] Ding, Wen and William Yurcik. (2006). Economics of Internet security outsourcing: simulation results based on the Schneier model. Workshop on the Economics of Securing the Information Infrastructure (WESII), Washington DC.

[64] Acquisti, Alessandro and Jens Grossklags. (2004). Privacy and rationality: Preliminary evidence from pilot data. Third Workshop on the Economics of Information Security, Minneapolis, Mn.

[65] John, Leslie, K., Alessandro Acquisti and George Loewenstein. (2009). The best of strangers: Context dependent willingness to divulge personal information.

[66] Afroz, S., Islam, A.C., Santell, J., Chapin, A. and Greenstadt, R. (2013). How privacy flaws affect consumer perception. pp. 10–17. *In*: Socio-Technical Aspects in Security and Trust (STAST), Third Workshop on, IEEE.

CHAPTER 15

Teaching Johnny to Thwart Phishing Attacks
Incorporating the Role of Self-efficacy into a Game Application

Nalin A.G. Arachchilage[1] *and Mumtaz Abdul Hameed*[2,*]

1. Introduction

In March 2016, John Podesta, Chairman of the Hillary Clinton presidential campaign, received a phishing e-mail with a subject line, "*Someone has your password*" (see Fig. 1). The e-mail greeted Podesta, "Hi John," and then read, "someone just used your password to try to sign into your Google account john.podesta@gmail.com." Then it provided a time stamp (Saturday, 19 March, 8:34:30 UTC) and an IP address (134.249.139.239) in the location (Ukraine), where someone used his password. The e-mail also stated, "Google stopped this sign-in attempt." It then offered a link luring him to change his password immediately. Bit.ly provides Web address shortening service (i.e., this is heavily used by Twitter users), which can make users easier to share. So, hackers created a Bit.ly account called Fancy Bear, belonging to a group of Russian hackers, to bypass the Web browser phishing e-mail filtering system to make the attack successful. Unfortunately, Podesta clicked on the link and disclosed his credentials to the hackers. The hacking group, which created the Bit.ly account linked to a domain under the control of Fancy Bear, failed to make the account private. Therefore, this shows that "hackers are human too and they make mistakes (i.e., weakest link)."

[1] School of Engineering and Information Technology (SEIT), University of New South Wales, Australia; nalin.asanka@adfa.edu.au
[2] Technovation Consulting and Training Pvt. Ltd., 1/33, Chandhani Magu, Male'. 20-03. Maldives.
* Corresponding author: mumtazabdulhameed@gmail.com

```
> *From:* Google <no-reply@accounts.googlemail.com>
> *Date:* March 19, 2016 at 4:34:30 AM EDT
> *To:* ████████a@gmail.com
> *Subject:* "Someone has your password"
>
> Someone has your password
> Hi John
>
> Someone just used your password to try to sign in to your Google Account
> ████████@gmail.com.
>
> Details:
> Saturday, 19 March, 8:34:30 UTC
> IP Address: 134.249.139.239
> Location: Ukraine
>
> Google stopped this sign-in attempt. You should change your password
> immediately.
>
> CHANGE PASSWORD <https://bit.ly/1PibSU0>
```

Fig. 1: The John Podesta e-mails released by WikiLeaks [1].

Phishing (identity theft) is particularly dangerous to computer users [2–5]. It synthesizes social engineering techniques (i.e., the art of human hacking) along with the technical subterfuge to make the attack successful [6]. Phishing aims to steal sensitive information, such as username, password and online banking details from its victims [2]. Mass-marketing perpetrators commit identity theft to contact, solicit and obtain money, funds, or other items of value from victims [7]. Online Mass-Marketing Fraud (MMF) is a serious, complex and often very organized crime, which exploits mass communication techniques (e.g., email, instance messaging service, spams, social networking website) to steal people's money.

Automated tools (e.g., anti-phishing, anti-virus and anti-spyware) have been developed and used to alert users of potential fraudulent e-mails and Websites [8, 9]. However, these tools are not entirely reliable in detecting and preventing people from online phishing attacks, for example, even the best anti-phishing tools missed over 20 per cent of phishing Websites [10, 11] because the 'humans' are the weakest link in cyber security [2]. It is impossible to completely avoid the end-user [12]; for example, in personal computer use, one mitigating approach for cyber security is to educate the end-user in security prevention [65]. Educational researchers and industry experts talk about well-designed user security education being effective [2, 3, 13, 10, 14, 15]. However, we know to our cost no one talks about how to better design-security education for end-users. Therefore, the aim of this research proposal is to focus on designing an innovative and gamified approach to educate individuals about phishing attacks. The study asks how one can integrate 'self-efficacy', which has a co-relation with the user's knowledge into an anti-phishing educational game to thwart phishing attacks?

We initially focus on identifying how people's 'self-efficacy' enhances their phishing threat-avoidance behavior. One of the main reasons would appear to be a lack of user knowledge to prevent phishing attacks. Based on the literature, we then attempt to identify the elements that influence (in this case, either conceptual or procedural knowledge or their interaction effect) to enhance people's phishing prevention behavior through their motivation. Based on the literature, we will also attempt to identify whether conceptual knowledge or procedural knowledge has a positive effect on computer users' self-efficacy in relation to thwarting phishing attacks. Furthermore, the current research work integrates either procedural or conceptual (or the interaction effect of both procedural and conceptual) knowledge into an anti-phishing educational game to educate people better on how to protect themselves against phishing attacks. We believe this will certainly improve users' ability to thwart phishing attacks.

2. Related Work

Educational games and simulations have become increasingly acceptable as an enormous and powerful teaching tool that may result in an 'instructional revolution' [16–19]. The main reason is that game-based education and the use of virtual environment allow users to learn through experience while leading them to approach problem solving through critical thinking [2, 20] describes Serious Games as "a mental contest, played with a computer in accordance with specific rules." [21] proposed the following thematic classification based upon Zyda's definition of Serious Games— Military Games, Government Games, Educational Games, Corporate Games, Health-care Games, Political and Religious Games. Nevertheless, serious games are quite useful as an effective teaching medium [22] because it enables users to learn in an interactive and attractive manner [23]. There is a considerable amount of published studies in the literature describing the role of games in the educational context. Bellotti, et al. [24] designed a Massive Multiplayer Online Game (MMOG) for high-school called 'SeaGame' to promote best practices in sea-related behaviors, such as sailing or beach surveillance. The main focus of the 'SeaGame' was to embed the educational content into the gaming context in a meaningful, homogeneous and compelling whole, where the player can enjoy learning while having fun. Authors have concluded that this type of games helps people to improve their best practices in behavior.

In addition, game-based education is further useful in motivating players to change their behavior. Gustafsson and Bang [25] designed a pervasive mobile-based game called 'Power Agent' to educate teenagers and their families to reduce energy consumption in their homes. They attempted to transform the home environment and its devices into a learning arena for hands-on experience with electricity usage. The results suggested that the game concept was more efficient in engaging teenagers and their families to reduce their energy consumption during the game sessions. Furthermore, serious games can be effective, not only for changing people's behavior, but also for developing their logical thinking to solve mathematical problems. [26] designed 'Wu's Castle' through an empirical investigation, which is a 2D (2-dimensional) role-playing game where students developed their C++ coding skills, such as loops, arrays, algorithms and logical thinking and problem solving. The results showed that 'Wu's Castle' is more effective than a traditional programming assignment for learning on how to solve problems on arrays and loops. One of the most well-established facts within the serious games research is the ability to enhance players' motivation towards learning as they are able to retain their attention and keep engaged and immersed in games [2, 27–30]. Other interesting facts of educational games are: (a) The provision of immersive gaming environments that can be freely explored by players and promote self-directed learning [31, 32]. (b) The immediate feedback process with perception of progress [33, 29, 28]. (c) Their relation to constructivist theories and support of learning in the augmented environment [28, 34].

The design of serious games is a double-edged sword. When its power is properly harnessed to serve good purposes, it has tremendous potential to improve human performance. However, when it is exploited for violation purposes, it can pose a huge threat to individuals and society. Therefore, the designing of educational games is not an easy task and there are no all-purpose solutions [35]. The notion that game-based education offers the opportunity to embed learning in a natural environment, has repeatedly emerged in the research literature [36, 2, 3, 15, 10, 6, 27, 37].

A number of educational games have been designed and developed to protect computer users and to assert security issues in the cyberspace [17], for example, some educational games teach information assurance concepts, whereas others teach pure entertainment with no basis in information assurance principles or reality [66]. However, there is little research on engagement in the virtual world that also combines the human aspect of security [10]. Therefore, it is worth investigating further on game-based learning in order to protect computer users from malicious IT threats, such as MMFs.

Even though there are usability experts who claim that user education and training does not work [10], other researchers have revealed that well-designed end-user education could be a recommended approach to combating against cyber-attacks, such as MMFs [38, 14, 15, 39, 40]. In line with [41], also [14] and other researchers argue that current security education on malicious IT threats offers little protection to end users, who access potentially malicious websites [15, 10, 13].

Another reason for ineffectiveness of current security education for phishing prevention is because security-education providers assume that users are keen to avoid risks and thus likely to adopt behavior that might protect them. [14] claimed that security education should consider the drivers of end-user behavior rather than warning users of dangers. Therefore, well-designed security education (i.e., user-centered security education) should develop the threat perception where users are aware that such a threat is present in the cyberspace. It should also encourage users to enhance avoidance behavior through motivation to protect them from malicious IT threats.

Literature revealed that well-designed games focusing on education could be helpful in learning, even when used without assistance. The 'Anti-phishing Phil' game developed by [10] reported results that confirm that games educate people about phishing and other security attacks in a more effective way than other educational approaches, such as reading anti-phishing tutorial or reading existing online training materials. [2] developed a mobile game prototype to teach people how to thwart phishing attacks. Their mobile game design aimed to enhance the user's avoidance behavior through motivation to protect themselves against phishing threats. The designed mobile game was somewhat effective in teaching people how to thwart phishing attacks as the study results showed a significant improvement in participants' phishing threat-avoidance behavior in their post-test assessment. Furthermore, the study findings suggested that participants' threat perception, safeguard effectiveness, self-efficacy, perceived severity and perceived susceptibility elements positively impact threat avoidance behavior, whereas safeguard cost had a negative impact on it.

[42] have developed a board game that contributes to enhance users' awareness of online phishing scams. Their findings reveal that after playing the game, participants

had a better understanding of phishing scams and learnt how to better protect themselves. [43] designed and evaluated an embedded training e-mail system that teaches people to protect themselves from phishing attacks during their normal use of e-mail. The authors conducted lab experiments contrasting the effectiveness of standard security notices about phishing with two embedded training designs they had developed. They found that embedded training works better than the current practice of sending security notices.

[44] evaluated the impact of end-user education in differentiating phishing e-mails from legitimate ones. They provided an overview of phishing education, targeting on context aware attacks and introduced a new strategy for end-user education by combining phishing IQ tests and classroom discussions. The technique involved displayed both legitimate and fraudulent e-mails to users and asked them to identify the phishing attempts from the authentic e-mails. The study concluded that users identified phishing e-mails correctly after having the phishing IQ test and classroom discussions. Users also acknowledged the usefulness of the IQ test and classroom discussions. Researchers at Indiana University also conducted a similar study of 1,700 students in which they collected Websites frequently visited by students and either sent them phishing messages or spoofed their e-mail addresses [45].

Tseng et al. [46] also developed a game to teach users about phishing, based on the content of the Website. The authors proposed the phishing attack frame hierarchy to describe stereotype features of phishing-attack techniques. The inheritance and instantiation properties of the frame model allowed them to extend the original phishing pages to increase game contents. Finally, the authors developed an anti-phishing educational game to evaluate the effectiveness of the proposed frame hierarchy. The evaluation results showed that most of the lecturers and experts were satisfied with this proposed system.

Previous research has revealed that technology alone is insufficient to ensure critical IT security issues. So far, there has been little work on end user behavior of performing security and preventing users from attacks which are imperative to cope with phishing attacks [2, 13, 3, 47–50, 23, 51]. Many discussions have terminated with the conclusion of "if we could only remove the user from the system, we would be able to make it secure" [52]. Where it is not possible to completely eliminate the user, for example, in home use, the best possible approach for computer security is to educate the user in security prevention [2, 13, 3, 14, 53]. Previous research has revealed well-designed user-security education can be effective [10, 15, 43, 54]. This could be Web-based training materials, contextual training and embedded training to improve users' ability to avoid phishing attacks.

Therefore, this research focuses on investigating how one can educate the people better in order to protect themselves from phishing attacks. To address the problem, this research attempts to understand how people's 'self-efficacy' enhances their phishing threat-avoidance behavior. The research work reported in this paper then discusses how one can integrate people's 'self-efficacy' into an anti-phishing educational game design in order to better educate themselves against phishing attacks.

3. Methodology

The aim of the proposed game design is to integrate people's 'self-efficacy' into an anti-phishing educational-game design in order to educate themselves better to thwart phishing attacks. Self-efficacy has a corelation with individuals' knowledge [55]; for example, uses are more confident to take relevant actions to thwart phishing attacks, when they are knowledgeable of phishing threats. [56] has revealed that one's knowledge can be influenced by learning procedural and conceptual knowledge. [57] has argued that conceptual knowledge is close to the idea of 'know that' and procedural knowledge 'know how' in which, both the ideas are imperative to educate one to thwart phishing attacks. Furthermore, his research work describes that such conceptual knowledge permits one to explain why, hence the difference of 'know how' and 'know why'. Additionally, [56] stated that the two ideas of conceptual and procedural knowledge are frequently treated individually, with their relationship being disregarded.

Therefore, in this research, the elements derived from a theoretical model [13] will be used to incorporate into the game design to thwart phishing attacks. The theoretical model [13] (Fig. 2) examines if conceptual knowledge or procedural knowledge affects computer users' self-efficacy in thwarting phishing attacks. Their findings revealed that the interaction effect of both procedural and conceptual knowledge will positively influence self-efficacy, which contributes to enhancing computer users' phishing threat-avoidance behavior (through their motivation).

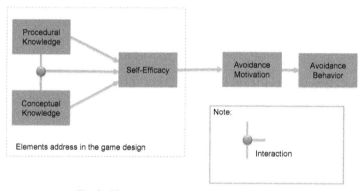

Fig. 2: Elements address in the game design [13].

4. Integrating 'Self-efficacy' into a Game Design

To explore the viability of using a game to thwart phishing attacks based on 'self-efficacy', a prototype designed was proposed. We developed a story addressing 'self-efficacy' (in this case, both procedural and conceptual knowledge as well as its interaction effect) of phishing URLs within a game design context.

4.1 Story

The game story was created based on a scenario of a character of a small and big fish, both living in a big pond. The game-player role-plays the character of the small fish, which wants to eat worms in order to become a big fish. Worms were randomly generated in the game design. However, the small fish should be very careful of phishers, who may try to trick him with fake worms. This represents phishing attacks by developing threat perception. The other character is the big fish, which is an experienced and grown-up fish in the pond. The proposed mobile game-design prototype contains two sections: teaching the concept of phishing URLs and phishing e-mails. The proposed game design teaches how one can differentiate legitimate URLs (Uniform Resource Locators) from fraudulent ones.

4.2 Storyboarding

The game scenario focused on designing an educational game for users to thwart phishing attacks. Therefore, the current study attempts to design a storyboard based on the game scenario as the initial step for the game.

A storyboard is a short graphical representation of a narrative, which can be used for a variety of tasks [58]. The exercise of creating storyboards has drawn from a long history with particular communities, such as those for developing films, television program segments and cartoon animations [59]. Storyboarding is a common technique in Human Computer Interaction (HCI) to demonstrate the system interfaces and contexts of use [58]. One of the best ways to vizualize the game is 'storyboarding' it to create a sequence of drawings that shows different scenes and goals [59]. Storyboard sketches are used in this study to do brainstorming and getting down the general flow of the game on a paper. Stories are rich and the fleshed-out descriptions of settings, people, motivation, activities, goals and values are presented in a coherent and causally connected way.

As interactive computing moves off the desktop to mobile devices, storyboards should demonstrate not only the details related to the specific interface-design context but also the high concepts surrounding the user's motivation and emotion during the systems use [58]. This is because the lack of the user's motivation can cause less satisfaction in the system use. This situation leads the user to engage the system, which also directly causes not to produce the expected level of output by the system. Storyboards can illustrate a visual scenario of how an application feature works in designing new technologies. The designer creates an information design scenario that specifies the representation of a task's objects and actions that may help the end-user perceive, interpret and make sense of the proposed functionalities. The interaction design scenario specifies how the user would interact with the system to perform new activities. Therefore, in the current study, the process of creating a story can hold the game player's attention to the factors necessary to make an effective solution.

The story-based design methods can be beneficial in two specific ways [60]. First, use of case descriptions is important in understanding how technology reshapes user activities. Second, the story can be created before the system is built and its impacts are figured out. Thus, it has been shown that designers often use storyboards to visualise the depiction of information.

There are a number of existing research utilising storyboarding tools, such as Silk [61], DENIM [62] and DEMAIS [42] to create interaction scenarios to convey how the end-user interacts with the envisioned system. However, designers brainstorm individually their design alternatives while capturing their ideas in quick sketches using ink pen or pencil and papers [58] because this is the easiest and quickest way to visualise the idea instead of using drawing tools, such as Microsoft PowerPoint and Microsoft Paint, or much more advanced graphics tools, such as Adobe Photoshop and

Adobe Illustrator. In addition, paper prototypes are low-fidelity prototypes to suggest changes. A prototype design can be either high fidelity or low-fidelity prototype. There are some important trade off here in developing high-fidelity prototypes and low-fidelity prototypes. The high fidelity prototype does look and feel more real to the user. Consequently, users may be more reluctant to critique the interface. Therefore, a quick and dirty-paper prototype is more powerful to design storyboards [63]. Finally, quick and dirty-paper prototypes are considered more powerful to use for designing storyboard. Therefore, the game was initially sketched in a storyboard, using ink pen, post-it notes and papers based on the above-mentioned story and which is shown in Fig. 3 [58].

Fig. 3: The game storyboard design.

4.3 Mechanism

Funny stories are a great tonic for maintaining users' attraction in eLearning [64]. Storytelling techniques are used to grab attention, which also exaggerates the interesting aspect of reality. Stories can be based on personal experiences or famous tales or could also be aimed to build a storyline that associates content units, inspire or reinforce.

The game is based on a scenario of a character of a small fish and 'his' teacher who live in a big pond. The more appropriate, realistic and content in the story, the better the chances of triggering the users. The main character of the game is the small fish, who wants to eat worms to become a big fish. The game player role-plays as a small fish. However, he should be careful of phishers that try to trick him with fake worms. This represents phishing attacks by developing threat perception. Each worm is associated with a Website address and the Unified Resource Locator (URL) appears as a dialog box. The small fish's job is to eat all the real worms which associate which legitimate Website addresses and reject fake worms which associate with fake Website addresses before the time is up. This attempts to develop the severity and susceptibility of the phishing threat in the game design.

The other character is the small fish's teacher, which is a mature and experienced fish in the pond. If the worm associated with the URL is suspicious and if it is difficult to identify, the small fish can go to his teacher and request for help. The teacher could help him by giving some tips on how to identify bad worms. For example, "Website addresses associating with numbers in the front are generally scams", or "a company name followed by a hyphen in a URL is generally a scam". Whenever the small fish requests help from the teacher, the time score will be reduced by a certain amount (in this case by 100 seconds) as a payback for the safeguard measure. This attempts to address the safeguard effectiveness and the cost needs to pay for the safeguard in the game design.

The game prototype design consists of a total of 10 URLs to randomly display worms, including five good worms (associated with legitimate URLs) and five fake worms (associated with phishing URLs). If the user correctly identifies all good worms while avoiding all fake worms by looking at URLs, then he will gain 10 points (in this case each attempt scores 1 point). If the user falsely identifies good worms or fake worms, each attempt loses one life out of the total lives remaining to complete the game. If the user seeks help from the big fish (in this case small fishaˆAˇZ´s teacher) each attempt loses 100 seconds out of the total remaining time to complete the game, which is 600 seconds.

4.4 Game Design

The game was designed, based on the aforementioned storyboard mechanism (Fig. 3). Each worm is associated with an URL, which appears as a dialog box. The small fish's job is to eat all the real worms, which are associated with legitimate URLs while preventing eating of fake worms, which are associated with fraudulent URLs before time is up. If the phishing URL is correctly identified, then the game player is prompted to identify which part of the URL indicates phishing (Fig. 4). This determines whether or not the game player has understood the conceptual knowledge of the phishing URL. At this point in time, the score of the game player will encourage him/her in order to proceed with the game. Nevertheless, if the phishing URL is incorrectly identified, then the game player will get real time feedback, saying why their decision was wrong with an appropriate example, such as "Legitimate websites usually do not have numbers at the beginning of their URLs" or http://187.52.91.111/.www.hsbc.co.uk. Therefore, this attempts to teach the player about conceptual knowledge of phishing URLs within the game-design context.

Furthermore, one can argue that presenting the player with different types of URLs to identify if it is phishing or legitimate (i.e., procedural knowledge), then asking them to identify which part of the URL is phishing (i.e., conceptual knowledge) would positively impact his/her self-efficacy to enhance the phishing-threat avoidance.

The proposed game design is presented at different levels, such as beginner, intermediate and advance. When the player moves from the beginner to advanced

Please identify which part of the URL indicates phishing?

http://187.52.91.111 **/.www.hsbc.co.uk**".

Fig. 4: The game player is prompted to identify which part of the URL indicates phishing.

level, complexity of the combination of URLs (i.e., procedural knowledge) dramatically increases while considerably decreasing the time period to complete the game. Procedural knowledge of phishing prevention will be then addressed in the game design, as the complexity of the URLs is presented to the player through the design. The overall game design focuses on designing an innovative and gamified approach to integrate both conceptual and procedural knowledge effects on the game player's (i.e., computer user) self-efficacy to thwart phishing attacks.

If the worm associated with the URL is suspicious or if it is difficult to identify/ guess the small fish (in this case the game player) can go to the big fish and request for help. The big fish will then provide some tips on how to recognize bad worms associated with fraudulent URLs. For example, "a company name followed by a hyphen in an URL is generally a scam" or "Website addresses associated with numbers in the front are generally scams". This will again aid to develop the player's 'self-efficacy' to combat against phishing attacks. Whenever the small fish asks for some help from the big fish, the time left will be reduced by a certain amount (e.g., 100 seconds) as a payback for the safeguarding measure.

5. Conclusion and Future Work

This research focuses on designing an innovative and gamified application to integrate both conceptual and procedural knowledge effects on the game player's (i.e., computer user) self-efficacy to combat against phishing attacks. The study asks how one can integrate 'self-efficacy' into an anti-phishing educational gaming tool that teaches people to thwart phishing attacks? The game design teaches how one can differentiate legitimate URLs (Uniform Resource Locators) from fraudulent ones. The elements derived from a theoretical model [13] are incorporated into the proposed game design to thwart phishing attacks. The theoretical model [13] reveals the interaction effects of both conceptual knowledge or procedural knowledge that will positively affect self-efficacy, which contributes to enhancing the computer user's phishing threat-avoidance behavior (through motivation). Furthermore, as future research, we attempt to implement this game design and empirically investigate through users in order to understand how their knowledge (the interaction effect of both conceptual and procedural) will positively affect self-efficacy, which eventually contributes to enhancing their phishing threat-avoidance behaviour.

The transformative nature of research work is in presenting the game-based awareness design for individuals to protect oneself against phishing crimes. The proposed game design that integrated the user's 'self-efficacy' will enhance the individual's phishing threat-avoidance behaviour. This concept is based on the notion that not only can a computer game provide anti-phishing education, but also provide a better learning environment, because it motivates the user to remain attentive by providing immediate feedback. This can also be considered an appropriate way to reach individuals in society where the message of phishing-threat awareness is vital, making a considerable contribution to enabling the cyberspace to be a safe environment for everyone.

The proposed game is designed to encourage users to enhance their avoidance behavior through motivation to protect themselves from phishing attacks. Moreover, we will record how users employ their strategies to differentiate phishing attacks from legitimate ones through the game and then develop a threat model, understanding how cybercriminals leverage their attacks within the organization through human exploitation.

References

[1] Krawchenko, K. (2016, October). The phishing e-mail that hacked the account of John Podesta. CBS News.

[2] Arachchilage, N.A.G., Love, S. and Beznosov, K. (2016). Phishing threat avoidance behaviour: An empirical investigation. Computers in Human Behavior 60: 185–197.

[3] Arachchilage, N.A.G. and Love, S. (2013). A game design framework for avoiding phishing attacks. Computers in Human Behavior 29(3): 706–714.

[4] Almomani, A., Gupta, B.B., Atawneh, S., Meulenberg, A. and Almomani, E. (2013). A survey of phishing email filtering techniques. IEEE Communications Surveys and Tutorials 15(4): 2070–2090.

[5] Gupta, B.B., Tewari, A., Jain, A.K. and Agrawal, D.P. (2017). Fighting against phishing attacks: State of the art and future challenges. Neural Computing and Applications 28(12): 3629–3654.

[6] Gupta, B., Arachchilage, N.A. and Psannis, K.E. (2017). Defending against phishing attacks: Taxonomy of methods, current issues and future directions. Telecommunication Systems, 1–21.

[7] Whitty, M.T. (2015). Mass-marketing fraud: A growing concern. IEEE Security & Privacy 13(4): 84–87.

[8] Jain, A.K. and Gupta, B.B. (2017). Phishing detection: Analysis of visual similarity-based approaches. Security and Communication Networks, 2017.

[9] Almomani, A., Gupta, B.B., Wan, T.C., Altaher, A. and Manickam, S. (2013). Phishing dynamic evolving neural fuzzy framework for online detection zero-day phishing e-mail. Indian Journal of Science and Technology 6(1): 3960–3964.

[10] Sheng, S., Magnien, B., Kumaraguru, P., Acquisti, A., Cranor, L.F., Hong, J. and Nunge, E. (2007). Anti-phishing phil: The design and evaluation of a game that teaches people not to fall for phish. pp. 88–99. *In*: Proceedings of the 3rd Symposium on Usable Privacy and Security.

[11] Dhamija, R., Tygar, J.D. and Hearst, M. (2006). Why phishing works. pp. 581–90. *In*: Proceedings of the Sigchi Conference on Human Factors in Computing Systems.

[12] Arachchilage, N.A.G., Namiluko, C. and Martin, A. (2013). A taxonomy for securely sharing information among others in a trust domain. pp. 296–304. *In*: Internet Technology and Secured Transactions (icitst), 8th International Conference for.

[13] Arachchilage, N.A.G. and Love, S. (2014). Security awareness of computer users: A phishing threat avoidance perspective. Computers in Human Behavior 38: 304–312.

[14] Kirlappos, I. and Sasse, M.A. (2012). Security education against phishing: A modest proposal for a major rethink. IEEE Security & Privacy 10(2): 24–32.

[15] Kumaraguru, P., Sheng, S., Acquisti, A., Cranor, L.F. and Hong, J. (2008). Lessons from a real world evaluation of anti-phishing training. pp. 1–12. *In*: Ecrime Researchers Summit.

[16] Walls, R. (2012). Using computer games to teach social studies' (unpublished doctoral dissertation), http://www.diva-portal.org/smash/record.jsf?pid=diva2University, Disciplinary Domain of Humanities and Social Sciences, Faculty of Educational Sciences, Department of Education.

[17] Cone, B.D., Irvine, C.E., Thompson, M.F. and Nguyen, T.D. (2007). A video game for cyber security training and awareness. Computers and Security 26(1): 63–72.

[18] Foreman, J. (2004). Video game studies and the emerging instructional revolution. Innovate: Journal of Online Education 1(1).

[19] Arachchilage, G. and Asanka, N. (2012). Security awareness of computer users: A game-based learning approach (unpublished doctoral dissertation), Brunel University, School of Information Systems, Computing and Mathematics.

[20] Zyda, M. (2005). From visual simulation to virtual reality to games. Computer 38(9): 25–32.

[21] Michael, D.R. and Chen, S.L. (2005). Serious Games: Games that Educate, Train and Inform, Muska & Lipman/Premier-Trade.

[22] Froschauer, J., Seidel, I., Ga¨rtner, M., Berger, H. and Merkl, D. (2010). Design and evaluation of a serious game for immersive cultural training. pp. 253–260. *In*: Virtual Systems and Multimedia (vsmm), 2010 16th International Conference on.

[23] Arachchilage, N.A.G. and Cole, M. (2011). Design a mobile game for home computer users to prevent from phishing attacks'. pp. 485–489. *In*: Information Society (i-society), 2011 International Conference on.

[24] Bellotti, F., Berta, R., Gloria, A.D. and Primavera, L. (2009). Enhancing the educational value of video games. Computers in Entertainment (CIE) 7(2): 23.

[25] Gustafsson, A., Katzeff, C. and Bang, M. (2009). Evaluation of a pervasive game for domestic energy engagement among teenagers. Computers in Entertainment (CIE) 7(4): 54.

[26] Eagle, M. and Barnes, T. (2008). Wu's castle: Teaching arrays and loops in a game. pp. 245–249. *In*: ACM Sigcse Bulletin, Vol. 40.

[27] Gee, J.P. (2003). What video games have to teach us about learning and literacy. Computers in Entertainment (CIE) 1(1): 20–20.

[28] Fotouhi-Ghazvini, F., Earnshaw, R., Robison, D and Excell, P. (2009). The mobo city: A mobile game package for technical language learning. International Journal of Interactive Mobile Technologies 3(2).

[29] Gunter, G.A., Kenny, R.F. and Vick, E.H. (2008). Taking educational games seriously: Using the retain model to design endogenous fantasy into stand-alone educational games. Educational Technology Research and Development 56(5-6): 511–537.

[30] Malone, T.W. (1981). Towards a theory of intrinsically motivating instruction. Cognitive Science 5(4): 333–369.

[31] Oblinger, D. (2004). The next generation of educational engagement. Journal of Interactive Media in Education, 2004(1).

[32] Squire, K. (2003). Video games in education. *In*: International Journal of Intelligent Simulations and Gaming.

[33] De Freitas, S. and Oliver, M. (2006). How can exploratory learning with games and simulations within the curriculum be most effectively evaluated? Computers & Education 46(3): 249–264.

[34] Prensky, M. (2001). Digital Game-based Learning.

[35] Moreno-Ger, P., Burgos, D., Mart´ınez-Ortiz, I., Sierra, J.L. and Ferna´ndez-Manj´on, B. (2008). Educational game design for online education. Computers in Human Behavior 24(6): 2530–2540.

[36] Andrews, G., Woodruff, E., MacKinnon, K.A. and Yoon, S. (2003). Concept development for kindergarten children through health simulation. Journal of Computer Assisted Learning 19(2): 209–219.

[37] Roschelle, J. and Pea, R. (2002). A walk on the wild side: How wireless handhelds may change computer-supported collaborative learning. International Journal of Cognition and Technology 1(1): 145–168.

[38] Allen, M. (2006). Social Engineering: A Means to Violate a Computer System, SANS Institute, InfoSec Reading Room.

[39] Schechter, S.E., Dhamija, R., Ozment, A. and Fischer, I. (2007). The emperor's new security indicators. pp. 51–65. *In*: Security and Privacy, sp'07. IEEE Symposium on.

[40] Timko, D. (2008). The social engineering threat. Information Systems Security Association Journal.

[41] Herley, C. (2009). So long, and no thanks for the externalities: The rational rejection of security advice by users. pp. 133–144. *In*: Proceedings of the 2009 Workshop on New Security Paradigms Workshop.

[42] Bailey, B.P., Konstan, J.A. and Carlis, J.V. (2001). Demais: Designing multimedia applications with interactive storyboards. pp. 241–250. *In*: Proceedings of the Ninth ACM International Conference on Multimedia.

[43] Kumaraguru, P., Rhee, Y., Acquisti, A., Cranor, L.F., Hong, J. and Nunge, E. (2007). Protecting people from phishing: The design and evaluation of an embedded training email system. pp. 905–914. *In*: Proceedings of the Sigchi Conference on Human Factors in Computing Systems.

[44] Robila, S.A. and Ragucci, J.W. (2006). Don't be a phish: Steps in user education. pp. 237–241. *In*: Acm Sigcse Bulletin, Vol. 38.

[45] Jagatic, T.N., Johnson, N.A., Jakobsson, M. and Menczer, F. (2007). Social phishing. Communications of the ACM 50(10): 94–100.

[46] Tseng, S.-S., Chen, K.-Y., Lee, T.-J. and Weng, J.-F. (2011). Automatic content generation for anti-phishing education game. pp. 6390–6394. *In*: Electrical and Control Engineering (ICECE), International Conference on.

[47] Liang, H. and Xue, Y. (2009). Avoidance of information technology threats: A theoretical perspective. MIS Quarterly, 71–90.

[48] Liang, H. and Xue, Y. (2010). Understanding security behaviors in personal computer usage: A threat avoidance perspective. Journal of the Association for Information Systems 11(7): 394.

[49] Workman, M., Bommer, W.H. and Straub, D. (2008). Security lapses and the omission of information security measures: A threat control model and empirical test. Computers in Human Behavior 24(6): 2799–2816.

[50] Aytes, K. and Connolly, T. (2004). Computer security and risky computing practices: A rational choice perspective. Journal of Organizational and End User Computing (JOEUC) 16(3): 22–40.

[51] Yan, Z., Robertson, T., Yan, R., Park, S.Y., Bordoff, S., Chen, Q. and Sprissler, E. (2018). Finding the weakest links in the weakest link: How well do undergraduate students make cybersecurity judgment? Computers in Human Behavior 84: 375–382.

[52] Gorling, S. (2006). The Myth of User Education.

[53] Schneier, B. (2000). Semantic attacks: The third wave of network attacks. Crypto-Gram Newsletter, 14.

[54] Le Compte, A., Elizondo, D. and Watson, T. (2015). A renewed approach to serious games for cyber security. pp. 203–216. *In*: Cyber Conflict: Architectures in Cyberspace (cycon), 7th International Conference on.

[55] Hu, W. (2010). Self-efficacy and individual knowledge sharing. pp. 401–404. *In*: Information Management, Innovation Management and Industrial Engineering (iciii), International Conference on, Vol. 2.

[56] McCormick, R. (1997). Conceptual and procedural knowledge. pp. 141–159. *In*: Shaping Concepts of Technology, Springer.

[57] Plant, M. (1994). How is science useful to technology? Design and Technology in the Secondary Curriculum: A Book of Readings, The Open University, Milton Keynes, 96–108.

[58] Truong, K.N., Hayes, G.R. and Abowd, G.D. (2006). Storyboarding: An empirical determination of best practices and effective guidelines. pp. 12–21. *In*: Proceedings of the 6th Conference on Designing Interactive Systems.

[59] Gruen, D. (2000). Storyboarding for design: An overview of the process. Lotus Research— Accessed 3(11): 10.

[60] Rosson, M.B. and Carroll, J.M. (2009). Scenario-based design. Human-Computer Interaction, Boca Raton, FL, 145–162.

[61] Landay, J.A. and Myers, B.A. (1996). Sketching storyboards to illustrate interface behaviors. pp. 193–194. *In*: Conference Companion on Human Factors in Computing Systems.

[62] Lin, J., Newman, M.W., Hong, J.I. and Landay, J.A. (2000). Denim: finding a tighter fit between tools and practice for website design. pp. 510–517. *In*: Proceedings of the Sigchi Conference on Human Factors in Computing Systems.

[63] Sefelin, R., Tscheligi, M. and Giller, V. (2003). Paper prototyping—what is it good for? Acomparison of paper-and computer-based low-fidelity prototyping. pp. 778–779. *In*: Chi'03 Extended Abstracts on Human Factors in Computing Systems.

[64] Taran, C. (2005). Motivation techniques in e-learning. pp. 617–619. *In*: Advanced Learning Technologies, icalt 2005, Fifth IEEE International Conference on.

[65] Arachchilage, N.A.G. (2011). Designing a mobile game for home computer users to protect against phishing attacks. Internatioal Journal for e-Learning Security (IJeLS) 1/2: 8.

[66] Baslyman, M. and Chiasson, S. (2016). Smells phishy? An educational game about online phishing scams. pp. 1–11. *In*: Electronic Crime Research (e-crime), apwg symposium on.

Index

9 780367 780272